Flexible Shells

Theory and Applications

Edited by
E. L. Axelrad and F. A. Emmerling

With 107 Figures

Springer-Verlag
Berlin Heidelberg New York Tokyo 1984

Prof. Dr. E. L. AXELRAD
Prof. Dr.-Ing. F. A. EMMERLING

Institut für Mechanik
Hochschule der Bundeswehr München
Fachbereich Luft- und Raumfahrttechnik
Werner-Heisenberg-Weg 39
D-8014 Neubiberg

Library of Congress Cataloging in Publication Data
Main entry under title:
Flexible shells.
"Euromech-Colloquium Nr. 165" -- Prelim. p.
1. Shells (Engineering) -- Congresses.
2. Elastic plates and shells -- Congresses.
I. Axelrad, E. L. (Ernest L.),
II. Emmerling, F. A. (Franz Alfons).
TA660.S5F56 1984 624.1'7762 84-5541

Softcover reprint of the hardcover 1st edition 1984

ISBN-13: 978-3-642-48015-7 e-ISBN-13: 978-3-642-48013-3
DOI: 10.1007/ 978-3-642-48013-3

Preface

Euromech-Colloquium Nr. 165

The shell-theory development has changed its emphasis during the last two decades. Nonlinear problems have become its main motive.

But the analysis was until recently predominantly devoted to shells designed for strength and stiffness. Nonlinearity is here relevant to buckling, to intensively varyable stress states. These are (with exception of some limit cases) covered by the quasi-shallow shell theory.

The emphasis of the nonlinear analysis begins to shift further – to shells which are *designed* for and actually *capable* of large elastic displacements. These shells, used in industry for over a century, have been recently termed *flexible shells.*

The European Mechanics Colloquium 165. was concerned with the theory of elastic shells in connection with its applications to these shells. The Colloquium was intended to discuss:

1. The formulations of the nonlinear shell theory, different in the generality of kinematic hypothesis, and in the choice of dependent variables.
2. The specialization of the shell theory for the class of shells and the respective elastic stress states assuring flexibility.
3. Possibilities to deal with the complications of the buckling analysis of flexible shells, caused by the precritial perturbations of their shape and stress state.
4. Methods of solution appropriate for the nonlinear flexible-shell problems.
5. Applications of the theory.

There were 71 participants the sessions were presided over (in that order) by E. Reissner, J. G. Simmonds, W. T. Koiter, R. C. Tennyson, F. A. Emmerling, E. Ramm, E. L. Axelrad. Particular thanks for the aid in the composition of this volume are due to Professors W. T. Koiter, E. Reissner, J. G. Simmonds and E. Stein.

The Continuum Mechanics and in particular the solution of flexible-shell problems owes very much to Professor E. Reissner who has become 70 in 1983. The participants of the Colloquium have used the opportunity to express to Eric Reissner their admiration and best wishes.

Munich, January 1984

E. L. Axelrad
F. A. Emmerling

List of Contributors

Axelrad, E. L.
Probst-Heinrich-Straße 4
D-8000 München 40 – FRG

Bernadou, M.
Institut National de Recherche
en Informatique et en Automatique
Domaine de Voluceau-Rocquencourt
B.P. 105
F-78153 Le Chesnay Cedex – France

Emmerling, F. A.
Hochschule der Bundeswehr München
Institut für Mechanik
Werner-Heisenberg-Weg 39
D-8014 Neubiberg – FRG

Hübner, W.
Hochschule der Bundeswehr München
Institut für Mechanik
Werner-Heisenberg-Weg 39
D-8014 Neubiberg – FRG

Kröplin, B.
Universität Dortmund
Abteilung Bauwesen, Konstruktiver Ingenieurbau
August-Schmidt-Straße
D-4600 Dortmund 50 – FRG

Lloyd, D. W.
University of Leeds
Department of Textile Industries
GB-Leeds LS2 9JT – England

Nash, W. A.; Shaaban, S. H.; Watawala, L.; Lee, S. C.
Dept. of Civil Engineering
Marston Hall
University of Massachusetts
Amherst Massachusetts 01003 – USA

Oliver, J.; Oñate, E.
Universitat Politècnica de Barcelona
ETS. d'Enginyers de Camins, Canals i Ports
Jordi Girona Salgado, 31
E-Barcelona – 34 – Spain

Pietraszkiewicz, W.
Institute of Fluid-Flow Machinery
of the Polish Academy of Sciences
ul. Gen. J. Fiszera 14
PL-80-952 Gdańsk 6 – Polen

Ramm, E.
Universität Stuttgart
Institut für Baustatik
Pfaffenwaldring 7
D-7000 Stuttgart 80 – FRG

Reissner, E.
University of California, San Diego
Department of Applied Mechanics
and Engineering Sciences
Mail Code B-010
La Jolla, California 92093 – USA

Simmonds, J. G.
Department of Applied Mathematics and Computer Science
University of Virginia
Thornton Hall, Charlottesville
Virginia 22901 – USA

Schmidt, R.
Institut für Mechanik
Ruhr-Universität Bochum
Universitätsstraße 150
D-4630 Bochum 1 – FRG

Stein, E.; Wagner, W.; Lambertz, K.-H.
Universität Hannover
Institut für Baumechanik und Numerische Mathematik
Callinstraße 32
D-3000 Hannover 1 – FRG

Stumpf, H.
Lehrstuhl für Mechanik II
Ruhr-Universität Bochum
Universitätsstraße 150 I A3
D-4630 Bochum 1 – FRG

Scheidl, R.; Troger, H.
Institut für Mechanik
Technische Universität Wien
Karlsplatz 13
A-1040 Wien – Austria

Wan, F. Y. M.
Applied Mathematics, FS-20 University of Washington
Seattle, Washington 98195 – USA

Table of Contents

The Nonlinear Thermodynamical Theory of Shells: Descent from 3-Dimensions without Thickness Expansions

James G. Simmonds

Department of Applied Mathematics and Computer Science
University of Virginia
Charlottesville, Virginia 22901

Introduction

A glance at the current engineering literature on shells reveals two strong, interacting trends: the inexorable rise and spread of the finite element method and the pressures of economics that are demanding light, efficient structures for automobiles, air- and spacecraft.

It seems obvious that as the applications of shell theory grow, its foundations should be secured proportionally. This is the first aim of this brief overview: to show that a nonlinear, thermodynamic theory of shells can be derived from 3-dimensional continuum mechanics in a natural, clean, and comprehensive way. We show that <u>all</u> approximations can be thrown into a <u>postulated</u>, 2-dimensional form of the First Law of Thermodynamics. A second aim is to indicate some open questions and to convince the reader, in this day of mind-boggling computer codes, that there is an important place for analytic work. What is summarized here is developed at length in a recent monograph by Libai and me [1].

Derivation of the Shell Equations

The derivation has these desirable features:

a. The 2-dimensional, impulse-integral form of the equations of motion and The Second Law of Thermodynamics (Clausius-Duhem inequality) for a shell follow naturally and <u>exactly</u> from their 3-dimensional counterparts.

b. One obtains unique, concrete definitions of shell variables such as stress resultants and couples, deformed position, spin, and entropy resultant in terms of weighted integrals of 3-dimensional quantities through the thickness.

c. There are no series expansions in the thickness direction.

d. There is no Kirchhoff hypothesis in the kinematics.

e. Two-dimensional strain measures fall out automatically from The Mecahnical Work Identity.

f. All approximations appear in The First Law.

An interesting by-product of the descent from 3-dimensions is that, contrary to what one finds in the literature on thermal stresses in shells, the 2-dimensional temperature field that emerges is not a thickness average.

The derivation of the impulse-integral form of the equations of motion for a shell starts from those of a 3-dimensional body:

$$\int_{t_1}^{t_2}\left(\int_{\partial V} \mathbf{S}dA + \int_V \mathbf{f}dV\right) = \int_V \rho\dot{\mathbf{x}}dV\Big|_{t_1}^{t_2} \qquad (1)$$

$$\int_{t_1}^{t_2}\left(\int_{\partial V} \mathbf{x}\times\mathbf{S}dA + \int_V \mathbf{x}\times\mathbf{f}dV\right) = \int_V \mathbf{x}\times\rho\dot{\mathbf{x}}dV\Big|_{t_1}^{t_2} . \qquad (2)$$

Here (t_1,t_2) is an arbitrary interval of time and V denotes any volume in the reference shape of the body. V is assumed to have a piecewise smooth boundary ∂V and dA and dV are, respectively, differential elements of area and volume. $\mathbf{S}dA$ and $\mathbf{f}dV$ are contact and body forces acting across and throughout the deformed images of dA and dV, and ρ is the initial mass density at a particle P with initial position $\mathbf{X}$. We assume that $\mathbf{x}$, the position of P at time t, is given by an invertible smooth function of the form $\mathbf{x} = \hat{\mathbf{x}}(\mathbf{X},t)$; $\dot{\mathbf{x}}\equiv\partial\hat{\mathbf{x}}/\partial t$. The stress vector $\mathbf{S}$ may be expressed as

$$\mathbf{S} = \mathbf{N}\cdot\underline{\underline{S}} = \underline{\underline{S}}^T\cdot\mathbf{N}, \qquad (3)$$

where $\underline{\underline{S}}$ is the nominal stress tensor and $\underline{\underline{S}}^T$ is the first Piola-Kirchhoff stress tensor.

A <u>shell</u> is a body whose initial shape may be given the parametric form

$$\mathbf{X} = \hat{\mathbf{X}}(\sigma^\alpha,\zeta) = \hat{\boldsymbol{\xi}}(\sigma^\alpha)+\zeta\hat{\mathbf{n}}(\sigma^\alpha),\ \sigma^\alpha \varepsilon D,\ \hat{H}_-(\sigma^\alpha) \le \zeta \le \hat{H}_+(\sigma^\alpha), \quad (4)$$

where $\mathbf{n}$ is a unit normal to S. If $\zeta=0$, we are on the <u>reference surface</u> S.

To derive shell equations, let Σ, with piecewise smooth boundary $\partial\Sigma$, be generated by taking $\sigma^\alpha \varepsilon \Delta \subset D$ and let V be the corresponding volume generated by (4). By using some elementary differential geometry and integrating with respect to ζ, we obtain the <u>global</u> equations of motion of a shell:

$$\int_{t_1}^{t_2} \left(\int_{\partial\Sigma} \mathbf{N}^\alpha \nu_\alpha ds + \int_\Sigma \mathbf{p} d\Sigma\right)dt = \int_\Sigma m\dot{\mathbf{y}} d\Sigma \Big|_{t_1}^{t_2} \quad (5)$$

$$\int_{t_1}^{t_2} \left[\int_{\partial\Sigma} (\mathbf{y}\times\mathbf{N}^\alpha + \mathbf{M}^\alpha)\nu_\alpha ds + \int_\Sigma (\mathbf{y}\times\mathbf{p} + \boldsymbol{\ell})d\Sigma\right]dt$$

$$= \int_\Sigma (\mathbf{y}\times m\dot{\mathbf{y}} + I\boldsymbol{\omega})d\Sigma \Big|_{t_1}^{t_2}, \quad (6)$$

where

$$\mathbf{N}^\alpha = \int_-^+ \mathbf{S}^\alpha \mu d\zeta, \quad \mathbf{M}^\alpha = \int_-^+ \boldsymbol{\eta}\times\mathbf{S}^\alpha \mu d\zeta \quad (7)$$

are stress resultants and couples,

$$\mathbf{p} = \int_-^+ \mathbf{f}\mu d\zeta + [\mu(\mathbf{S}^3 - H,_\alpha \mathbf{S}^\alpha)]_-^+ \quad (8)$$

$$\boldsymbol{\ell} = \int_-^+ \boldsymbol{\eta}\times\mathbf{f}\mu d\zeta + [\mu\boldsymbol{\eta}\times(\mathbf{S}^3 - H,_\alpha \mathbf{S}^\alpha)]_-^+ \quad (9)$$

are distributed surface forces and couples per unit area of S,

$$m = \int_-^+ \rho\mu d\zeta, \qquad I = \int_-^+ \rho\zeta^2\mu d\zeta \quad (10)$$

are the mass/area and transverse moment of inertia of S,

$$\mathbf{y} = \int_-^+ \rho\mathbf{x}\mu d\zeta/m, \qquad \boldsymbol{\omega} = \int_-^+ \boldsymbol{\eta}\times\rho\dot{\boldsymbol{\eta}}\mu d\zeta/I \quad (11)$$

are a weighted position and spin of the deformed shell, and

$$\mu = 1 - 2\zeta M + \zeta^2 K. \quad (12)$$

where M and K are the mean and Gaussian curvatures at a point of S. In (5)-(11), $\int_-^+$ is short for $\int_{H_-}^{H_+}$ and $\boldsymbol{\eta} = \mathbf{x} - \mathbf{y}$.

As (5) and (6) must hold for all $t_2 > t_1$ and all Σ, it follows that if the integrands are continuous and if there are no impulsive or concentrated forces, then the following local equations of motion must hold.

$$\mathbf{N}^\alpha|_\alpha + \mathbf{p} = m\ddot{\mathbf{y}} \tag{13}$$

$$\mathbf{M}^\alpha|_\alpha + \mathbf{y}_\alpha \times \mathbf{N}^\alpha + \boldsymbol{\ell} = I\dot{\boldsymbol{\omega}}, \tag{14}$$

where $\mathbf{y}_\alpha \equiv \partial \mathbf{y}/\partial \sigma^\alpha$, and a vertical bar denotes covariant differentiation based on the metric of the undeformed reference surface.

The Mechanical Work Identity (MWI)

If we dot (13) with $\dot{\mathbf{y}}$ and (14) with $\boldsymbol{\omega}$, add, integrate over an arbitrary piece Σ of the reference surface, and apply the divergence theorem, we obtain

$$\int_{t_1}^{t_2} \mathcal{W} dt \equiv \mathcal{K}|_{t_1}^{t_2} + \int_{t_1}^{t_2} \mathcal{D} dt. \tag{15}$$

where

$$\mathcal{W} = \int_{\partial\Sigma} (\mathbf{N}^\alpha \cdot \dot{\mathbf{y}} + \mathbf{M}^\alpha \cdot \boldsymbol{\omega}) v_\alpha ds + \int_\Sigma (\mathbf{p} \cdot \dot{\mathbf{y}} + \boldsymbol{\ell} \cdot \boldsymbol{\omega}) d\Sigma \tag{16}$$

is the apparent mechanical power.

$$\mathcal{K} = \frac{1}{2} \int_\Sigma (m\dot{\mathbf{y}} \cdot \dot{\mathbf{y}} + I\boldsymbol{\omega} \cdot \boldsymbol{\omega}) d\Sigma \tag{17}$$

is the kinetic energy, and

$$\mathcal{D} = \int_\Sigma [\mathbf{N}^\alpha \cdot (\dot{\mathbf{y}}_\alpha - \boldsymbol{\omega} \times \mathbf{y}_\alpha) + \mathbf{M}^\alpha \cdot \boldsymbol{\omega},_\alpha] d\Sigma \equiv \int_\Sigma \sigma \cdot \dot{\varepsilon} d\Sigma \tag{18}$$

is the deformation power.

Strains

Let $\{\mathbf{a}_i\} \equiv \{\mathbf{a}_\alpha, \mathbf{n}\}$, $i = 1,2,3$, denote the standard triad of base vectors and unit normal on S, and let $\{\mathbf{A}_i\} \equiv \{\mathbf{A}_\alpha, \mathbf{N}\}$ be a triad that coincides with $\{\mathbf{a}_i\}$ at t=0 and has spin $\boldsymbol{\omega}$ relative to $\{\mathbf{a}_i\}$. Then, as shown in [1], we may introduce a stretching $\mathbf{E}_\alpha$ and a bending $\mathbf{K}_\alpha$ such that

$$\dot{\mathbf{y}}_\alpha - \boldsymbol{\omega} \times \mathbf{y}_\alpha = \mathbf{E}^*_\alpha, \qquad \boldsymbol{\omega},_\alpha = \mathbf{K}^*_\alpha, \tag{19}$$

where the *-derivative is the time-rate-of-change relative to the frame $\{\mathbf{A}_i\}$. The stretching and bending satisfy the local compatibility conditions

$$\varepsilon^{\alpha\beta}(\mathbf{K}_\beta \times \mathbf{A}_\alpha + \mathbf{E}_{\alpha|\beta}) = \mathbf{0}, \quad \varepsilon^{\alpha\beta}(\mathbf{K}_{\alpha|\beta} + \frac{1}{2}\mathbf{K}_\alpha \times \mathbf{K}_\beta) = \mathbf{0}, \tag{20}$$

where $\varepsilon^{\alpha\beta}$ are the contravariant components of the permutation tensor.

In terms of components relative to the basis $\{\mathbf{A}_i\}$, defined by Eqs. (3.59) and (3.61) of [1], the deformation power density takes the form

$$\sigma\cdot\dot{\varepsilon} = \bar{N}^{\alpha\beta}\dot{\gamma}_{\alpha\beta} + \underline{N\dot{\Psi}} + \underline{Q^\alpha\dot{E}_\alpha} + \bar{M}^{\alpha\beta}\dot{R}_{\alpha\beta} + \underline{M^\alpha\dot{K}_\alpha} + \underline{\bar{M}\dot{B}}, \tag{21}$$

where

$$N^{\alpha\beta} = \bar{N}^{\alpha\beta} + \varepsilon^{\alpha\beta}\bar{N} \quad , \quad \bar{N}^{\alpha\beta} = \bar{N}^{\beta\alpha} \tag{22}$$

$$M^{\alpha\beta} = \bar{M}^{\alpha\beta} + \varepsilon^{\alpha\beta}\bar{M} \quad , \quad \bar{M}^{\alpha\beta} = \bar{M}^{\beta\alpha} \tag{23}$$

$$E_{\alpha\beta} = \gamma_{\alpha\beta} + \varepsilon_{\alpha\beta}\Psi \quad , \quad \gamma_{\alpha\beta} = \gamma_{\beta\alpha} \tag{24}$$

$$K_{\alpha\beta} = R_{\alpha\beta} + \varepsilon_{\alpha\beta}B \quad , \quad R_{\alpha\beta} = R_{\beta\alpha}. \tag{25}$$

In the classical or <u>first-approximation</u> theory of shells, as set forth by Koiter [2] and Sanders [3], the MWI, with the underlined terms in (21) omitted, is taken as axiomatic and equations of equilibrium are derived from it.

<u>Heating of a Shell</u>

The net heating of a piece of a body that initially occupies a region V may be expressed as

$$\mathcal{Q} = \int_{\partial V} v\,dA + \int_V r\,dV, \tag{26}$$

where vdA is the heat <u>influx</u> across the deformed image of dA and rdV is the rate of heat production of the particles initially in dV. Applying the same procedures that led from the 3-dimensional equations of motion, (1) and (2), to the shell equations of motion, (5) and (6), and noting that

$$v = -\mathbf{q}\cdot\mathbf{N}, \qquad \mathbf{q} = q^\alpha \mathbf{X}_{,\alpha} + q^3\mathbf{N} \tag{27}$$

$$dA = \{[(1-2HM)^2-H^2K]a^{\alpha\beta}+2H(1-HM)b^{\alpha\beta}]H,_{\alpha}H,_{\beta}+\mu^2\}_{\pm}^{\frac{1}{2}}d\Sigma \equiv \tilde{\mu}_{\pm}d\Sigma, \quad (28)$$

we infer from (26) that for a shell-like body with Σ as a piece of its reference surface,

$$\mathcal{Q} = \int_{\partial\Sigma} h^{\alpha}\nu_{\alpha}ds + \int_{\Sigma} s d\Sigma, \quad (29)$$

where

$$h^{\alpha} = -\int_{-}^{+} q^{\alpha}\mu d\zeta, \qquad s = (v\tilde{\mu})_{-} + (v\tilde{\mu})_{+} + \int_{-}^{+} r\mu d\zeta. \quad (30)$$

We call h^{α} the surface contravariant components of the heat <u>influx</u> resultant and s the <u>surface heating</u>. We can also write the second of Eqs. (30) in a form analogous to (8) by noting that $v\tilde{\mu} = -\mu(q^3 - H,_{\alpha}q^{\alpha})$.

<u>The Clausius-Duhem Inequality for a Shell</u>

For an arbitrary piece of a body of initial volume V, this form of the 2nd Law of Thermodynamics reads

$$\int_V \eta dV\Big|_{t_1}^{t_2} \geq \int_{t_1}^{t_2} \left(\int_{\partial V} \frac{v}{\theta}\, dA + \int_V \frac{r}{\theta}\, dV\right)dt, \quad (31)$$

where ηdV is the entropy of the particles initially in dV and $\theta>0$ is the absolute temperature.

If V is generated by a piece Σ of the reference surface of a shell-like body, then (31) implies that

$$\int_{\Sigma} \iota d\Sigma\Big|_{t_1}^{t_2} \geq \int_{t_1}^{t_2} \left(\int_{\partial\Sigma} p^{\alpha}\nu_{\alpha}ds + \int_{\Sigma} v d\Sigma\right)dt, \quad (32)$$

where

$$\iota = \int_{-}^{+} \eta\mu d\zeta \qquad \text{and} \qquad p^{\alpha} = \int_{-}^{+} q^{\alpha}\theta^{-1}\mu d\zeta \quad (33)$$

are the <u>entropy resultant</u> and the surface contravariant components of the <u>entropy flux vector</u>, and

$$v = \left[\frac{v\tilde{\mu}}{\theta}\right]_{-} + \left[\frac{v\tilde{\mu}}{\theta}\right]_{+} + \int_{-}^{+} \frac{r}{\theta}\,\mu d\zeta \quad (34)$$

$$\equiv \frac{s}{T} + \beta$$

is the <u>entropy supply</u>. Here,

$$\frac{1}{T} \equiv \frac{1}{2} [\frac{1}{\theta_+} + \frac{1}{\theta_-}] \qquad (35)$$

is an effective temperature mean and

$$\beta \equiv \frac{1}{2} [\frac{1}{\theta_+} - \frac{1}{\theta_-}] [(\nu\tilde{\mu})_+ - (\nu\tilde{\mu})_-] + \int_-^+ [\frac{1}{\theta} - \frac{1}{T}] r\mu d\zeta \qquad (36)$$

is an (unknown) measure of the supply of entropy due to the nonuniform distribution of temperature through the thickness. (If θ=T, β=0.) Note that temperature enters (32) explicitly, via v, only through its values, θ_+, θ_-, on the upper and lower faces of the shell.

The First Law of Thermodynamics for Shells

So far

(a) all our shell equations have been exact, and

(b) mechanical and thermal effects have been uncoupled.

In our approach to shell theory, the First Law is used to couple, in an approximate way, exact mechanical and thermal quantities. Namely, we postulate that there exists an internal energy $\mathcal{U}$ such that

$$\int_{t_1}^{t_2} (\mathcal{Q} + \mathcal{W})dt = [\mathcal{K} + \mathcal{U}]_{t_1}^{t_2}, \qquad (37)$$

or, in view of the MWI, (15),

$$\int_{t_1}^{t_2} (\mathcal{Q} + \mathcal{D})dt = \mathcal{U} |_{t_1}^{t_2}. \qquad (38)$$

Thermoelastic Shells

Let $\Lambda \equiv (\varepsilon, T, \beta)$, $\varepsilon = (E_{\alpha\beta}, R_{\alpha\beta}, E_\alpha, \Psi, K_\alpha, B)$, denote a generalized state variable. We define a shell to be thermoelastic if there exists an internal energy per unit area of S that depends on Λ and its gradient $\Lambda,_\gamma$ such that

$$\mathcal{U} = \int_\Sigma u d\Sigma. \qquad (39)$$

If the various fields involved are sufficiently smooth, then (32) and (38), with $\mathcal{D}$, $\mathcal{Q}$ and $\mathcal{U}$ given by (18), (29), and (39), imply the local equations

$$\dot{\iota} \geq p^{\alpha}|_{\alpha} + v \tag{40}$$

$$h^{\alpha}|_{\alpha} + s + \sigma\cdot\dot{\varepsilon} = \dot{u}. \tag{41}$$

It is convenient to express (41) in another form. Adding $-\iota\dot{T} - g\dot{\beta}$ to both sides of (41), where $g = \partial u/\partial\beta$, we may write the resulting equation as

$$h^{\alpha}|_{\alpha} + s + \Xi\cdot\dot{\Lambda} = \dot{\Phi} + \dot{\iota}T + \dot{g}\beta, \tag{42}$$

where

$$\Phi = u - \iota T - g\beta = \hat{\Phi}(\Lambda,\Lambda,_{\gamma}) \tag{43}$$

is the <u>free energy density</u> and

$$\Xi = (\sigma,\ -\iota,\ -g). \tag{44}$$

Setting

$$p^{\alpha} = T^{-1}h^{\alpha} + \gamma^{\alpha}, \tag{45}$$

where γ^{α} measures the deviation of the entropy flux vector from its expected value if θ=T, and substituting (40) and (45) into (42), we obtain

$$\Xi\cdot\dot{\Lambda} - \dot{\Phi} \geq -T^{-1}h^{\alpha}T,_{\alpha} + T\gamma^{\alpha}|_{\alpha} + T\beta + g\beta. \tag{46}$$

We now introduce constitutive laws, assuming that Φ, Ξ, h^{α}, and γ^{α} are functions of Λ and $\Lambda,_{\gamma}$ only. This enables us to wring a number of consequences from (46).

First, setting $\tilde{\Xi} \equiv (\sigma,T,0)$ and $\tilde{\Phi} = \Phi + g\beta$, a modified free-energy density, and noting that the surface covariant derivative of γ^{α} has the form

$$\gamma^{\alpha}|_{\alpha} = \gamma^{\alpha}_{\Lambda}\cdot\Lambda,_{\alpha} + \gamma^{\alpha}_{\Lambda,_{\beta}}\cdot\Lambda_{\beta}|_{\alpha}, \tag{47}$$

we may rewrite (46) as

$$(\tilde{\Xi}-\tilde{\Phi}_{\Lambda})\cdot\dot{\Lambda}-\tilde{\Phi}_{\Lambda,_{\gamma}}\cdot\dot{\Lambda},_{\gamma}\geq -T^{-1}h^{\alpha}T,_{\alpha}+T\beta+T(\gamma^{\alpha}_{\Lambda}\cdot\Lambda,_{\alpha}+\gamma^{\alpha}_{\Lambda,_{\beta}}\cdot\Lambda_{\beta}|_{\alpha}). \tag{48}$$

At any value of (σ^α, t), the elements of the set

$$\Upsilon = \{\Lambda, \Lambda,_\gamma, \dot{\Lambda}, \dot{\Lambda},_\gamma, \Lambda_\gamma|_\delta\} \tag{49}$$

may be assigned arbitrary values, provided that **p**, ℓ and s are chosen so that the equations of motion, (13) and (14), and the equation of energy balance, (41), are satisfied. By the assumed form of the constitutive laws, the coefficients of the last three elements of Υ as well as the remaining terms in (48) are <u>independent</u> of these quantities. Hence if (48) is to be inviolate, then we must have $\tilde{\Xi} = \tilde{\Phi}_\Lambda$, $\tilde{\Phi}_{\Lambda,_\gamma} = 0$ and $\gamma^\alpha_{\Lambda,_\gamma} = 0$. The first two conditions imply that $\tilde{\Phi} = \Phi(\varepsilon, T)$, so that so simple a phenomenon as the bending of a non-uniformly heated plate is excluded. Moreover, unless $\gamma^\alpha \equiv 0$, γ^α must depend on $\Lambda,_\gamma$ if it is to transform properly under a change of surface coordinates.

To avoid these unwanted conclusions, we assume that there is a relation between $\gamma^\alpha|_\alpha + \dot{g}\beta$ and the constitutive arguments Λ and $\Lambda,_\gamma$ of the form

$$T\gamma^\alpha|_\alpha + \dot{g}\beta = D(\Lambda, \Lambda,_\gamma) \equiv C(\Lambda, \Lambda,_\gamma) - T\beta, \tag{50}$$

where C is a constitutive function that must be determined by experiment. Applying the same arguments to (46) as we applied to (48), we now conclude that

$$\Xi = \Phi_\Lambda \quad , \quad \Phi_{\Lambda,_\gamma} = 0 \tag{51}$$

$$T^{-1} h^\alpha T,_\alpha \geq C. \tag{52}$$

Thus Φ does not depend on gradients of strain or temperature. Further restrictions on the possible forms of the constitutive laws for Φ, h^α, γ^α, and C follow from dimensional analysis and invariance requirements. See [1], [4], and [5] for details in special cases.

<u>Comments and Open Questions</u>

The first part of this overview was devoted to deriving the equations of the nonlinear, thermoelastic theory of shells. In this second part, we comment on different forms of the field equations and their solutions.

1. <u>The displacement form of the field equations is ill-conditioned in (near) inextensional bending</u>. This may be seen already in the linear, static theory of shells. Love's equations for a circular cylindrical shell [6, pp. 574] provide a striking example. As explained in [7], if h is the thickness of the shell and a is the radius of its midsurface, then approximating the term $1+(1/12)(h/a)^2$ by 1--unavoidable on any computer if the shell is sufficiently thin--can lead to a relative error of O(1) in inextensional bending problems.

2. <u>For small strains, there are robust forms of the field equations that avoid ill-conditioning in all cases</u>. Such forms, where the unknowns occur in static-geometric pairs, are given by Simmonds & Danielson [8], Reissner [9], and Koiter & Simmonds [10].

3. <u>When strains are large, what are the robust forms of the field equations</u>? The problem here is that the stress resultants are nonlinear functions of the strains. Even if these stress-strain relations could be inverted, the dual unknown approach that yields satisfactory results with small strains would lead to partial differential equations nonlinear in their highest derivatives.

4. <u>What does the Principle of Virtual Work tell us about the boundary conditions of shell theory</u>? Clearly, the constitutive laws must enter the picture. For example, when these laws are those of a membrane, then, in the linear theory at least, the governing partial differential equations are elliptic, parabolic, or hyperbolic at a point of the reference surface as the Gaussian curvature there is positive, zero, or negative. Each type of equation seems to demand a different type of boundary condition. Or, if plastic flow occurs in a plate, how does one justify the Kirchhoff boundary conditions of classical plate theory?

5. <u>Do higher order shell theories predict, say, stress resultants more accurately than does classical shell theory in a neighborhood of concentrated loads or edges</u>?

6. Should we worry about finite element codes in the hands of non-experts in shell theory? How can we trust codes in which the errors may grow as the shell grows thinner? Will it ever be possible to "certify" codes or at least to write programs that will warn the user when the output (as a consequence of the input) is suspect?

References

1. Libai, A., and Simmonds, J. G., "Nonlinear Elastic Shell Theory," in Advances in Applied Mechanics (Hutchinson & Wu, eds.), Vol. 23, Academic Press, 1983, pp. 271-371.

2. Koiter, W. T., "A Consistent First Approximation in the General Theory of Thin Elastic Shells," in The Theory of Thin Elastic Shells (Proc. I.U.T.A.M. Sympos., Delft, 1959, Koiter, ed.), North-Holland, 1960, pp. 12-33.

3. Sanders, J. L., Jr., "Nonlinear Theories for Thin Shells," Q. Appl. Math., 21:21-36, 1963.

4. Green, A. E., and Naghdi, P. M., "On Thermal Effects in the Theory of Shells," Proc. Roy. Soc. London, Ser. A, 365:161-190, 1979.

5. Simmonds, J. G., "The Strain Energy Density of Rubber-Like Shells," Int. J. Solids & Structures (to appear).

6. Love, A. E. H., A Treaties on the Mathematical Theory of Elasticity, 3rd edition, Cambridge Univ. Press, 1920.

7. Simmonds, J. G., "A Set of Simple, Accurate Equations for Circular Cylindrical Elastic Shells," Int. J. Solids & Structures, 2:525-541, 1966.

8. Simmonds, J. G., and Danielson, D. A., "Nonlinear Shell Theory with Finite Rotation and Stress-Function Vectors," J. Appl. Mechs., 39:1085-1090, 1972.

9. Reissner, E., "Linear and Nonlinear Theory of Shells," in Thin-Shell Structures (Sympos. Thin Shell Structures, Pasadena, 1972, Fung & Sechler, eds.), Prentice-Hall, 1974, pp. 29-44.

10. Koiter, W. T., and Simmonds, J. G., "Foundations of Shell Theory," in Proc. 13th Int. Cong. Theor. & Appl. Mechs. (Moscow, 1972, Becker & Mikhailov, eds.), Springer-Verlag, 1973, pp. 150-176.

Acknowledgement

This work was supported by the National Science Foundation under Grant CEE-8117103.

On the Derivation of the Differential Equations of Linear Shallow Shell Theory

E. REISSNER

Department of Applied Mechanics and Engineering Sciences
University of California, San Diego
La Jolla, California 92093

Abstract

Given that derivations of linear shallow shell theory depend on (1) geometric shallowness assumptions and (2) static shallowness assumptions it is the purpose of the present discussion to verify the less obvious validity of the second set of assumptions through the application of an iterative procedure. The result of this justification includes the observation that additional steps of the iterative procedure will be meaningful only in conjunction with simultaneous iterative improvements in regard to the geometric shallowness assumption.

Introduction

In what follows we respond to the invitation of the organizers of the Euromech Colloquium on Flexible Shells to contribute to the Proceedings of the Colloquium by a "state-of-the-art review on relevant developments of the theory", with a note on the derivation of the differential equations of linear shallow shell theory. It had originally been thought that the work which is described in what follows should have been carried further. Inasmuch as it now appears that changes in the author's interests make it unlikely that these desirable extensions will in actuality be undertaken by him, and inasmuch as the present interim results may be of some interest by themselves they are offered as they existed in manuscript form, prior to the Colloquium. We have added a number of references so as to facilitate an appreciation of the origin of the equations which represent the point of departure for the present considerations.

Formulation of the Problem

We are concerned with the differential equations

$$D\nabla^4 w + L(K) = p\ , \quad B\nabla^4 K - L(w) = 0 \tag{1a,b}$$

for the transverse deflection w and the Airy stress function K of a uniform isotropic shallow shell, with middle surface equation $z = z(x_1,x_2)$ and with the operator L defined by

$$L(\ ) = z_{,11}(\ )_{,22} + z_{,22}(\ )_{,11} - 2z_{,12}(\ )_{,12}\ . \tag{2}$$

It is known that the derivation of (1) from formulations of general linear shell theory depends on two basic assumptions. The first of these is the geometric shallowness assumption

$$\mathrm{Max}\left| z_{,i}z_{,j} \right| << 1\ . \tag{3}$$

The second basic assumption is of a less obvious nature and consists, in essence, of the statement that it is appropriate for shallow shells to neglect transverse shear stress resultant terms in the equations of tangential force equilibrium and to neglect stress couple terms in the normal component equation of moment equilibrium. In order to justify this second assumption, it is often said that its appropriateness may be verified a posteriori by seeing that the results obtained on the basis of a neglect of these terms are of such nature as to be, in fact, consistent with the original assumption. The purpose of what follows is to show that it is possible to carry out this verification once for all, for a fairly general class of cases.

We depart from known statements of the differential equations of equilibrium and compatibility of the linear theory of shells. It is readily shown that the geometric assumption (3) means that the cartesian base plane coordinates x_i are also orthogonal cartesian coordinates on the middle surface of the shell, with the radii of curvature R_{ij} of this surface given by

$$R_{ij}^{-1} = z_{,ij}\ . \tag{4}$$

In what follows we will, for simplicity's sake, limit ourselves to case for which the R_{ij} are constants.

With equations (3) and (4) the differential equations of force equilibrium for tangential stress resultants N_{ij} and transverse resultants Q_i are

$$N_{11,1} + N_{21,1} + \frac{Q_1}{R_{11}} + \frac{Q_2}{R_{12}} + p_1 = 0 \,, \tag{5a}$$

$$N_{12,1} + N_{22,2} + \frac{Q_1}{R_{21}} + \frac{Q_2}{R_{22}} + p_2 = 0 \,, \tag{5b}$$

$$Q_{1,1} + Q_{2,2} - \frac{N_{11}}{R_{11}} - \frac{N_{22}}{R_{22}} - \frac{N_{12} + N_{21}}{R_{12}} + p = 0 \,. \tag{5c}$$

The associated system of moment equilibrium equations, involving the stress couples M_{ij}, is

$$Q_1 = M_{11,1} + M_{21,2} \,, \qquad Q_2 = M_{12,1} + M_{22,2} \,, \tag{6a,b}$$

$$N_{12} - N_{21} + \frac{M_{22} - M_{11}}{R_{12}} + \frac{M_{12}}{R_{11}} - \frac{M_{21}}{R_{22}} = 0 \,. \tag{6c}$$

We note that equations (5) and (6) may be considered to be direct consequences of a system of equations first stated in [1], upon making use of the shallowness assumptions (3). More general considerations, suggesting the derivation of (5) and (6) as a step in a parametric expansion procedure applied to a system of shell equations in oblique-coordinate form may be found in [4].

The system (5) and (6) is associated with a dual system of compatibility equations for force and moment strains ε_{ij}, λ_i and κ_{ij}, of the form

$$\kappa_{22,1} - \kappa_{12,2} + \frac{\lambda_1}{R_{21}} - \frac{\lambda_2}{R_{11}} = 0 \,, \tag{7a}$$

$$\kappa_{21,1} - \kappa_{11,2} - \frac{\lambda_1}{R_{22}} + \frac{\lambda_2}{R_{12}} = 0 \,, \tag{7b}$$

$$\lambda_{2,1} - \lambda_{1,2} + \frac{\kappa_{11}}{R_{22}} + \frac{\kappa_{22}}{R_{11}} - \frac{\kappa_{12} + \kappa_{21}}{R_{12}} = 0 \,, \tag{7c}$$

and

$$\lambda_1 = \varepsilon_{21,1} - \varepsilon_{11,2} \ , \qquad \lambda_2 = \varepsilon_{22,1} - \varepsilon_{12,2} \ , \tag{8a,b}$$

$$\kappa_{21} - \kappa_{12} + \frac{\varepsilon_{12}}{R_{22}} - \frac{\varepsilon_{21}}{R_{11}} + \frac{\varepsilon_{11} - \varepsilon_{22}}{R_{12}} = 0 \ . \tag{8c}$$

with the moment strain components λ_i having a formal rather than physical significance in the present context. Equations (7) and (8) are derived most easily as direct consequences of more general results for nonlinear theory in [6], upon again using (3). Intermediate results which may be said to lead to (7) and (8) begin with variationally consistent strain displacement relations in [2]. A vectorial version of equations which imply (7) and (8) is given in [3], and a component version, with $1/R_{12} = 0$, is derived as a consequence of three-dimensional theory in [5].

Equations (5) to (8) are supplemented by constitutive equations which are here taken in the form

$$\varepsilon_{11} = B(N_{11} - \nu N_{22}) \ , \qquad \varepsilon_{22} = B(N_{22} - \nu N_{11}) \ , \tag{9a,b}$$

$$\varepsilon_{12} = \varepsilon_{21} = \frac{1}{2}\,(1 + \nu)B(N_{12} + N_{21}) \ , \tag{9c}$$

$$M_{11} = D(\kappa_{11} + \nu\kappa_{22}) \ , \qquad M_{22} = D(\kappa_{22} + \nu\kappa_{11}) \ . \tag{10a,b}$$

$$M_{12} = M_{21} = \frac{1}{2}\,(1 - \nu)D(\kappa_{12} + \kappa_{21}) \ . \tag{10a}$$

Conventional Shallow Shell Theory.

We neglect the terms with Q_i in (5a,b), and the terms with M_{ij} in (6c)$^+$. We assume for simplicity's sake, absence of surface load terms p_i and have then that the abbreviated equations

$^+$ For the reduce equations which ensue as a consequence of these stipulations reference may be made to [7].

(5a,b) and (6c) imply validity of the Airy stress function representation

$$N_{11} = K_{,22}, \quad N_{22} = K_{,11}, \quad N_{12} = N_{21} = -K_{,12} \ . \tag{11}$$

Neglect of the Q_i and M_{ij}-terms in (5a,b) and (6c) is equivalent, via appropriate virtual work considerations, to neglect of the λ_i and ε_{ij}-terms in (7a,b) and (8c). The abbreviated equations (7a,b) and (8c) imply validity of the strain displacement relations

$$\kappa_{11} = -w_{,11}, \quad \kappa_{22} = -w_{,22}, \quad \kappa_{12} = \kappa_{21} = -w_{,12} \ , \tag{12}$$

with the minus signs having been introduced for reasons which, in the present context, are of no concern.

Having equations (11) and (12) we have, from (9) and (10), expressions for the ε_{ij} in terms of K and for the $\bar{M}_{ij}$ in terms of w. We next use (6a,b) to express the Q_i in terms of w, and (8a,b) to express the λ_i in terms of K and obtain, for the case that D and B are independent of x_1 and x_2

$$Q_i = -D(\nabla^2 w)_{,i}, \quad \lambda_1 = -B(\nabla^2 K)_{,2}, \quad \lambda_2 = B(\nabla^2 K)_{,1} \ . \tag{13}$$

With equations (13), and with (11) and (12), we finally obtain, from the remaining equilibrium equation (5c) and the remaining compatibility equation (8c), the fourth order differential equations (1a,b) for the determination of w and K.

Corrections to Conventional Shallow Shell Theory.

We now do not neglect the terms with Q_i in (5a,b) and the terms with M_{ij} in (6c). Instead, we assume that these terms may be taken, approximately, in the form in which they appear in the conventional theory. This means that we now consider, in place of the abbreviated equation $N_{11,1} + N_{21,2} = N_{12,1} + N_{22,2} = N_{12} - N_{21} = 0$, equations of the form

$$N_{11,1} + N_{21,2} = D\left[\frac{(\nabla^2 w)_{,1}}{R_{11}} + \frac{(\nabla^2 w)_{,2}}{R_{12}}\right], \qquad (14a)$$

$$N_{12,1} + N_{22,2} = D\left[\frac{(\nabla^2 w)_{,1}}{R_{12}} + \frac{(\nabla^2 w)_{,2}}{R_{22}}\right], \qquad (14b)$$

$$N_{12} - N_{21} = (1-\nu)D\left[\frac{w_{,22} - w_{,11}}{R_{12}} + \frac{w_{,12}}{R_{11}} - \frac{w_{,12}}{R_{22}}\right], \qquad (14c)$$

as point of departure.

From equations (14) there is obtained, in modification of the conventional Airy stress function representation,

$$N_{11} = K_{,22} + D\left[\frac{\nabla^2 w}{R_{11}} - (1-\nu)\left(\frac{w_{,12}}{R_{12}} + \frac{w_{,22}}{R_{22}}\right)\right], \qquad (15a)$$

$$N_{21} = -K_{,21} + D\left[\frac{\nabla^2 w}{R_{12}} + (1-\nu)\left(\frac{w_{,11}}{R_{12}} + \frac{w_{,21}}{R_{22}}\right)\right], \qquad (15b)$$

$$N_{12} = -K_{,12} + D\left[\frac{\nabla^2 w}{R_{12}} + (1-\nu)\left(\frac{w_{,12}}{R_{11}} + \frac{w_{,22}}{R_{21}}\right)\right], \qquad (15c)$$

$$N_{22} = K_{,11} + D\left[\frac{\nabla^2 w}{R_{22}} - (1-\nu)\left(\frac{w_{,11}}{R_{11}} + \frac{w_{,21}}{R_{12}}\right)\right]. \qquad (15d)$$

As regards the form of these equations, we note in particular the fact that the Q_i-terms in (5a,b) as well as the M_{ij}-terms in (6c) result in modifications of the conventional representation (11) for the N_{ij} of one and the same order of magnitude, with the terms coming from the moment equilibrium equation (6c) being distinguished from the terms coming from the force equations (5a,b) by the factor $1 - \nu$ in front of them.

Corresponding to the representations (15) for the N_{ij}, we find from the compatibility equations (7a,b) and (8c) as modifications of the bending strain representations (12),

$$\kappa_{11} = -w_{,11} + B\left[\frac{\nabla^2 K}{R_{22}} - (1+\nu)\left(\frac{K_{,21}}{R_{12}} + \frac{K_{,11}}{R_{11}}\right)\right], \tag{16a}$$

$$\kappa_{21} = -w_{,21} - B\left[\frac{\nabla^2 K}{R_{12}} + (1+\nu)\left(\frac{K_{,22}}{R_{12}} + \frac{K_{,12}}{R_{11}}\right)\right], \tag{16b}$$

$$\kappa_{12} = -w_{,12} - B\left[\frac{\nabla^2 K}{R_{12}} + (1+\nu)\left(\frac{K_{,11}}{R_{12}} + \frac{K_{,21}}{R_{22}}\right)\right], \tag{16c}$$

$$\kappa_{22} = -w_{,22} + B\left[\frac{\nabla^2 K}{R_{11}} - (1+\nu)\left(\frac{K_{,12}}{R_{12}} + \frac{K_{,22}}{R_{22}}\right)\right]. \tag{16d}$$

Having equations (15) and (16) we may now, as before, obtain expressions for the ε_{ij} and M_{ij} in terms of w and K from equations (9) to (10) and then, from (6a,b) and (8a,b), expressions for the Q_i and λ_i. With these we then derive from the remaining force equilibrium equation (5c) and the remaining moment strain compatibility equation (7c) the two differential equations for w and K which take the place of equations (1a,b) of the conventional theory.

In carrying out this procedure we may forego some of the detailed calculations by the following observation. It is readily seen that the expressions for Q_i and λ_i come out to be of the form

$$\frac{Q_1}{D} = -\left[\nabla^2 w + O\left(\frac{B}{R}\,\nabla^2 K\right)\right]_{,1}, \quad \frac{\lambda_1}{B} = -\left[\nabla^2 K + O\left(\frac{D}{R}\,\nabla^2 w\right)\right]_{,2}. \tag{17}$$

etc. Upon substituting (17), in conjunction with (15) and (16), in equations (5c) and (7c) we obtain as differential equations for w and K

$$D\nabla^4 w + O(BDR^{-1}\nabla^4 K) + L(K)$$

$$+ D\left[\left(\frac{1}{R_{11}} + \frac{1}{R_{22}}\right)^2 + (3-\nu)\left(\frac{1}{R_{12}^2} - \frac{1}{R_{11}R_{22}}\right)\right]\nabla^2 w = p\ , \tag{18a}$$

$$B\nabla^4 K + O(BDR^{-1}\nabla^4 w) - L(w)$$

$$+ B\left[\left(\frac{1}{R_{11}} + \frac{1}{R_{22}}\right)^2 + (3+\nu)\left(\frac{1}{R_{12}^2} - \frac{1}{R_{11}R_{22}}\right)\right]\nabla^2 K = 0 \ . \qquad (18b)$$

Having refrained from making explicit the form of the second terms on the left of (18a) and (18b), we begin by showing that the order of magnitude of these second terms is in fact such that they are negligible. For this purpose we take account of the material constant order of magnitude relation $BD = O(h^2)$ +.

With a as the representative distance over which significant changes of K and w occur, this then gives as the order of magnitude of the second terms in (18a,b)

$$BDR^{-1}\nabla^4(K,w) = O[(h^2/a^2)L(K,w)] \ . \qquad (19)$$

Since the negligibility of terms of relative order h^2/a^2 is one of the basic assumptions of shell theory, for shallow as well as for non-shallow shells, we do have in fact that the second terms on the left of (18a,b) are negligible.

It remains to appraise the fourth terms on the left of (18a,b), for which we may note the remarkable way in which the curvature radii R_{ij} appear ++. Evidently, the fourth terms

---

+For the example of a shell with properties uniform in thickness direction we have $B = 1/(Eh)$ and $D = Eh^3/[12(1 - \nu^2)]$ so that $BD = h^2/[12(1 - \nu^2)]$.

++It seems worthy of particular note that had we considered the Q_i-terms in (5a,b) but had omitted the M_{ij}-terms in (6c), with corresponding assumptions for (7a,b) and (8c), the form of these fourth terms would have been the same as in (18a,b) except that instead of the factors 3 - ν there would have been one and the same factor 2 so that in effect, the M-terms in (6c) are as important, or as unimportant, as the Q-terms in (5a,b).

in (18a,b), when compared to the first terms are of relative order a^2/R^2. Accordingly, these fourth terms are certainly negligible in the event that a^2 is small compared to R^2, as is for instance the case when $a^2 = O(h\,R)$.

It remains to consider the possibility that a is of the same order of magnitude as R. For this case fourth and first terms in (18a,b) are of the same order of magnitude. However, it is readily shown that now both these terms are of the order of magnitude h^2/R^2 relative to the third terms, so that again it is appropriate to disregard the fourth terms in (18a,b).

The question may be asked whether it would be rational to use the fourth terms in (18a,b) for the purpose of determining meaningful corrections to results obtained through use of the conventional shallow shell equations. That this is, in general, not so follows from a consideration of the order of magnitude of the geometrical terms $z_{,i}z_{,j}$ which at the outset were neglected compared to unity. To be specific, consider a shallow spherical shell with middle surface equation $z = H - r^2/(2R)$, and with the edge radius a of the shell given by $H - a^2/(2R)=0$. The maximum value of the slope square is now $(z_{,r}(a))^2 = a^2/R^2$, which is just the relative order of magnitude which the fourth terms in (18a,b) may have in one of the two cases considered above. Accordingly, retention of these fourth terms would not lead to meaningful corrections for the results of conventional shallow shell theory.

References

1. Knowles, J.K. and Reissner, E.: A Derivation of the Equations of Shell Theory for General Orthogonal Coordinates, J. Math. and Phy. 35, 351-358, 1956.

2. Reissner, E.: Variational Considerations for Elastic Beams and Shells, Proc. Amer. Soc. Civil Engrg. 8 (EM) 23-57, 1962.

3. Reissner, E.: On the Foundations of the Theory of Elastic Shells, Proc. 11th Intern. Congr. Appl. Mech. (München 1964) pp. 20-30.

4. Reissner, E.: On Oblique Coordinates and Shallowness in Shell Theory, Adv. in Math. 4, 264-276, 1970.

5. Reissner, E.: On Consistent First Approximations in the General Linear Theory of Thin Shells, Ingenieur-Archiv 40, 402-419, 1971.

6. Reissner, E.: Linear and Nonlinear Theory of Shells, Thin Shell Structures (Sechler Anniversary Volume) pp. 29-44, 1974.

7. Reissner, E. and Wan, F.Y.M.: On the Equations of Linear Shallow Shell Theory, Studies Appl. Math. 48, 133-145, 1969.

Geometrically Nonlinear Theory and Incremental Analysis of Thin Shells

E. STEIN, W. WAGNER and K.-H. LAMBERTZ

Universität Hannover, Institut für Baumechanik
und Numerische Mechanik, Germany

1. Introduction

Within the last years, a wide variety of complicated geometrically and physically nonlinear shell problems was approximately solved using mainly the Finite Element Method (FEM). Complex engineering structures need large systems of nonlinear equations for incremental load paths. Own investigations, using the SHEBA-element with 63 DOF's /4/, /8/, have shown that the limit capacity even of large and fast scalar computers (in our case a Cyber 76/73 configuration) may be exceeded in the case of critical and postcritical calculations.

So a certain trend can be observed to use nonlinear elements with rather small numbers of primary kinematic variables (displacements and rotations). Moreover, sophisticated solution algorithms were developed, see BATOZ /21/ and CRISFIELD /22/ e.g., so that reasonable results could be achieved with less computation effort than before. Shear deformations are also included. The 'shear-locking' effect can be avoided by using selected reduced integration, see KANOKNUKULCHAI /18/ e.g. or by using special interpolation functions, see TESSLER /19/ e.g. In the plane DKT-element, see BATHE, HO /14/, special interpolation functions for the slopes of the boundaries fulfill the normal-hypothesis (KIRCHHOFF-LOVE) only in discrete edge points but not in the inner domain of an element. This element behaves rather efficient for geometrically and physically nonlinear folded plate computations as shown by KROG /9/.

Concerning an appropriate bending theory of thin shells, the consistent moderate rotation concept is used, see PIETRASZKIEWICZ /3/, BERG /4/ and /8/. The principle of virtual work is formulated in an updated incremental version. It is the starting point for FE-discretizations with the DKT-element. Some illustrative examples show the quality of the element and the efficiency of the solution strategy.

2. Coordinates of Shell Space

In fig. 1, the coordinates for describing the undeformed state (initial configuration (i.c.)) and for the deformed state (current configuration (c.c.)) are represented.

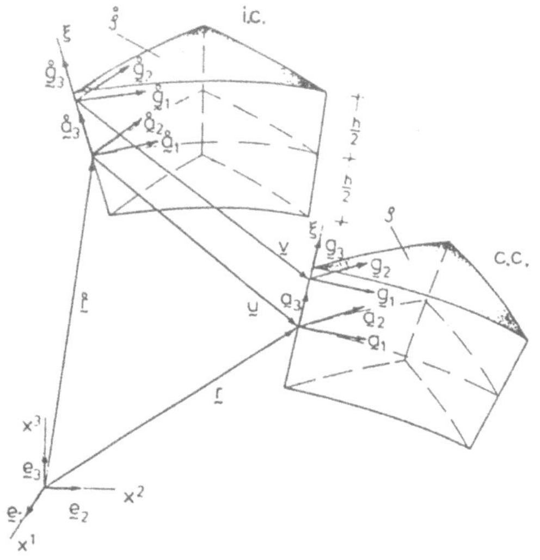

Fig. 2.1 Geometry of a thin shell in the initial configuration (i.c.) and in the current configuration (c.c.)

Quantities in the i.c. are marked by an upper circle o. A point P of shell space in the c.c. is described by

$$\underset{\sim}{p} = \underset{\sim}{r} + \xi \underset{\sim}{a}_3 \quad . \tag{2.1}$$

The tangential vectors $\underset{\sim}{a}_\alpha$ and the unit normal vector $\underset{\sim}{a}_3$ of the middle surface (MS) are

$$\underset{\sim}{a}_\alpha = \underset{\sim}{r}_{,\alpha} \quad , \qquad \underset{\sim}{a}_3 = \frac{\underset{\sim}{a}_1 \times \underset{\sim}{a}_2}{|\underset{\sim}{a}_1 \times \underset{\sim}{a}_2|} \quad . \tag{2.2, 2.3}$$

Then, the tangential vectors and the unit normal vector in shell space are

$$\underset{\sim}{g}_\alpha = \underset{\sim}{a}_\alpha + \xi \underset{\sim}{a}_{3,\alpha} \quad , \qquad \underset{\sim}{g}_3 = \underset{\sim}{a}_3 \quad . \tag{2.4}$$

The curvature tensor is defined by the tangential vectors as

$$\underset{\sim}{B} = b_{\alpha\beta}\, \underset{\sim}{a}^{\alpha} \otimes \underset{\sim}{a}^{\beta} = -(\underset{\sim}{a}_{3,\alpha} \cdot \underset{\sim}{a}_{\beta})\, \underset{\sim}{a}^{\alpha} \otimes \underset{\sim}{a}^{\beta}$$

$$= -\underset{\sim}{a}_{3,\beta} \otimes \underset{\sim}{a}^{\beta} = -\operatorname{grad} \underset{\sim}{a}_3 \quad , \tag{2.5}$$

so that the tangential vectors in shell space can be presented in the form

$$\underset{\sim}{g}_{\alpha} = (\underset{=}{1} - \xi \underset{\sim}{B})\, \underset{\sim}{a}_{\alpha} = \underset{=}{g}\, \underset{\sim}{a}_{\alpha} \quad , \tag{2.6}$$

with the unit tensor $\underset{=}{1} = \underset{\sim}{a}_{\alpha} \otimes \underset{\sim}{a}^{\alpha} + \underset{\sim}{a}_3 \otimes \underset{\sim}{a}_3$, (2.7)

and the shell shifter $\underset{=}{g} = (\underset{=}{1} - \xi \underset{\sim}{B})$.

3. Kinematics in Shell Space

Postulating thin shells with smooth curvatures, we introduce the well-known KIRCHHOFF-LOVE hypothesis.

The material deformation gradient $\underset{\sim}{F}$ in shell space is defined for the introduced convected coordinates as

$$\underset{\sim}{F} = \underset{\sim}{g}_i \otimes \overset{\circ}{\underset{\sim}{g}}{}^i \quad . \tag{3.1}$$

According to PIETRASZKIEWICZ /3/, the material deformation gradient $\underset{\sim}{G}$ of the MS is given by

$$\underset{\sim}{G} = \underset{\sim}{F}(\xi = 0) = \underset{\sim}{a}_i \otimes \overset{\circ}{\underset{\sim}{a}}{}^i = \underset{\sim}{a}_{\alpha} \otimes \overset{\circ}{\underset{\sim}{a}}{}^{\alpha} + \underset{\sim}{a}_3 \otimes \overset{\circ}{\underset{\sim}{a}}_3 \quad . \tag{3.2}$$

So, $\underset{\sim}{F}$ can be represented by $\underset{\sim}{G}$ and the shifter $\underset{=}{g}$ as

$$\underset{\sim}{F} = \underset{=}{g}\, \underset{\sim}{G}\, \overset{\circ}{\underset{=}{g}}{}^{-1} \quad . \tag{3.3}$$

The GREEN strain tensor in shell space is given as

$$\underset{\sim}{E} = \frac{1}{2}(\underset{\sim}{F}^T \underset{\sim}{F} - \overset{\circ}{\underset{=}{1}}) = \frac{1}{2}(g_{ij} - \overset{\circ}{g}_{ij})\, \overset{\circ}{\underset{\sim}{g}}{}^i \otimes \overset{\circ}{\underset{\sim}{g}}{}^j \quad , \tag{3.4, 3.5}$$

and with (3.3)

$$\underset{\sim}{E} = \overset{\circ}{\underset{=}{g}}{}^{T-1} \frac{1}{2}(\underset{\sim}{G}^T \underset{=}{g}^T \underset{=}{g}\, \underset{\sim}{G} - \overset{\circ}{\underset{=}{g}}{}^T \overset{\circ}{\underset{=}{g}})\, \overset{\circ}{\underset{=}{g}}{}^{-1} \quad . \tag{3.6}$$

So, the GREEN strain tensor can be represented in the form, see /3/

$$\underset{\sim}{E} = \overset{\circ}{\underset{=}{g}}{}^{T-1} (\underset{=}{\gamma} + \xi \underset{=}{\varkappa} + \xi^2 \underset{=}{\nu})\, \overset{\circ}{\underset{=}{g}}{}^{-1} \quad , \tag{3.7}$$

with

$$\underset{=}{\gamma} = \frac{1}{2}(\underset{\sim}{G}^T \underset{\sim}{G} - \overset{\circ}{\underset{=}{1}}) , \quad \underset{=}{\varkappa} = -(\underset{\sim}{G}^T \underset{\sim}{B} \underset{\sim}{G} - \overset{\circ}{\underset{\sim}{B}}) , \tag{3.8, 3.9}$$

and

$$\underset{=}{\nu} = \frac{1}{2}(\underset{\sim}{G}^T \underset{\sim}{B}^T \underset{\sim}{B} \underset{\sim}{G} - \overset{\circ}{\underset{\sim}{B}}{}^T \overset{\circ}{\underset{\sim}{B}}) . \tag{3.1o}$$

Now the strains are expressed by the displacements. The displacement vector of points in the middle surface is introduced as

$$\underset{\sim}{u} = u^\alpha \overset{\circ}{\underset{\sim}{a}}_\alpha + u_3 \overset{\circ}{\underset{\sim}{a}}_3 . \tag{3.11}$$

The displacement gradient is defined as (compare equ. 2.5)

$$\underset{\sim}{H} = \text{Grad}\, \underset{\sim}{u} = \underset{\sim}{u}_{,\alpha} \otimes \overset{\circ}{\underset{\sim}{a}}{}^\alpha . \tag{3.12}$$

According to fig. 2.1 it holds

$$\underset{\sim}{u} = \underset{\sim}{r} - \overset{\circ}{\underset{\sim}{r}} , \qquad \underset{\sim}{\beta} = \underset{\sim}{a}_3 - \overset{\circ}{\underset{\sim}{a}}_3 , \tag{3.13, 3.14}$$

so that the displacement of a point in shell space is given by

$$\underset{\sim}{v} = \underset{\sim}{u} + \xi \underset{\sim}{\beta} . \tag{3.15}$$

For equs. (3.8) - (3.1o) follows

$$\underset{=}{\gamma} = \frac{1}{2}(\underset{\sim}{u}_{,\alpha} \cdot \overset{\circ}{\underset{\sim}{a}}_\beta + \underset{\sim}{u}_{,\beta} \cdot \overset{\circ}{\underset{\sim}{a}}_\alpha + \underset{\sim}{u}_{,\alpha} \cdot \underset{\sim}{u}_{,\beta}) \overset{\circ}{\underset{\sim}{a}}{}^\alpha \otimes \overset{\circ}{\underset{\sim}{a}}{}^\beta , \tag{3.16}$$

$$\underset{=}{\varkappa} = (\underset{\sim}{u}_{,\alpha} \cdot \overset{\circ}{\underset{\sim}{a}}_{3,\beta} + \underset{\sim}{\beta}_{,\beta} \cdot \overset{\circ}{\underset{\sim}{a}}_\alpha + \underset{\sim}{u}_{,\alpha} \cdot \underset{\sim}{\beta}_{,\beta}) \overset{\circ}{\underset{\sim}{a}}{}^\alpha \otimes \overset{\circ}{\underset{\sim}{a}}{}^\beta , \tag{3.17}$$

$$\underset{=}{\nu} = \frac{1}{2}(\underset{\sim}{\beta}_{,\alpha} \cdot \overset{\circ}{\underset{\sim}{a}}_{3,\beta} + \underset{\sim}{\beta}_{,\beta} \cdot \overset{\circ}{\underset{\sim}{a}}_{3,\alpha} + \underset{\sim}{\beta}_{,\alpha} \cdot \underset{\sim}{\beta}_{,\beta}) \overset{\circ}{\underset{\sim}{a}}{}^\alpha \otimes \overset{\circ}{\underset{\sim}{a}}{}^\beta . \tag{3.18}$$

With the displacement gradient defined in equ. (3.12) we get the strains of the middle surface in symbolic tensor notation

$$\underset{=}{\gamma} = \frac{1}{2}(\underset{\sim}{H}^T \overset{\circ}{\underset{=}{1}}_2 + \overset{\circ}{\underset{=}{1}}_2 \underset{\sim}{H} + \underset{\sim}{H}^T \underset{\sim}{H}) , \quad \overset{\circ}{\underset{=}{1}}_2 = \overset{\circ}{\underset{\sim}{a}}_\alpha \otimes \overset{\circ}{\underset{\sim}{a}}{}^\alpha , \tag{3.19}$$

and the changes of curvature

$$\underset{=}{\varkappa} = \underset{\sim}{H}^T \text{Grad}\, \overset{\circ}{\underset{\sim}{a}}_3 + \overset{\circ}{\underset{=}{1}}_2 \text{Grad}\, \underset{\sim}{\beta} + \underset{\sim}{H}^T \text{Grad}\, \underset{\sim}{\beta}$$

$$= \underset{\sim}{H}^T \text{Grad}\, \underset{\sim}{a}_3 + \overset{\circ}{\underset{=}{1}}_2 \text{Grad}\, \underset{\sim}{\beta} . \tag{3.2o}$$

Following KOITER /1/ and PIETRASZKIEWICZ /3/, we introduce

$$\Phi^{\beta}_{\cdot\alpha} = \mathring{\underset{\sim}{a}}^{\beta} \cdot \underset{\sim}{u}_{,\alpha} \quad ; \quad \Phi_{\beta\alpha} = \mathring{\underset{\sim}{a}}_{\beta} \cdot \underset{\sim}{u}_{,\alpha} \tag{3.21}$$

$$\Phi_{3\alpha} = \mathring{\underset{\sim}{a}}_{3} \cdot \underset{\sim}{u}_{,\alpha}$$

$$\beta^{\beta}_{\cdot\alpha} = \mathring{\underset{\sim}{a}}^{\beta} \cdot \underset{\sim}{\beta}_{,\alpha} \quad ; \quad \beta_{\beta\alpha} = \mathring{\underset{\sim}{a}}_{\beta} \cdot \underset{\sim}{\beta}_{,\alpha} \tag{3.22}$$

$$\beta_{3\alpha} = \mathring{\underset{\sim}{a}}_{3} \cdot \underset{\sim}{\beta}_{,\alpha} \quad .$$

Introducing the equs. (3.21) and (3.22) into equs. (3.16) and (3.17) finally results in the strain tensors

$$\underset{=}{\gamma} = \frac{1}{2}\left(\Phi_{\beta\alpha} + \Phi_{\alpha\beta} + \Phi^{\lambda}_{\cdot\alpha}\, \Phi_{\lambda\beta} + \Phi_{3\alpha}\, \Phi_{3\beta}\right) \mathring{\underset{\sim}{a}}^{\alpha} \otimes \mathring{\underset{\sim}{a}}^{\beta} \tag{3.23}$$

and

$$\underset{=}{\varkappa} = \left(- \mathring{b}^{\lambda}_{\beta}\, \Phi_{\lambda\alpha} + \beta_{\alpha\beta} + \Phi^{\lambda}_{\cdot\alpha}\, \beta_{\lambda\beta} + \Phi_{3\alpha}\, \beta_{3\beta}\right) \mathring{\underset{\sim}{a}}^{\alpha} \otimes \mathring{\underset{\sim}{a}}^{\beta} . \tag{3.24}$$

These formulas are valid for small strains (the strain energy is a quadratic functional of the strains) without restrictions with respect to the size of rotations. PIETRASZKIEWICZ proposed in /3/ a classification of different levels of consistent nonlinear shell theories in dependence of a rotation vector. Within a 'moderate rotation theory' the approximated strains of the MS follow from (3.23) as

$$\underset{=}{\gamma} = \frac{1}{2}\left(\Phi_{\beta\alpha} + \Phi_{\alpha\beta} + \Phi_{3\alpha}\, \Phi_{3\beta}\right) \mathring{\underset{\sim}{a}}^{\alpha} \otimes \mathring{\underset{\sim}{a}}^{\beta} \tag{3.25}$$

and similar to equ. (3.19)

$$\underset{=}{\gamma} = \frac{1}{2}\left(\underset{\sim}{H}^{T}\, \mathring{\underset{=}{1}}_{2} + \mathring{\underset{=}{1}}_{2}\, \underset{\sim}{H} + \tilde{\underset{\sim}{H}}^{T}\, \tilde{\underset{\sim}{H}} \right) , \tag{3.26}$$

with

$$\tilde{\underset{\sim}{H}} = (\underset{\sim}{u}_{,\alpha})_{n} \otimes \mathring{\underset{\sim}{a}}^{\alpha} \quad , \quad (\underset{\sim}{u}_{,\alpha})_{n} = \Phi_{3\alpha}\, \mathring{\underset{\sim}{a}}_{3} \quad . \qquad \begin{matrix}(3.27a)\\(3.27b)\end{matrix}$$

The difference vector $\underset{\sim}{\beta}$, compare (3.14), can be represented in the linearized form, see NAGHDI /6/

$$\underset{\sim}{\beta} = \left(- \underset{\sim}{u}_{,\alpha} \cdot \mathring{\underset{\sim}{a}}_{3}\right) \mathring{\underset{\sim}{a}}^{\alpha} = - \left(\mathring{\underset{\sim}{a}}^{\alpha} \otimes \underset{\sim}{u}_{,\alpha}\right) \mathring{\underset{\sim}{a}}_{3} \quad . \tag{3.28}$$

Corresponding to equ. (3.21), $\underset{\sim}{\beta}$ can be written as

$$\underset{\sim}{\beta} = - \Phi_{3\alpha} \, \overset{\circ}{\underset{\sim}{a}}{}^{\alpha} . \tag{3.29}$$

So, the normal component of the difference vector $\underset{\sim}{\beta}$ is neglected. The tangential components of the gradient of $\underset{\sim}{\beta}$ are

$$\beta_{\alpha\beta} = -\Phi_{3\alpha|\beta} \quad . \tag{3.30}$$

Using this notation, we get the linear approximation of the change-of-curvature tensor

$$\underset{=}{\varkappa} = - (\Phi_{3\alpha|\beta} + \overset{\circ}{b}{}^{\lambda}_{\beta} \; \Phi_{\lambda\alpha}) \, \overset{\circ}{\underset{\sim}{a}}{}^{\alpha} \otimes \overset{\circ}{\underset{\sim}{a}}{}^{\beta} . \tag{3.31}$$

Then, with displacements the strain tensor of the middle surface gets the final form

$$\begin{aligned} \underset{=}{\gamma} = \frac{1}{2} (& u_{\alpha|\beta} + u_{\beta|\alpha} - 2 \overset{\circ}{b}_{\alpha\beta} u_3 + u_{3,\alpha} u_{3,\beta} + \overset{\circ}{b}{}^{\lambda}_{\beta} u_{\lambda} u_{3,\alpha} \\ & + \overset{\circ}{b}{}^{\lambda}_{\alpha} u_{\lambda} u_{3,\beta} + \overset{\circ}{b}{}^{\lambda}_{\alpha} b^{\mu}_{\beta} u_{\lambda} u_{\mu}) \, \overset{\circ}{\underset{\sim}{a}}{}^{\alpha} \otimes \overset{\circ}{\underset{\sim}{a}}{}^{\beta} , \end{aligned} \tag{3.32}$$

and the change-of-curvature tensor of the middle surface can be represented as

$$\begin{aligned} \underset{=}{\varkappa} = - (& u_{3|\alpha\beta} + \overset{\circ}{b}{}^{\lambda}_{\alpha|\beta} u_{\lambda} + \overset{\circ}{b}{}^{\lambda}_{\alpha} u_{\lambda|\beta} + \overset{\circ}{b}{}^{\lambda}_{\beta} u_{\lambda|\alpha} \\ & - \overset{\circ}{b}{}^{\lambda}_{\beta} \overset{\circ}{b}_{\lambda\alpha} u_3) \, \overset{\circ}{\underset{\sim}{a}}{}^{\alpha} \otimes \overset{\circ}{\underset{\sim}{a}}{}^{\beta} . \end{aligned} \tag{3.33}$$

For plates in bending the strain tensors become

$$\begin{aligned} \underset{=}{\gamma} &= \frac{1}{2} (u_{\alpha|\beta} + u_{\beta|\alpha} + u_{3,\alpha} u_{3,\beta}) \, \overset{\circ}{\underset{\sim}{a}}{}^{\alpha} \otimes \overset{\circ}{\underset{\sim}{a}}{}^{\beta} , \\ \underset{=}{\varkappa} &= - (u_{3|\alpha\beta}) \, \overset{\circ}{\underset{\sim}{a}}{}^{\alpha} \otimes \overset{\circ}{\underset{\sim}{a}}{}^{\beta} . \end{aligned} \tag{3.34}$$

4. Statical equilibrium conditions

In /4/ BERG defines a section force tensor

$$\underset{\sim}{Q} = Q^{\alpha\beta} \, \overset{\circ}{\underset{\sim}{a}}_{\alpha} \otimes \overset{\circ}{\underset{\sim}{a}}_{\beta} + Q^{3\beta} \, \overset{\circ}{\underset{\sim}{a}}_{3} \otimes \overset{\circ}{\underset{\sim}{a}}_{\beta} \tag{4.1}$$

and gets - in conformity with the three dimensional theory - the equilibrium conditions for forces with respect to the undeformed configuration (LAGRANGIAN description)

$$\mathrm{Div}\, \underset{\sim}{G}\underset{\sim}{Q} + \mathring{\rho}\, \underset{\sim}{b} = \underset{\sim}{0} \quad . \tag{4.2}$$

Introducing a 2nd generalized section force tensor

$$\underset{\sim}{M} = M^{\alpha\beta}\, \underset{\sim}{\mathring{a}}_{\alpha} \otimes \underset{\sim}{\mathring{a}}_{\beta} \tag{4.3}$$

the equilibrium condition for forces and moments can be expressed in a single equation, see BERG /4/

$$\mathrm{Div}\, \{ \underset{\sim}{G}\, \underset{\sim}{N} + \mathrm{Grad}\, \underset{\sim}{a}_3 \underset{\sim}{M}^T + (\underset{\sim}{\mathring{a}}_3 \otimes \underset{\sim}{G}^{-1}\, \mathrm{Div}\, \underset{\sim}{G}\, \underset{\sim}{M})\, \underset{=}{\mathring{1}}_2 \} + \mathring{\rho}\, \underset{\sim}{b} = 0 \; . \tag{4.4}$$

In (4.4) the symmetric tangential force tensor $\underset{\sim}{N}$ is defined as

$$\underset{\sim}{N} = N^{\alpha\beta}\, \underset{\sim}{\mathring{a}}_{\alpha} \otimes \underset{\sim}{\mathring{a}}_{\beta} \tag{4.5}$$

with the components

$$N^{\alpha\beta} = Q^{\alpha\beta} + b^{\alpha}_{\lambda}\, M^{\beta\lambda} \quad . \tag{4.6}$$

5. Constitutive equations for thin elastic shells

We restrict the algorithms to homogeneous and isotropic - so-called hyper elastic materials. Then, there exists a unique relation between stresses and strains in every material point. Restricting to small strains, one can introduce a linear dependence of the 2nd PIOLA-KIRCHHOFF stress tensor $\underset{\sim}{S}$ form the GREEN strain tensor $\underset{\sim}{E}$ in symbolic notation

$$\underset{\sim}{S} = \frac{C}{1-\nu^2}\, \underset{\sim}{\mathbb{H}}\, [\underset{\sim}{E}] \; . \tag{5.1}$$

Integration over the shell thickness and estimating the orders of magnitude of all terms within the restriction to small strains leads to the wellknown first approximation

$$\underset{\sim}{N} = \frac{C\, h}{1-\nu^2}\, \underset{\sim}{\mathbb{H}}\, [\underset{=}{\gamma}] \quad , \tag{5.2}$$

$$\underset{\sim}{M} = \frac{C\,h^3}{12(1-\nu^2)}\ \underset{\sim}{\mathbb{H}}\,[\underset{=}{\varkappa}] \quad . \tag{5.3}$$

These material equs. (5.2) and (5.3) are very practicable because of their simple structure. Their derivation was given and discussed by KOITER /11/, NAGHDI /6/, PIETRASZKIEWICZ /3/, BASAR and KRÄTZIG /12/ and others.

6. The principle of virtual work in updated LAGRANGE formulation

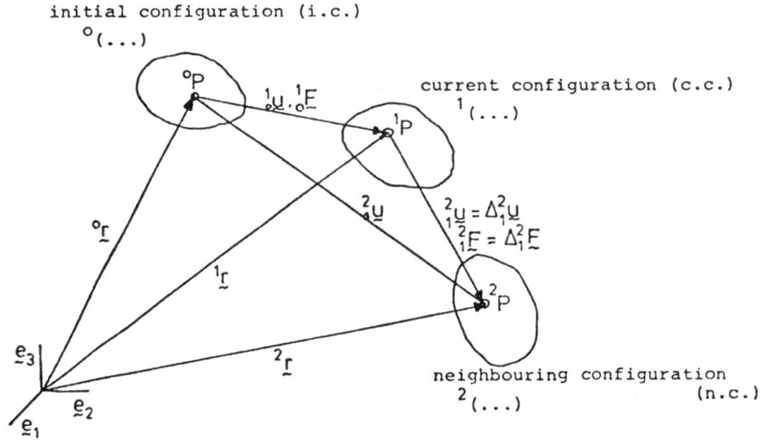

Fig. 6.1 Configurations of a deformable solid, e.g. a shell

In the following an incremental principle of virtual work is given in an updated LAGRANGIAN form as derived by BATHE, RAMM, WILSON /1o/ and KROG /9/.

For the neighboured configuration (n.c.) of a deformable solid continuum this principle can be formulated as

$$\int_{{}^1B} {}^2_1\underset{\sim}{S} \cdot \delta^2_1\underset{\sim}{E}\; d\overset{1}{v} - \int_{\partial^0 B_\sigma} {}^2_0\bar{\underset{\sim}{t}} \cdot \delta^2_1\underset{\sim}{u}\; d\overset{\circ}{v} - \int_{{}^0B} {}^2_0\bar{\underset{\sim}{f}} \cdot \delta^2_1\underset{\sim}{u}\; d\overset{\circ}{a} = 0 \; . \tag{6.1}$$

Postulating conservative loads the virtual work of the given surface forces is calculated with respect to the initial (undeformed) surfaces.

In equ. (6.1) ${}^{2}_{1}\mathbf{S}$ is the 2nd P.-K. stress tensor in the n.c. referred to the metric of the c.c., and ${}^{2}_{1}\mathbf{E}$ is the GREEN strain tensor in the n.c. referred to the c.c..

With the assumption of small desplacement-, strain- and stress increments, an additive decomposition of the 2nd P.-K.-stress tensor in

$$ {}^{2}_{1}\mathbf{S} = {}^{1}_{1}\mathbf{T} + \Delta^{2}_{1}\mathbf{S} \tag{6.2} $$

is possible with

$$ {}^{2}_{1}\mathbf{u} = \Delta^{2}_{1}\mathbf{u} \quad , \qquad {}^{2}_{1}\mathbf{E} = \Delta^{2}_{1}\mathbf{E} \quad . \tag{6.3} $$

The increment of the GREEN strain tensor is also split into a linear and a nonlinear part

$$ \Delta^{2}_{1}\mathbf{E} = \Delta^{2}_{1}\mathbf{E}^{lin} + \Delta^{2}_{1}\mathbf{E}^{nl} \quad . \tag{6.4} $$

Then, the incremental form of the principle of virtual work gets the form

$$ \int_{{}^{1}\mathcal{B}} \Delta^{2}_{1}\mathbf{S} \cdot \delta\Delta^{2}_{1}\mathbf{E}\, d\overset{1}{v} + \int_{{}^{1}\mathcal{B}} {}^{1}_{1}\mathbf{T} \cdot \delta\Delta^{2}_{1}\mathbf{E}^{nl} d\overset{1}{v} = \int_{\partial^{\circ}\mathcal{B}_{\sigma}} {}^{2}_{0}\bar{\mathbf{t}} \cdot \delta\Delta^{2}_{1}\mathbf{u}\, d\overset{\circ}{a} $$

$$ + \int_{{}^{\circ}\mathcal{B}} {}^{2}_{0}\bar{\mathbf{f}} \cdot \delta\Delta^{2}_{1}\mathbf{u}\, d\overset{\circ}{v} - \int_{{}^{1}\mathcal{B}} {}^{1}_{1}\mathbf{T} \cdot \delta\Delta^{2}_{1}\mathbf{E}^{lin} d\overset{1}{v} \quad . \tag{6.5} $$

For shells, we arrive at the following version of the incremental principle of virtual work

$$ \int_{{}^{1}\mathcal{M}} (\Delta^{2}_{1}\mathbf{N} \cdot \delta\Delta^{2}_{1}\boldsymbol{\gamma} + \Delta^{2}_{1}\mathbf{M} \cdot \delta\Delta^{2}_{1}\boldsymbol{\varkappa})\, d\overset{1}{a} + \int_{{}^{1}\mathcal{M}} ({}^{1}_{1}\mathbf{N} \cdot \delta\Delta\, {}^{2}_{1}\boldsymbol{\gamma}^{nl} $$

$$ + {}^{1}_{1}\mathbf{M} \cdot \delta\Delta^{2}_{1}\boldsymbol{\varkappa}^{nl})\, d\overset{1}{a} = \int_{\partial^{\circ}\mathcal{M}_{\sigma}} {}^{2}_{0}\bar{\mathbf{t}} \cdot \delta\Delta^{2}_{1}\mathbf{u}\, d\overset{\circ}{a} + \int_{{}^{\circ}\mathcal{M}} {}^{2}_{0}\bar{\mathbf{f}} \cdot \delta\Delta^{2}_{1}\mathbf{u}\, d\overset{\circ}{a} $$

$$ - \int_{{}^{1}\mathcal{M}} ({}^{1}_{1}\mathbf{N} \cdot \delta\Delta^{2}_{1}\boldsymbol{\gamma}^{lin} + {}^{1}_{1}\mathbf{M} \cdot \delta\Delta^{2}_{1}\boldsymbol{\varkappa}^{lin})\, d\overset{1}{a} \quad . \tag{6.6} $$

The term $\int_{^1\mathcal{M}} {}^1_1M\, \delta\Delta^2_1 \underline{\underline{\varkappa}}^{nl} d\overset{1}{a}$ vanishes in the frame of a moderate-rotation-theory because the 2nd strain tensor contains - as outlined before - only linear parts. Under the presumption of adequate small increments of the loads, the incremental strains are small enough to allow only linear terms of the incremental strain tensor. So, a post-iteration is avoidable for sufficiently small increments and we get the incremental principle in the form

$$\int_{^1\mathcal{M}} (\Delta^2_1\underset{\sim}{N} \cdot \delta^2_1\Delta\underline{\underline{\gamma}}^{lin} + \Delta^2_1\underset{\sim}{M} \cdot \delta^2_1\Delta\underline{\underline{\varkappa}}^{lin})\, d\overset{1}{a} + \int_{^1\mathcal{M}} {}^1_1\underset{\sim}{N} \cdot \delta^2_1\Delta\underline{\underline{\gamma}}^{nl}\, d\overset{1}{a} \tag{6.7}$$

$$= \int_{\partial^\circ\mathcal{M}_\sigma} {}^2_0\underset{\sim}{\bar{t}} \cdot \delta\Delta^2_1\underset{\sim}{u}\, da + \int_{^\circ\mathcal{M}} {}^2_0\underset{\sim}{\bar{f}} \cdot \delta\Delta^2_1\underset{\sim}{u}\, d\overset{\circ}{a} - \int_{^1\mathcal{M}} ({}^1_1\underset{\sim}{N} \cdot \delta\Delta^2_1\underline{\underline{\gamma}}^{lin} + {}^2_1\underset{\sim}{M} \cdot \delta\Delta^2_1\underline{\underline{\varkappa}}^{lin})\, d\overset{1}{a}\,.$$

With equ. (3.25)

$$\begin{aligned} \underline{\underline{\gamma}} &= \left[\frac{1}{2}(\phi_{\beta\alpha} + \phi_{\alpha\beta}) + \frac{1}{2}\phi_{3\alpha}\phi_{3\beta}\right] \overset{\circ}{\underset{\sim}{a}}{}^{\alpha} \otimes \overset{\circ}{\underset{\sim}{a}}{}^{\beta} \\ &= \left[\Theta_{\alpha\beta} + \frac{1}{2}\phi_{3\alpha}\phi_{3\beta}\right] \overset{\circ}{\underset{\sim}{a}}{}^{\alpha} \otimes \overset{\circ}{\underset{\sim}{a}}{}^{\beta} \end{aligned} \tag{6.8}$$

we get

$$\delta\Delta^2_1\underline{\underline{\gamma}} = \left[\delta\Delta^2_1\Theta_{\alpha\beta} + \frac{1}{2}\delta\Delta^2_1\phi_{3\alpha}\Delta^2_1\phi_{3\beta} + \frac{1}{2}\Delta^2_1\phi_{3\alpha}\delta\Delta^2_1\phi_{3\beta}\right] \overset{\circ}{\underset{\sim}{a}}{}^{\alpha} \otimes \overset{\circ}{\underset{\sim}{a}}{}^{\beta}\,. \tag{6.9}$$

Here, the linear part is

$$\delta\Delta^2_1\underline{\underline{\gamma}}^{lin} = \delta\Delta^2_1\Theta_{\alpha\beta}\, \overset{\circ}{\underset{\sim}{a}}{}^{\alpha} \otimes \overset{\circ}{\underset{\sim}{a}}{}^{\beta}\,, \tag{6.1o}$$

and the nonlinear part is

$$\delta\Delta^2_1\underline{\underline{\gamma}}^{nl} = \frac{1}{2}(\delta\Delta^2_1\phi_{3\alpha}\Delta^2_1\phi_{3\beta} + \Delta^2_1\phi_{3\alpha}\delta\Delta^2_1\phi_{3\beta})\, \overset{\circ}{\underset{\sim}{a}}{}^{\alpha} \otimes \overset{\circ}{\underset{\sim}{a}}{}^{\beta}\,. \tag{6.11}$$

Introducing the constitutive equations (5.2) and (5.3) into the work principle equ. (6.7) we get the variational equation for small strains, moderate rotations and small load increments.

$$\int_{{}^1\mathcal{M}} \Big\{ \frac{C\,h}{1-\nu^2} \underset{\sim}{\mathbb{H}} [\Delta_1^2 \underline{\underline{\gamma}}^{lin}] \cdot \delta\Delta_1^2 \underline{\underline{\gamma}}^{lin} + \frac{C\,h^3}{12(1-\nu^2)} \underset{\sim}{\mathbb{H}} [\Delta_1^2 \underline{\underline{\varkappa}}^{lin}] \cdot \delta\Delta_1^2 \underline{\underline{\varkappa}}^{lin} \Big\} d\overset{1}{a}$$

$$+ \int_{{}^1\mathcal{M}} {}_1^1\underset{\sim}{N} \cdot \delta\Delta_1^2 \underline{\underline{\gamma}}^{nl} d\overset{1}{a} = \int_{\partial^\circ\mathcal{M}_\sigma} {}_0^2\bar{\underset{\sim}{t}} \cdot \delta\Delta_1^2 \underset{\sim}{u}\, d\overset{\circ}{a} + \int_{{}^\circ\mathcal{M}} {}_0^2\bar{\underset{\sim}{f}} \cdot \delta\Delta_1^2 \underset{\sim}{u}\, d\overset{\circ}{a}$$

$$- \int_{{}^1\mathcal{M}} ({}_1^1\underset{\sim}{N} \;\; \delta\Delta_1^2 \underline{\underline{\gamma}}^{lin} + {}_1^1\underset{\sim}{M} \;\; \delta\Delta_1^2 \underline{\underline{\varkappa}}^{lin})\, d\overset{1}{a} \quad . \qquad (6.12)$$

7. The DKT-Shell-Element

7.1 General description

The examples, given in this paper, were calculated using the plane triangular DKT-Element (DKT means Discrete KIRCHHOFF Theory). It was investigated by BATHE, HO /14/ and BATOZ, BATHE, HO /13/ especially for plates and has proved to be very efficient so far. As kinematic freedoms of the corresponding plane triangular folded plate element, the 6 primar DOF's are considered in each of the 3 corner points, so that the element has 18 primar kinematic variables.

The stiffness matrix of an element is given as a linear superposition of the following three independent parts:

a) Membrane stiffness matrix $\underline{K}_M$
b) Bending stiffness matrix $\underline{K}_B$
c) Formal rotational stiffness matrix $\underline{K}_{\vartheta z}$ for rotations with respect to the normal direction of the element.

The geometry of the element is shown in fig. 7.1.

7.2 Membrane stiffness matrix

The element membrane stiffness matrix $\underline{K}_M$ is simply the constant strain plane stress stiffness matrix of a three node element /15/, /16/.

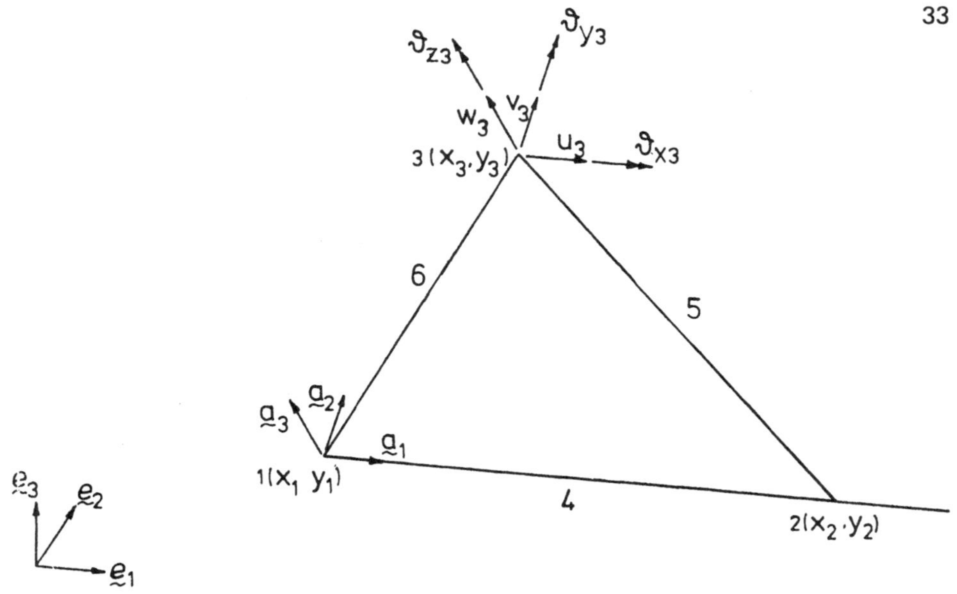

Fig. 7.1 Geometry of the triangular DKT-element

7.3 Bending stiffness matrix

Starting from the plate bending theory with shear deformations, see /13/, quadratic interpolation functions are introduced for the slopes β_x and β_y as

$$\beta_x = \sum_{i=1}^{6} N_i\ \beta_{xi} \quad ; \quad \beta_y = \sum_{i=1}^{6} N_i\ \beta_{yi} \quad , \tag{7.1a,b}$$

with N_i - Quadratic shape functions in homogeneous coordinates /15/.

β_{xi}, β_{yi} - slopes in the corner points 1, 2, 3 and in the midside points 4, 5, 6 at the boundaries.

The vector of the kinematic variables in the three corner points is, see fig. 7.1

$$\underline{\hat{u}}_B = \{ w_1\ \vartheta_{x1}\ \vartheta_{y1}\ w_2\ \vartheta_{x2}\ \vartheta_{y2}\ w_3\ \vartheta_{x3}\ \vartheta_{y3} \} \tag{7.2}$$

In order to get a unique representation for the nodal displacement vector, equ. (7.2), from the interpolation functions, equ. (7.1a,b), the following assumptions are introduced:

- The KIRCHHOFF hypothesis holds
  a) in the corner nodes 1, 2, 3

$$\beta_x + w_{,x} = 0 \quad , \quad \beta_y - w_{,y} = 0 \quad . \tag{7.3a,b}$$

  b) in the midside nodes 4, 5, 6

$$\beta_{SK} + w_{,SK} = 0 \quad . \tag{7.4}$$

- The variation of w along the element boundary is cubic, and so variation of the slope is quadratic. The slope of midside node k between the corner nodes i and j (with the side length l_{ij}) is given by

$$w_{,SK} = -\frac{3}{2l_{ij}}\, w_i - \frac{1}{4}\, w_{,Si} + \frac{3}{2l_{ij}}\, w_S - \frac{1}{4}\, w_{,Sj} . \tag{7.5}$$

- A linear variation of the normal slope is imposed along the boundary (linear twisting)

$$\beta_{nk} = \frac{1}{2}\,(\beta_{ni} + \beta_{nj}) \quad , \tag{7.6}$$

  and therefore the element is fully geometrical compatible.

With respect to these assumptions the following remarks should be made:

a) By the assumption of a cubic polynomial for w along the edges, the relations between the slopes and the deflections according to equs. (7.1)-(7.6) are uniquely determined.
b) The deflection w is not defined in the inner element domain (only at the element edges).
c) As β_s and $w_{,s}$ coincide in 3 points of each element edge and as both have a quadratic course, the KIRCHHOFF hypothesis $w_{,s} + \beta_s = 0$ is fulfilled along the whole edges.
d) The element is fully geometrical compatible, i.e. the deflections w and the slopes β_s and β_n are continuous along the edges.

With the direction cosines of the side-normals, simple transformations between the slopes β_x, β_y and β_n, β_s and correspondingly for $w_{,n}$, $w_{,s}$ and ϑ_x, ϑ_y can be done /13/. So, in total, β_x and β_y are expressed by the nodal displacement vector $\underline{\hat{u}}_B$, equ. (7.2),

$$\beta_y = \underline{H}_y^T(\xi,\eta)\,\underline{\hat{u}}_B \quad , \quad \beta_x = \underline{H}_x^T(\xi,\eta)\,\underline{\hat{u}}_B \,. \tag{7.7a,b}$$

The row vectors $\underline{H}_x$ and $\underline{H}_y$ can be seen as new shape functions (compare equs. (7.1) /13/, /14/. They express the slopes in the element domain by the kinematic variables of the corner points. Choosing an adequate local coordinate system, yields (with equs. (7.7)) the strain-displacement transformation matrix $\underline{B}_B$. Introducing the strain-displacement relations into the principle of virtual work, equ. (6.12), yields the following expressions

$$^1\underline{K}_M = \int_M {}^1\underline{B}_M^T\,\underline{C}_M\,{}^1\underline{B}_M\,d^1A \quad , \quad ^1\underline{K}_B = \int_M {}^1\underline{B}_B^T\,\underline{C}_B\,{}^1\underline{B}_B\,d^1A \,, \tag{7.8a, 7.8b}$$

$$\underline{C}_M = \frac{C}{(1-\nu^2)}\begin{bmatrix} 1 & \nu & 0 \\ & 1 & 0 \\ \text{sym.} & & \frac{1-\nu}{2} \end{bmatrix} \,, \quad C_B = \frac{h^3}{12}\,\underline{C}_M \,. \tag{7.9}$$

The integration over the middle surface is performed numerically using 7 GAUSS-points.

7.4 Rotational stiffness with respect to the plate normal

With the stiffness matrices $\underline{K}_M$ and $\underline{K}_B$, no possibility of computing the rotation ϑ_z around the normal to the middle surface is given. Folded plate systems need the introduction of the kinematic variable ϑ_z.

In order to avoid nonregular systems of equations in the case of completely plane systems, a minimum rotational stiffness is adjoined to the variable ϑ_z; it causes a larger stiffness of the system.

Coinciding with own experiences, the proposed stiffness by BATHE/HO /14/,

$$\underline{K}_{\vartheta z} = 10^{-4} \min K_B \underline{I} \qquad (7.1o)$$

with min K_B : smallest main-diagonal-element of the bending-stiffness-matrix,

bewares the regularity and effects no essential stiffening effect.

7.5 The nonlinear part of the stiffness matrix

According to the chosen linear shape functions the nonlinear terms in the GREEN strain tensor (for the membrane part) are constant within the element, see /14/. The needed "initial" normal forces $\hat{\underline{N}}$ are known from configuration 1. So we get the relation

$$ {}^{2}_{1}\underline{K}_{NL} = \int_M {}^{2}_{1}\underline{B}^T_{NL} \, {}^{1}\hat{\underline{N}} \; {}^{2}_{1}\underline{B}_{NL} \, d^1A \quad . \qquad (7.11)$$

7.6 Correction terms on the right-hand-side

A fictitious load vector ${}^{1}_{o}\underline{t}_R$ is calculated from the section forces ${}^{1}_{1}\underline{N}$ and ${}^{1}_{1}\underline{M}$ as

$$ {}^{1}_{0}\underline{F}_M = \int_M {}^{1}\underline{B}^T_M \, {}^{1}_{1}\underline{N} \; d^1A \;, \qquad {}^{1}_{0}\underline{F}_B = \int_M {}^{1}\underline{B}^T_B \, {}^{1}_{1}\underline{M} \; d^1A \;, \qquad \begin{bmatrix} {}^{1}_{0}\underline{F}_M \\ {}^{1}_{0}\underline{F}_B \end{bmatrix} = {}^{1}_{0}\underline{t}_R \quad . \qquad (7.12a,b,c)$$

The real load vector ${}^{2}_{o}\bar{\underline{t}}$ is split into the part ${}^{1}_{o}\bar{\underline{t}}$ - which was already applied in the state 1 - and the incremental load vector ${}^{2}_{1}\bar{\underline{t}}$. From the difference

$$ {}^{1}_{0}\underline{t}_{uN} = {}^{1}_{0}\bar{\underline{t}} - {}^{1}_{0}\underline{t}_R \quad , \qquad (7.13)$$

we get an error vector of non-equilibrated forces which is applied as a load vector together with the incremental load vector in the next incremental step.

7.7 System of equations

In matrix notation, the work principle equ. (6.12) gets the form (introducing equs. (8.a,b), (7.11) and (7.13))

$$\delta\,{}_1^2\underline{V}^T \; [\, \underline{K}_o + {}_1^2\underline{K}_{NL} \,]\, {}_1^2\underline{V} \; = \; \delta\,{}_1^2\underline{V}^T \; [\, \Delta_1^2\underline{t} + {}_0^1\underline{t}_{UN} \,] \; , \tag{7.14}$$

with the nodal displacement vector

$$\underline{V} = \{\, u_1 \; u_2 \; u_3 \; v_1 \; v_2 \; v_3 \; w_1 \; \vartheta_{x1} \; \vartheta_{y1} \; w_2 \; \vartheta_{x2} \; \vartheta_{y2} \; w_3 \; \vartheta_{x3} \; \vartheta_{y3} \; \vartheta_{z1} \; \vartheta_{z2} \; \vartheta_{z3} \,\} \, . \tag{7.15}$$

7.8 FE-calculation of shells using facet-elements

The load carrying behaviour of a shell does not admit a generally separation of membrane and bending parts.

A sufficiently dense FE-mesh with plane elements which approximates the curvature of the shell as good as possible, leads to a useful and reliable approximation of the whole problem where membrane and bending terms can be calculated uncoupled. There is no upper or lower bound combined with the facet concept. This could be shown for numerous own examples using the DKT-element.

8. Examples

8.1 Cantilever beam loaded by a single moment

This example is commonly used in order to show the quality of a specific element because of the availability of an analytical solution /14/. Own FE-calculations show a rather good agreement with the results of the analytical solution.

8.2 Cylindrical shell with two point loads

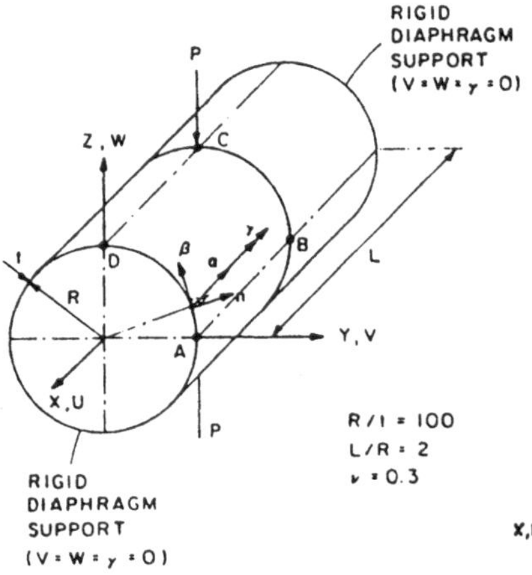

Fig. 8.1 Geometry of cylindrical shell

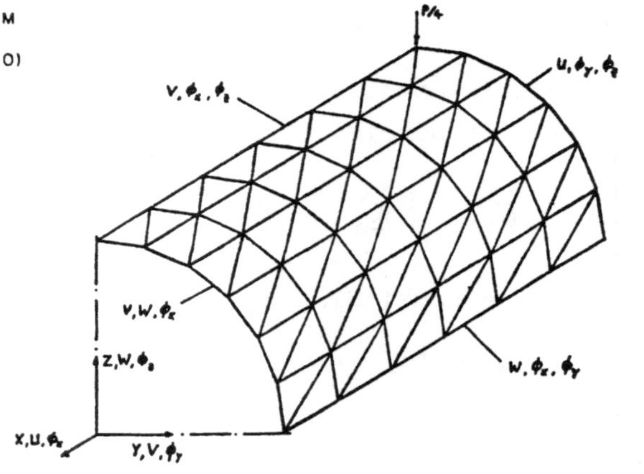

Fig. 8.2 FE-discretization 6x6 mesh

The cylindrical tube shown in fig. 8.1 is loaded by two opposite single forces normal to the surface. At the supports the shell is stiffened by diaphragms. An analytical solution with Fourier series was given by LINDBERG, OLSEN, COWPER /16/.

The FE-calculation was made for a 4x4- and a 6x6-discretization of the domain A B C D, i. e. using the symmetry properties of the system, see fig. 8.2.

Some significant resultants are represented in fig. 8.3a,b.

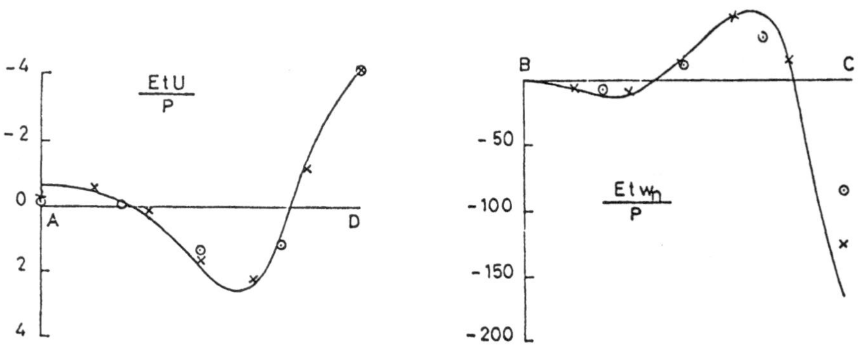

— analytical solution, ⊙ 4x4 mesh, x 6x6 mesh

Fig. 8.3a,b Normalized displacements along axes A-D and B-C

8.3 Clamped Cylindrical shell segment under uniform normal pressure

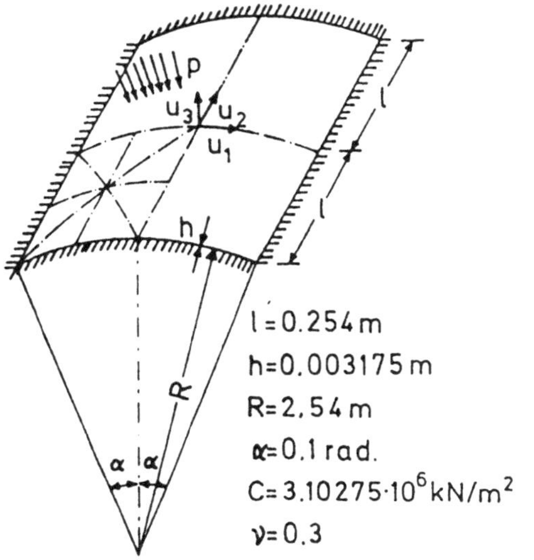

Fig. 8.4 Geometry of cylindrical shell segment

The FE-discretization was performed with 32 elements for a quarter of the shell. The load increments were chosen as o,1 kN/m^2 and o,o5 kN/m^2, respecticely. The results could be compared with former calculations using the 63-DOF-SHEBA-element, see STEIN/BERG/WAGNER /8/ and ARGYRIS /23/.

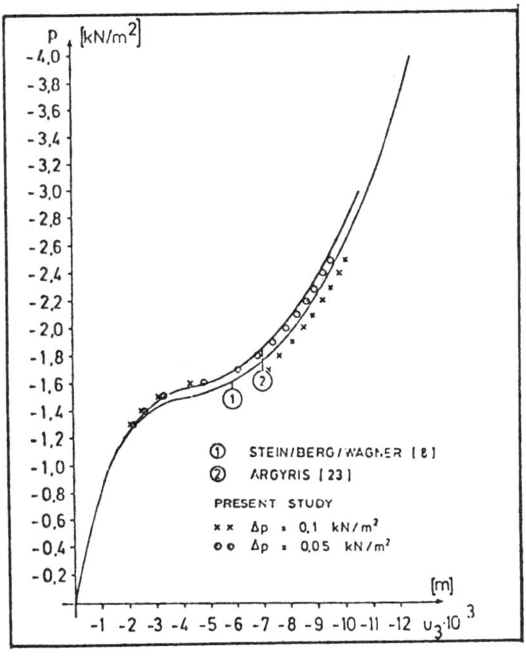

Fig. 8.5

Normal displacement of the midpoint under normal pressure

Fig. 8.5 shows that a rather good agreement could be achieved for sufficiently small load increments using the simple DKT-element with 18 DOF's.

8.4 Hinged spherical shell square segment under a concentrated apex load

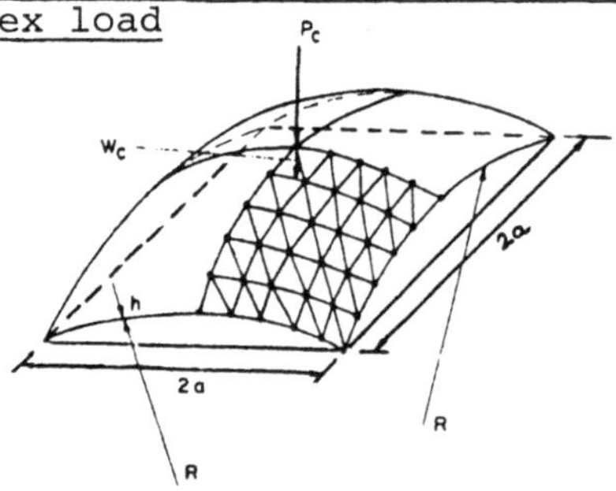

All edges hinged and immovable

E = 1oooo psi
ν = o,3
R = 1oo in
a = 3o,9o17 in
h = 3,9154 in

Fig. 8.6 Spherical shell subjected to apex load

The shell is hinged (all displacements zero) at all edges. The investigation concerns the snapping problem. A discretization for a quarter of the double symmetrical shell was chosen with a 4x4 mesh (32 DKT-elements) and a 5x5 mesh (5o DKT-elements). The solution of the nonlinear system of equations was performed using displacement control.

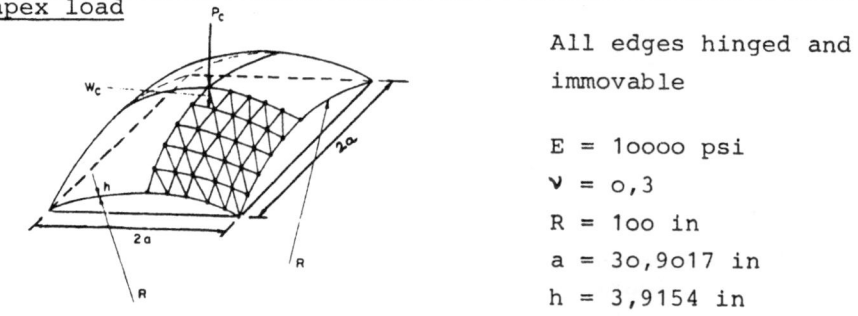

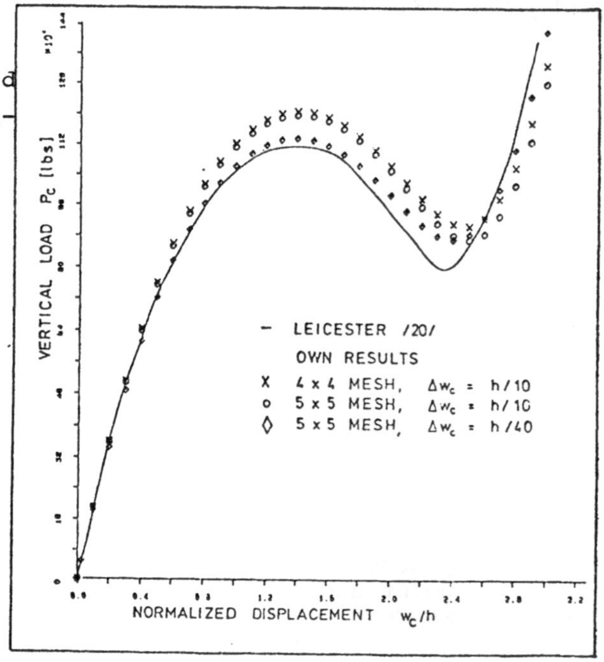

Fig. 8.7 Normal midpoint displacement under concentrated load

Fig. 8.7 shows comparisons of the results with those of LEICESTER /2o/.

It's evident that the plane DKT-element causes a stiffening effect which can be reduced slightly by refining the mesh.

The main reason for disagreements in the higher nonlinear range is to be seen in the fact that no iterations were applied, so that the displacement increments must be rather small.

9. Conclusions

In comparison with our former investigations using the 63-DOF-SHEBA-element, see /8/, the rather simple plane 18-DOF-element, containing the DKT representation for bending, was used for geometrical nonlinear computations.

The second modification to /8/ is the application of an updated Lagrangian representation without iterations for a single load or displacement increment. The errors in the equilibrium within a load step are calculated as unbalanced forces and considered in the next increment. Of course, an iteration is possible by choosing zero load increments.

In total the computation effort for the examples presented in this paper was considerably smaller compared with such needed in /8/.

Concerning the theoretical concept, the entirely symbolic tensor notation already used in /4/ and /8/ has proved to be very transparent in all parts, especially in formulating the incremental principle of virtual work as the starting point for the FEM.

The moderate rotation theory is an adequate concept together with an updated formulation in order to calculate highly nonlinear pre- and postcritical shell problems.

References

/1/ Koiter, W. T.: A consistent first approximation in the general theory of thin elastic shells, Proc. Symp. in the Theory of thin Elastic Shells, Delft, 1959.

/2/ John, F.: Estimates for the derivatives of the stresses in a thin shell and interior shell equations, Comp. Pure and applied Math. 18 (1965) 235-267.

/3/ Pietraszkiewicz, W.: Introducing to the non-linear theory of shells, Mitteilungen aus dem Institut für Mechanik, Nr. 1o, Ruhr-Universität Bochum, 1977.

/4/ Berg, A.: Beiträge zur geometrisch nichtlinearen Theorie und inkrementellen Finite-Element-Berechnung dünner elastischer Schalen, Forschungs- und Seminarberichte aus dem Bereich der Mechanik der Universität Hannover, Nr. 83/2, 1983.

/5/ Pietraszkiewicz, W.; Schmidt, R.: Variational principles in the geometrically non-linear theory of shells undergoing moderate rotations, Ing. Archiv 5o (1981) 187-2o1.

/6/ Nagdhi, P. M.: The theory of shells and plates, Handbuch der Physik, Band VIa/2, Springer Verlag, Berlin/Heidelberg/ New York, 1972.

/7/ Donnell, L. H.: A new theory for the buckling of thin cylinders under axial compression and bending, Trans. ASME 56 (1934) 795-8o6.

/8/ Stein, E.; Berg, A.; Wagner, W.: Different Levels of Nonlinear Shell Theory in Finite Element Stability Analysis, in Buckling of Shells, Proc. of a state-of-the-Art Colloquium, Springer Verlag, Berlin/Heidelberg/New York, 1982.

/9/ Krog, M.: Geometrisch nichtlineare FE-Berechnung von Faltwerken mit plastisch/viskoplastischem Deformationsverhalten, Forschungs- und Seminarberichte aus dem Bereich der Mechanik der Universität Hannover, in press.

/1o/ Bathe, K. J.; Ramm, E.; Wilson, E. L.: Finite element formulation for large deformation dynamic analysis, Int. J. Num. Meth. Engng. 9 (1975) 353-386.

/11/ Koiter, W. T.: On the stability of elastic equilibrium, ch.5, shells with finite deflections, Thesis Delft, H. J. Paris, Amsterdam (1945).

/12/ Basar, Y.; Krätzig, W. B.: Struktur konsistenter Grundgleichungen für das Beul- und Nachbeulverhalten allgemeiner Flächentragwerke, Der Stahlbau 46 (1977) 138-146.

/13/ Batoz, J. L.; Bathe, K. J.; Ho, L. W.: A study of three node triangular plate bending elements, Int. J. Num. Meth. Ingng. 15 (198o) 1771-1812.

/14/ Bathe, K. J.; Ho, L. W.: A simple and effective element for analysis of general shell structures, Comp. and Struct. 13 (1981) 673-681.

/15/ Gallagher, R. H.: Finite Element Analysis, Springer-Verlag, Berlin/Heidelberg/New York, 1976.

/16/ Zienkiewicz, O. C.; Methode der Finiten Elemente, Hanser-Verlag, München 1975.

/17/ Lindberg, G. M.; Olson, M. D.; Cowper, G. R.: New developments in the finite elemente analysis of shells, National Research Council of Canada, Quart. Bull. Div. Mech. Eng. and the National Aeronautical Establishment 4 (1969) 1-38.

/18/ Kanoknukulchai, W.: A simple and efficient finite element for general shell analysis, Int. J. Num. Meth. Engng 4 (1979), 179-2oo.

/19/ Tessler, A.: An efficient, conforming axisymmetric shell element including transverse shear and rotary inertia, Comp. & Struct. 15 (1982) 567-574.

/2o/ Leicester, R. H.: Finite deformations of shallow shells, Proc. ASCE 94 (EMG) (1968) 14o9-1423.

/21/ Batoz, J. L.; Dhatt, G.: Incremental Displacement Algorithms for Nonlinear Problems, Int. J. Num. Meth. Eng. 14 (1979) 1262-1267.

/22/ Crisfield, M. A.: A faster modified Newton-Raphson Iteration, Comp. Meth. in Appl. Mech. and Eng., 2o (1979) 267-278.

/23/ Argyris, J. H.: Entwicklung von Finiten Element Methoden für die Berechnung von Schalen allgemeiner Form, Vortrag anläßlich des Schlußkolloquiums des DFG-Schwerpunktes "Flächentragwerke im konstruktiven Ingenieurbau", 198o.

Flexible Shells

E.L. AXELRAD

Institut für Mechanik
HSBw München, FB LRT - DFG
Fèderal Republic of Germany

Summary

The nonlinear intrinsic equations of elastic thin shells are simplified by the assumption widely tested in the linear theory as the basis of the Novozhilov's complex equations. The obtained equations directly specialize themselves for the class of shells designed for large elastic displacements. Another aim of this contribution is to further the vector form of non-linear shell-theory. This form combines the graphic static-geometric duality with the short-cut way between the invariant and physical-component presentation.

1. Introduction

Shell structures serving to realize large elastic displacements by small strain (Fig.1) have been recently recognized to constitute a distinct class - *flexible shells* (FS). It is impossible to review here the efforts invested in the analysis of FS during several decades. Consider the main stages and scope of this development. Further references may be found in [14, 18].

All the FS problems treated before 1965 are in fact one-dimensional. Following the Kármán [1](1911) work on tube flexure, the boundary conditions were relaxed in the sense of St.-Venant. A similar approach may be perceived also in the H. Reissner (1912) concept of deformation of shells of revolution. Brazier (1927) extended the one-dimensional approach to non-linear tube flexure, Beskin [2](1945) - to non-axisymmetric linear tube problems.

E. Reissner [3,4](1949, 1950) introduced into this analysis

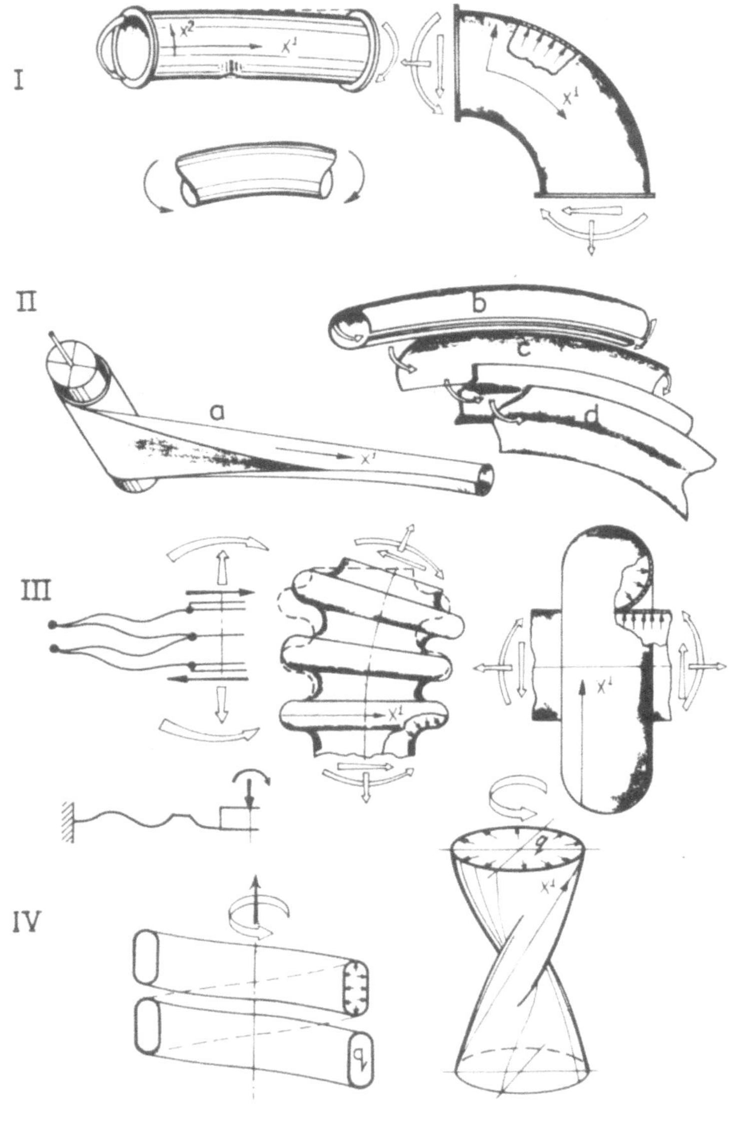

Fig.1. Four types of flexible shells

shell-theory methods. The H.Reissner-E.Meissner equations were extended to include: 1) the axisymmetric linear problem of curved-tube flexure and 2) the nonlinear deformation of shells of revolution.

The two ideas were combined in [6](1960) to derive equations encompassing nonlinear flexure of thin-walled tubes and open-sections beams with or without initial curvature. (Similar equations were independently obtained in [22], pp. 342, 343).

Thus, the one-dimensional FS problem could in principle be solved by 1964 (cf. [7,8] and further references in [18,19]).

A *two-dimensional* FS problem was first (as far as the present author is aware) solved in 1965 [9]. It was the nonlinear bending of a finite-length tube (Fig.1). The analysis was based on a specialized theory: The Wlassow's applied theory of cylinder shells was extended to double-curvature shells and to geometric nonlinearity. The later development of this FS theory is discussed in [14,18].

The reader has hardly failed to recall that in most of the shell-theory books there is not much said on the shell types being discussed. Indeed, the main concern have been shells designed for *strength*. These shells have to be *stiff*, flexibility is here a negative quality.

But a shift of emphasis is already recognizable. The new monograph of Libai and Simmonds [22] considers problems which are in fact those of *flexible* shells - it leaves out "The most widespread applications of elastic nonlinear shell theory... the prediction of stability and the analysis of postbuckling". These applications are covered by the Donnell-type quasi-shallow-shell theory generalized by Koiter [10](1966).

The Donnell-type theory effectively serves these purposes also for flexibile shells. But the main stress-state of FS is as far as possible from that of a stiff shell. It occupies the place between the domains of the two classical theories: The membrane stress states are *slowly* varying with respect to *both* surface coordinates [10]. The Donnell-type theory is applicable for stress states varying *intensively* with *both* coordinates. The main stress state of a FS is much less intensively variable with respect to *one* of the surface coordinates (x^1 in Fig.1).

The FS theory, arrived at by simplifying the general shell-theory according to the just formulated property [18,20], describes a membrane state in sections x^1 = const and retains all stress and strain resultants in sections x^2 = const. It reflects a *semi-membrane* or semi-momentless state *). The FS theory excludes the intensively varying with respect to x^1 component of the stressed state. In particular it is disentangled of the edge-effects at x^1 = const (which can be superimposed). This facilitates the numeric integration with the aid of shooting method.

But can a specialized FS theory be really useful, is it not becoming superfluous in the face of the general shell theory and the numeric procedures available? So far it was the case. Most FS problems have been investigated through the aid of a specialized theory (axisymmetric or two-dimensional). Moreover a connection between the general tensor-form theory and equations instrumental in solving FS problems has been established [17,22] only for the one-dimensional cases.

Naturally, the situation is changing. The computer allows to program almost any complicated problem. But this is not synonimous with obtaining a meaningful solution. The progress is accompanied by "computational empiricism", and it *increases* the need for theoretical analysis of problems [15].

---

*) An ideal limit-case of the semi-membrane deformation presents the plane strain of infinite cylindrical shells investigated by Koiter [17] and Libai, Simmonds [22] as an application of the nonlinear theory. It concerns, in fact, a perfect flexible-shell.

Most of the FS problems are far from being solved, however approximately. This determines the level of rigour and systematics aimed at in the following: the price in complications to be paid for going beyond the Kirchhoff-Love hypothesis and small strain is considered too high.

The reader's indulgence is asked for in connection with the (unusual in the recent time) path of reasoning. It starts with the simplest lines-of-curvature coordinates appropriate for most cases and stops by indicating the further step leading to invariant-form equations.

Only intrinsic equations are considered. Their advantages for nonlinear deformation are undeniable. The kinematic boundary conditions can also be optimally expressed through the intrinsic variables. As the actual shape is effectively computed for each loading stage, the differentiation with respect to the actual shape is preferred. This simplifies the presentation and displays a far-reaching static-geometric analogy of *nonlinear* theory; the calculations are clearer programmable (cf. [20]). If need be, a change to the use of undeformed metric is very direct.

2. Geometrical Relations

Consider geometry of a surface and its deformation in a vector form which assures the clarity of meaning of parameters and a simple structure of all relations.

Let $\mathbf{r}(x^\alpha), \alpha = 1,2$ denote the radius vector of a point of surface, x^α being the Gaussian coordinates of the point. The base vectors $\mathbf{a}_\alpha$, $\mathbf{t}_\alpha$, $\mathbf{n}$, $\mathbf{a}^\alpha$, $\mathbf{t}^\alpha$ and the metric tensor $a_{\alpha\beta}$, are defined by:

$$\mathbf{a}_\alpha = \mathbf{r}_{,\alpha} , \quad (\ )_{,\alpha} = \partial(\ )/\partial x^\alpha , \quad a_{\alpha\beta} = \mathbf{a}_\alpha \cdot \mathbf{a}_\beta ,$$

$$a_\alpha = \sqrt{a_{\alpha\alpha}} , \quad \mathbf{t}_\alpha = \mathbf{a}_\alpha / a_\alpha , \quad \mathbf{n} = \mathbf{a}_1 x \mathbf{a}_2 / \sqrt{a} ,$$

$$a = |\mathbf{a}_1 x \mathbf{a}_2|^2$$

$$a_{\alpha\beta}a^{\beta\lambda} = \delta_\alpha^\lambda \ (\delta_\alpha^\alpha = 1,\ \delta_2^1 = \delta_1^2 = 0),\ \mathbf{a}^\alpha = a^{\alpha\beta}\mathbf{a}_\beta,\ a^\alpha = \sqrt{a^{\alpha\alpha}}$$

$$\mathbf{t}^\beta = e^{\alpha\beta}\mathbf{n}\mathrm{x}\mathbf{t}_\alpha,\ \mathbf{n}\mathrm{x}\mathbf{t}^\alpha = e^{\alpha\beta}\mathbf{t}_\beta,\ (e^{\alpha\alpha} = 0,\ e^{12} = -e^{21} = 1) \tag{1}$$

The summation convention is implied here.

The characteristics of the deformed surface will be denoted by an asterisk: $\mathbf{r}^*$, $\mathbf{t}_\alpha^*$, $\cos\varphi^* = \mathbf{t}_1^*\cdot\mathbf{t}_2^*$ etc.

The curvature of the surface will be measured by the *curvature vectors* $\mathbf{k}_\alpha$ - the angular velocities of rotation of the tangent plane sliding along the coordinate lines x^α. The current position of the plane is identified by a unit vetor $\mathbf{n}^*$ normal to the surface. To measure its rotation around $\mathbf{n}^*$, the plane must be connected with a linear element of the surface. (The plane cannot be attached to more than one such element, for instance to $\mathbf{t}_\alpha a_\alpha dx^\alpha$, as the shear deformation changes the angle between them.) We choose for this an element $\mathbf{t}^*ds$, $\mathbf{t}^* = (\mathbf{t}_1^* + \mathbf{t}_2^*)/|\mathbf{t}_1^* + \mathbf{t}_2^*|$ bisecting the angle φ^* between the lines x^α.

The meaning of the curvature vectors becomes more lucid with their component representation. The following two presentations are identical to each other in the case of undeformed surface when $\mathbf{k}_\alpha^* = \mathbf{k}_\alpha$, $\mathbf{t}_\alpha^* = \mathbf{t}'_\alpha = \mathbf{t}_\alpha$, ...

$$\frac{\mathbf{k}_\alpha^*}{a_\alpha^*} = \frac{\mathbf{n}^*\mathrm{x}\mathbf{t}_\alpha^*}{R_\alpha^*} + \frac{\mathbf{n}^*\mathrm{x}\mathbf{t}^{*\beta}}{R_{\alpha\beta}^*} + \frac{\mathbf{n}^*}{\rho_\alpha^*}\ ,$$

$$\frac{\mathbf{k}_\alpha^*}{a_\alpha} = \frac{\mathbf{n}\mathrm{x}\mathbf{t}'_\alpha}{R'_\alpha} + \frac{\mathbf{n}\mathrm{x}\mathbf{t}'^\beta}{R'_{\alpha\beta}} + \frac{\mathbf{n}^*}{\rho'_\alpha} \quad (\beta \neq \alpha)\ . \tag{2}$$

The unit vetors $\mathbf{t}'_\alpha$ are defined as indicating the position attained by the vectors $\mathbf{t}_\alpha$ which, when rotating during the course of the shell deformation, are rigidly connected to the tangent plane, i.e. to the vectors $\mathbf{n}^*$ and $\mathbf{t}^*$. The difference between $\mathbf{t}'_\alpha$ and $\mathbf{t}_\alpha^*$ is determined solely by the shear angle $\gamma = \varphi - \varphi^*$. The vectors $\mathbf{t}'_\alpha$ remain orthogonal when the coordinate lines x^α are orthogonal before the deformation. (The use of such an orthogonal auxiliary basis was suggested by Simmonds and Danielson [11]).

The expressions of $\mathbf{t}'_\alpha$ for small strain ($|\gamma| \ll 1$) are:

$$\mathbf{t}'_\alpha = \mathbf{t}^*_\alpha - \frac{\gamma}{2}\,\mathbf{t}^{*\beta} \qquad (\beta \neq \alpha) \ . \tag{3}$$

The components of the curvature vectors as introduced by (2) are easily recognized to be the normal-section curvatures ($1/R_\alpha$), the twist ($1/R_{\alpha\beta}$) and the in-plane curvatures of the x^α-lines ($1/\rho_\alpha$).

The parameters $\mathbf{k}^*_\alpha$ render derivation formulas for all the basis vectors introduced. Consider an auxiliary unit vector $\mathbf{v}^*(x^\alpha)$ having at all points of the surface the same position with respect to $\mathbf{n}^*(x^\alpha)$ and $\mathbf{t}^*(x^\alpha)$.

The definition of $\mathbf{k}^*_\alpha$ means that $\mathbf{k}^*_\alpha dx^\alpha$ is the angle between $\mathbf{v}^*(x^\alpha)$ and the vector $\mathbf{v}^*(x^\alpha + dx^\alpha)$ at the adjacent point. This amounts to

$$\mathbf{v}^*(x^\alpha + dx^\alpha) = \mathbf{v}^*(x^\alpha) + \mathbf{k}^*_\alpha\, dx^\alpha \times \mathbf{v}^*(x^\alpha) \ .$$

Hence the derivatives of a unit vector $\mathbf{v}^*$ and its particular cases $\mathbf{n}^*$, $\mathbf{t}^*$ are:

$$\mathbf{v}^*_{,\alpha} = \mathbf{k}^*_\alpha \mathrm{x}\mathbf{v}^* \ , \quad \mathbf{n}^*_{,\alpha} = \mathbf{k}^*_\alpha \mathrm{x}\mathbf{n}^* \ , \quad \mathbf{t}^*_{,\alpha} = \mathbf{k}^*_\alpha \times \mathbf{t}^* \ . \tag{4}$$

The position of the vectors $\mathbf{t}'_\alpha$ and $\mathbf{t}^*_\alpha$ with respect to $\mathbf{t}^*$ varies inside the tangent plane with the angles $\varphi/2$ and $\varphi^*/2$, respectively. Taking account of this variation the derivation formulas for $\mathbf{t}'_\alpha$ and $\mathbf{t}^*_\alpha$ contain, compared to (4), additional terms:

$$\mathbf{t}'_{\alpha,\beta} = [\mathbf{k}^*_\beta + \tfrac{1}{2}(-1)^\alpha \varphi_{,\beta}\mathbf{n}^*]\ \mathrm{x}\mathbf{t}'_\alpha \ . \tag{5}$$

Formulas (4) lead to a compatibility equation between the two vector parameters $\mathbf{k}_\alpha^*$. Applying (4) to the equation $\mathbf{v}^*_{,21} = \mathbf{v}^*_{,12}$ and using the vector triple-product-expansion formula results in

$$\mathbf{k}^*_{1,2} - \mathbf{k}^*_{2,1} + \mathbf{k}^*_1 \times \mathbf{k}^*_2 = \mathbf{0} \quad . \tag{6}$$

This is equivalent to the three scalar Gauss-Codazzi equations.

With the derivation formula (5) the equation $\mathbf{r}_{,12} = \mathbf{r}_{,21}$ renders expressions for the geodetic curvatures $1/\rho_\alpha$. For $\varphi = \pi/2$:

$$a_1 a_2/\rho_1 = -a_{1,2} \ , \quad a_1 a_2/\rho_2 = a_{2,1} \quad . \tag{7}$$

3. Strain parameters, Compatibility equations

To describe the deformation of the surface, we introduce vector parameters $\boldsymbol{\chi}_\alpha$, $\boldsymbol{\varepsilon}_\alpha$ reflecting the change of the local-geometry characteristics from $\mathbf{k}_\alpha$, $\mathbf{a}_\alpha$ to $\mathbf{k}_\alpha^*$, $\mathbf{a}_\alpha^*$ in the simplest way possible:

$$\mathbf{k}_\alpha^* = \mathbf{k}_{\alpha R} + a_\alpha \boldsymbol{\chi}_\alpha , \quad \mathbf{a}_\alpha^* = \mathbf{a}_{\alpha R} + a_\alpha \boldsymbol{\varepsilon}_\alpha \quad . \tag{8}$$

The vectors $\mathbf{k}_{\alpha R}$, $\mathbf{a}_{\alpha R}$ must be defined in a way assuring the strain parameters $\boldsymbol{\chi}_\alpha$, $\boldsymbol{\varepsilon}_\alpha$ to be equal to zero at a point where the surface is not deformed. At such a point $\mathbf{k}_\alpha^* = \mathbf{k}_{\alpha R}$, $\mathbf{a}_\alpha^* = \mathbf{a}_{\alpha R}$ define the *initial* local shape - the same as $\mathbf{k}_\alpha$, $\mathbf{a}_\alpha$. Compared to $\mathbf{k}_\alpha$, $\mathbf{a}_\alpha$ the $\mathbf{k}_{\alpha R}$, $\mathbf{a}_{\alpha R}$, merely take into account the rotation of a locality of the surface. The components of $\mathbf{k}_{\alpha R}$, $\mathbf{a}_{\alpha R}$ in the "rotated" basis $\mathbf{n}^*$, $\mathbf{t}'_\alpha$ are identic to the components of $\mathbf{k}_\alpha$, $\mathbf{a}_\alpha$ with respect to $\mathbf{n}$, $\mathbf{t}_\alpha$:

$$\frac{\mathbf{k}_{\alpha R}}{a_\alpha} = \frac{\mathbf{n}^* \times \mathbf{t}'_\alpha}{R_\alpha} + \frac{\mathbf{n}^* \times \mathbf{t}'^{\beta}}{R_{\alpha\beta}} + \frac{\mathbf{n}^*}{\rho_\alpha} ,$$
$$\mathbf{a}_{\alpha R} = a_\alpha \mathbf{t}'_\alpha \quad (\beta \neq \alpha) \tag{9}$$

To derive compatibility equations between $\boldsymbol{\chi}_\alpha$, $\boldsymbol{\varepsilon}_\alpha$ consider the deformation of the surface at an arbitrary point $M(x^\alpha)$.

Without loss of generality, the analysis of strain may be simplified by taking as a reference for the displacement the tangent plane at the point M. This makes the rotated basis at the point M identical to the initial basis: $\mathbf{t}'_\alpha(M)$, $\mathbf{n}^*(M) = \mathbf{n}(M)$. Hence the *rotated* local-shape parameters, as defined in (9), remain at M equal to the *initial* ones:

$$\mathbf{k}_{\alpha R}(M) = \mathbf{k}_\alpha(M) \quad , \quad \mathbf{a}_{\alpha R} = \mathbf{a}_\alpha \ . \tag{10}$$

The application of the formulas (4), (5) with $\mathbf{k}^*_\beta$ according to (8), (10) results in

$$\begin{aligned} \mathbf{k}_{\alpha R,\beta} &= \mathbf{k}_{\alpha,\beta} + a_\beta \boldsymbol{\chi}_\beta \times \mathbf{k}_\alpha \ , \\ \mathbf{a}_{\alpha R,\beta} &= \mathbf{a}_{\alpha,\beta} + a_\beta \boldsymbol{\chi}_\beta \times \mathbf{a}_\alpha \ . \end{aligned} \tag{11}$$

We introduce the expressions (8), into the eqn. (6) and into $\mathbf{r}^*_{,12} = \mathbf{r}^*_{,21}$. Taking into account the derivation formulas (11), the relations (10), and finally the eqns (6) for the undeformed state and $\mathbf{r}_{,12} = \mathbf{r}_{,21}$ yields the compatibility equations [20] (cf./16/):

$$\begin{aligned} (a_2\boldsymbol{\chi}_2)_{,1} - (a_1\boldsymbol{\chi}_1)_{,2} &= a_1 a_2 \boldsymbol{\chi}_1 \times \boldsymbol{\chi}_2 \\ (a_2\boldsymbol{\varepsilon}_2)_{,1} - (a_1\boldsymbol{\varepsilon}_1)_{,2} + a_1 a_2 (\mathbf{t}'_1 \times \boldsymbol{\chi}_2 - \mathbf{t}'_2 \times \boldsymbol{\chi}_1) &= \mathbf{0} \end{aligned} \tag{12}$$

Having obtained the equations for an arbitrary point M, we have then changed the reference for displacements. This means a rigid-body rotation of the entire surface, which brings the vectors $\mathbf{t}_\alpha(M)$, $\mathbf{n}(M)$ into some positions $\mathbf{t}'_\alpha(M)$, $\mathbf{n}^*(M)$, giving the equations the general form (12).

The deformation parameters are most conveniently expressed in the rotated basis

$$\boldsymbol{\chi}_\alpha = \boldsymbol{\varkappa}_\alpha + \lambda_\alpha \mathbf{n}^* \ , \quad \boldsymbol{\varkappa}_\alpha = \mathbf{n}^* \times (\varkappa_\alpha \mathbf{t}'_\alpha + \tau_\alpha \mathbf{t}'^{\beta}) \ ,$$

$$\boldsymbol{\varepsilon}_\alpha = \varepsilon_\alpha \mathbf{t}'_\alpha + \frac{\gamma}{2} \mathbf{t}'^{\beta} \qquad (\beta \neq \alpha) \ . \tag{13}$$

The geometric meaning of the scalar parameters ε_α, γ, $\varkappa_\alpha$, τ_α, λ_α follows from the relations (8), (9) and (13). It does not differ substantially from the classic definition of the strain measures [12, 22] if the strain is small. In particular, ε_α is the relative extension along the x^α-line and

$$\frac{1}{R'_{\alpha\beta}} = \frac{1}{R_{\alpha\beta}} + \tau_\alpha \ , \quad \frac{1}{R'_\alpha} = \frac{1}{R_\alpha} + \varkappa_\alpha \ , \quad \frac{1}{\rho'_\alpha} = \frac{1}{\rho_\alpha} + \lambda_\alpha \ . \tag{14}$$

These parameters define through (2), (4), (5) the components of the derivatives and thus the six scalar equations following from (12). The numeric solutions with increment loading are well adapted to using at all stages parameters of the *deformed* structure, computed earlier.

For orthogonal coordinates the scalar equations equivalent to (12) coinside with the equations of E. Reissner's work [13] if the vectors $\boldsymbol{\varepsilon}_\alpha$, $\mathbf{M}_\alpha$ include the transverse components and if the position of the basis $\mathbf{t}'_\alpha$ in the tangential plane remains unspecified.

The metric and the second fundamental tensor of the surface can, of course, be expressed in terms of the vector parameters introduced in the foregoing. According to (1), (2), (4), (8), (9):

$$a^*_{\alpha\delta} = \mathbf{a}^*_\alpha \cdot \mathbf{a}^*_\delta \ , \ b^*_{\alpha\delta} = - \mathbf{n}^*_{,\alpha} \cdot \mathbf{a}_\delta \ , \ \mathbf{a}^*_\alpha = a_\alpha (\mathbf{t}^*_\alpha + \boldsymbol{\varepsilon}_\alpha) \ ,$$

$$b_{\alpha\alpha} = - \frac{a_{\alpha\alpha}}{R_\alpha} \ , \ b^\beta_\alpha = - \mathbf{n}_{,\alpha} \cdot \mathbf{a}^\beta = - \frac{a_\alpha a^\beta}{R_{\alpha\beta}} \qquad (\beta \neq \alpha) \ . \tag{15}$$

4. Equilibrium. Duality of Nonlinear Equations.

The well-known [13,23] vector equilibrium equations of an element of a deformed shell are

$$(a_2\mathbf{T}_1)_{,1} + (a_1\mathbf{T}_2)_{,2} = -\mathbf{q}\sqrt{a} \; ,$$
$$(a_2\mathbf{M}_1)_{,1} + (a_1\mathbf{M}_2)_{,2} + a_2 a_1^* \mathbf{t}_1^* \times \mathbf{T}_1 + a_1 a_2^* \mathbf{t}_2^* \times \mathbf{T}_2 = \mathbf{0} \; ; \tag{16}$$

where $\mathbf{T}_\alpha$, $\mathbf{M}_\alpha$ denote the stress resultants and couples related to the length of the normal section x^α = const before deformation. For virtually all cases of small strain, when $1 + |\varepsilon_\alpha| \ll 1$, the eqs. (16) may be simplified by assuming

$$a_\alpha^* \mathbf{t}_\alpha^* \times \mathbf{T}_\alpha = a_\alpha \mathbf{t}_\alpha' \times \mathbf{T}_\alpha \quad , \tag{17}$$

(or if the strains are not small, by transferring the terms $a_\beta(a_\alpha^*\mathbf{t}_\alpha^* - a_\alpha\mathbf{t}_\alpha') \times \mathbf{T}_\alpha$ to the right-hand side of the equations). The so simplified left-hand sides of the eqs. (16) are statically-geometrically dual to those of (12). The duality has been very useful in the linear theory (cf. references in [23]). One of its important products is the Novozhilov's simplification of the linear shell equations in complex form. Consider extension of this simplification to the nonlinear theory.

5. Complex Transformation.

We decompose now the vectors $\mathbf{T}_\alpha$, introducing notation for the normal, shear and transverse components of $\mathbf{T}_\alpha$ and for two components of $\mathbf{M}_\alpha$ (in a way analogous to (13)):

$$\mathbf{T}_\alpha = \mathbf{N}_\alpha + Q_\alpha \mathbf{n}^* \; , \quad \mathbf{N}_\alpha = N_\alpha \mathbf{t}'^\alpha + S_\alpha \mathbf{t}_\beta' \quad ,$$
$$\mathbf{M}_\alpha = \mathbf{n}^* \times (M_\alpha \mathbf{t}'^\alpha + H_\alpha \mathbf{t}_\beta') \quad . \tag{18}$$

We define *complex* vectorial and scalar parameters designating them with wavy-line supercript:

$$\tilde{\mathbf{N}}_\alpha = \mathbf{N}_\alpha + e_{\alpha\beta} iC\boldsymbol{\varkappa}_\beta , \quad \tilde{Q}_\alpha = Q_\alpha + iCe_{\alpha\beta}\lambda_\beta ,$$
$$\tilde{\mathbf{M}}_\alpha = M_\alpha + e_{\alpha\beta} iC\boldsymbol{\varepsilon}_\beta , \quad C = \sqrt{EhD} , \quad D = Eh^3/12(1 - \nu^2), \tag{19}$$
$$i = \sqrt{-1} .$$

The "complex stress resultants" and "moments" have comlex components. These are defined by (13), (18), (19):

$$\tilde{\mathbf{N}}_\alpha = \tilde{N}_\alpha \mathbf{t}'^\alpha + \tilde{S}_\alpha \mathbf{t}'_\beta , \quad \tilde{N}_\alpha = N_\alpha - iC\varkappa_\beta , \quad \tilde{S}_\alpha = S_\alpha + iC\tau_\beta,$$
$$\tilde{\mathbf{M}}_\alpha = \mathbf{n}^* \times (\tilde{M}_\alpha \mathbf{t}'^\alpha + \tilde{H}_\alpha \mathbf{t}'_\beta), \quad \tilde{M}_\alpha = M_\alpha + iC\varepsilon_\beta , \tag{20}$$
$$\tilde{H}_\alpha = H_\alpha - iC \frac{\gamma}{2} .$$

To obtain complex vector equations for $\tilde{\mathbf{N}}_\alpha$, $\tilde{Q}_\alpha$, $\tilde{\mathbf{M}}_\alpha$ it is sufficient to multiply the compatibility equations (12) by iC and add them to the equilibrium equations (16) with (17). This renders after decomposing in surface and transverse vectors according to (13), (18) the equations (summation with respect to $\alpha = 1,2$; $\beta \neq \alpha$):

$$(a_\beta \tilde{\mathbf{N}}_\alpha + a_\beta \tilde{Q}_\alpha \mathbf{n}^*)_{,\alpha} = -q\sqrt{a} +$$
$$+ iCa_1a_2(\boldsymbol{\varkappa}_1 \times \boldsymbol{\varkappa}_2 + e_{\alpha\beta}\lambda_\alpha \mathbf{n}^* \times \boldsymbol{\varkappa}_\beta) , \tag{21}$$

$$(a_\beta \tilde{\mathbf{M}}_\alpha)_{,\alpha} + a_1a_2\mathbf{t}'_\alpha \times (\tilde{\mathbf{N}}_\alpha + \tilde{Q}_\alpha \mathbf{n}^*) = \mathbf{0} . \tag{22}$$

The nonlinear terms with $\lambda_\alpha \boldsymbol{\varkappa}_\beta$ have mostly the order of magnitude of the strains, compared to $|\tilde{\mathbf{N}}_\alpha|$, and will be dropped.+)

+) The terms with $C\lambda_\alpha = e_{\alpha\beta} \mathrm{Im}\, Q_\beta$ can, however, be retained without basically influencing the following deliberations.

The equation (22) projected on $\mathbf{t}'^1$, $\mathbf{t}'^2$ allows to express the two complex parameters $\tilde{Q}_\alpha$ in terms of $\tilde{\mathbf{M}}_\alpha$; its projection on $\mathbf{n}^*$ provides one nondifferential relation between $\tilde{\mathbf{N}}_\alpha$ and $\tilde{\mathbf{M}}_\alpha$: Of course, each of these complex equations is equivalent to two real ones.

Equations (21), (22) represent the statics and the geometry of deformation. A full system must include also the constitutive relations.

6. Simplifications of Nonlinear Elastic-Shell Equations.

The comparatively compact form of the complex equations (21), (22) makes the possibility of simplifications easier to recognize. In simplifying the equations we will use the way leading in the linear approximation to the Novozhilov's complex equations ([23], pp. 56 - 57). The basic assumption will be

$$(1 + i\,\frac{h}{3R})\ [\tilde{\mathbf{N}}_\alpha\ \tilde{\mathbf{M}}_\alpha] \cong [\tilde{\mathbf{N}}_\alpha\ \tilde{\mathbf{M}}_\alpha]\ , \tag{23}$$

where 1/R denotes the maximum normal-section curvature.

The relations (23) mean that the membrane stresses in the direction x^α and the wall-bending or torsional stresses in the direction of the other coordinate x^β must be of not too different orders of magnitude:

$$\frac{h}{3R} \ll \left|\frac{N_\alpha/h}{E\varkappa_\beta h/2}\right|\ ,\ \left|\frac{S_\alpha/h}{E\tau_\alpha h}\right| \ll \frac{3R}{h} \tag{24}$$

The physical definitions of stress and strain and the relations between them are most naturally established with reference to *orthogonal coordinates*. We restrict now until further notice the choice of the coordinates: x^α-are (in the undeformed surface) lines of curvature.

Consider the six scalar equations equivalent to (21, (22). For the further transformation it is convenient to write the equations with the aid of operators defined by:

$$K(a,b,c,d) = \frac{(a_2 a)_{,1}}{a_1 a_2} - \frac{b}{\rho_2'} + \frac{(a_1 c)_{,2}}{a_1 a_2} - \frac{d}{\rho_1'} \quad . \tag{25}$$

The projections of the eqs (21), (22) written out with the aid of formulas (4), (5) where $\varphi_{,\beta} = 0$ and (25) are ($\beta \neq \alpha$):

$$\begin{aligned} &K(\tilde{N}_1,\tilde{N}_2,\tilde{S}_2,\tilde{S}_1) + \frac{1}{R_1'}\tilde{Q}_1 + \frac{1}{R_{21}'}\tilde{Q}_2 + q_1 = 0 \ , \\ &K(\tilde{S}_1,-\tilde{S}_2,\tilde{N}_2,-\tilde{N}_1) + \frac{1}{R_{12}'}\tilde{Q}_1 + \frac{1}{R_2'}\tilde{Q}_2 + q_2 = 0 \ , \end{aligned} \tag{26}$$

$$\sum_{\alpha=1}^{2}\left(\frac{\tilde{N}_\alpha}{R_\alpha'} + \frac{\tilde{S}_\alpha}{R_{\alpha\beta}'} - \frac{(\tilde{Q}_\alpha a_\beta)_{,\alpha}}{a_1 a_2}\right) - q_n + iC(\varkappa_1\varkappa_2 - \tau_1\tau_2) = 0, \tag{27}$$

$$\tilde{Q}_1 = K(\tilde{M}_1,\tilde{M}_2,\tilde{H}_2,\tilde{H}_1), \ Q_2 = K(\tilde{H}_1, -\tilde{H}_2, \tilde{M}_2, -\tilde{M}_1) \ , \tag{28}$$

$$\tilde{S}_1 - \frac{\tilde{H}_2}{R_2'} - \frac{\tilde{M}_1}{R_{12}'} = \tilde{S}_2 - \frac{\tilde{H}_1}{R_1'} - \frac{\tilde{M}_2}{R_{21}'} = \tilde{S} = S + iC\tau \ . \tag{29}$$

The eq. (29) introduce the new variable $\tilde{S}$ and its components S, $C\tau$.

We express now the $\tilde{Q}_\alpha$ with the aid of the elasticity relations [12, 17] which for the lines-of-curvature coordinates simplify to the form ([23] pp. 44):

$$\begin{aligned} &M_{(\alpha)} = D(\varkappa_\alpha + \nu\varkappa_\beta) \ , \quad H = (1 - \nu)D\tau \ , \\ &Eh\varepsilon_\beta = N_{(\beta)} - \nu N_{(\alpha)} \ , \quad Eh_\gamma = 2(1 + \nu)S \ . \end{aligned} \tag{30}$$

The quantities $M_{(\alpha)}$, H, $N_{(\alpha)}$, S are defined by the virtual work δV per unit area of the middle surface in a virtual deformation corresponding to the Kirchhoff-Love hypothesis:

$$\delta V = N_{(1)}\delta\varepsilon_1 + N_{(2)}\delta\varepsilon_2 + S\delta\gamma + M_{(1)}\delta\varkappa_1 + M_{(2)}\delta\varkappa_2 + 2H\delta\tau \ .$$

With the accuracy of the assumption (24) we have

$$M_{(\alpha)} = M_\alpha, \; N_{(\alpha)} = N_\alpha, \; H_\alpha = H \;, \; S_\alpha = S \;, \; \tau_\alpha = \tau \;. \qquad (31)$$

In what concerns S, τ these relations follow directly from (29), (24). Even more exactly $M_{(\alpha)} = M_\alpha$, $N_{(\alpha)} = N_\alpha$. - The difference is of the order of magnitude of the strain components in the shell.

Thus, for the analysis based on the assumption (23) the Love's "simplest approximation" of the elasticity relations is adequate. Relations (30) imply similar relations for the complex quantities $\tilde{M}_\alpha$, $\tilde{H}_\alpha$. We can write the elasticity relations (30) in the form ($\beta \neq \alpha$):

$$\tilde{M}_\alpha = ih'(\tilde{N}_\beta - \nu\bar{N}_\alpha), \; \tilde{H}_\alpha = - ih'(\tilde{S}_\alpha + \nu\bar{S}_\alpha) \;, \qquad (32)$$

where $\bar{N}_\alpha$, $\bar{S}_\alpha$ denote complex conjugates of $\tilde{N}_\alpha$, $\tilde{S}_\alpha$.
With (28), (32) we have expressions of $\tilde{Q}_\alpha$ in terms of $\tilde{N}_\alpha$, $\tilde{S}_\alpha = \tilde{S}$

$$\tilde{Q}_1 = ih'[K(\tilde{N}_2,\tilde{N}_1, - \tilde{S}_2, - \tilde{S}_1) - \nu K(\bar{N}_1,\bar{N}_2,\bar{S}_1,\bar{S}_2)] \;.$$

Expressing the first term of the right-hand side with the aid of the identity following from (25), (7)

$$K(b,a,-c,-d) + K(a,b,c,d) = \frac{1}{a_1}(a+b)_{,1} + (a+b)\lambda_2$$

we have

$$\tilde{Q}_1 = ih'\left\{\frac{1}{a_1}(\tilde{N}_1+\tilde{N}_2)_{,1} + (\tilde{N}_1+\tilde{N}_2)\lambda_2 - K(\tilde{N}_1,\tilde{N}_2,\tilde{S}_2,\tilde{S}_1) - \nu K(\bar{N}_1,\bar{N}_2,\bar{S}_1,\bar{S}_2)\right\} \qquad (a)$$

Inserting this into the first of equations (26) we recognize that according to the assumption (23) only the $\tilde{N}_1 + \tilde{N}_2$ terms of (a) must be retained. In the eqs. (26) the $\tilde{Q}_\alpha$-terms are adequately represented by

$$\tilde{Q}_\alpha = ih'[\frac{1}{a_\alpha}(\tilde{N}_1 + \tilde{N}_2)_{,\alpha} + (\tilde{N}_1 + \tilde{N}_2)\lambda_\beta] \tag{b}$$

The simplifications thus introduced include dropping in (26) of the parts of terms $\tilde{Q}_\alpha/R'_{\alpha\beta}$ which correspond to the quantities disregarded in (b) compared to (a). In the second of eqs. (26) the term $(h'/R'_{12})K(\tilde{N}_1,\tilde{N}_2,\tilde{S}_2,\tilde{S}_1)$ is neglected compared to $K(\tilde{S}_1,-\tilde{S}_2,\tilde{N}_2-\tilde{N}_1)$. For the lines of curvature $h'/R'_{\alpha\beta} \sim \tau h/3$, which has the order of magnitude of the shear strain. It is consequent to neglect such quantities while using the Hooke's law. On the same ground must be dropped the second term in (b): According to (19) $\lambda_\beta = \text{Im}Q_\alpha/C$ and with (30), (31): $h'|(N_1 + N_2)\lambda_\beta|$ is of the order of magnitude of $|(\varepsilon_1 + \varepsilon_2)\tilde{Q}_\alpha|$ which is negligible compared to $|\tilde{Q}_\alpha|$ in the same eqn.

Formula for $\tilde{Q}_1$ accurate in the sense of (23) also for the eq. (27) is obtained by introducing into (a) the expression of $K(\tilde{N}_1,\tilde{N}_2,\tilde{S}_2,\tilde{S}_1)$ and of its complex conjugate following from (26), (b). The respective formula for $\tilde{Q}_2$ is obtained similarly. Thus applying the criterium (23) we have

$$\tilde{Q}_\alpha = ih'\left\{\frac{1}{a_\alpha}(\tilde{N}_1 + \tilde{N}_2)_{,\alpha} + (1 + \nu)q_\alpha\right\} . \tag{33}$$

Substituting this into (26), (27) we obtain three complex equations which together with (31) constitute an extension of the Novozhilov's equations ([23], p. 57) to arbitrarily large displacements (the strains remaining small).

The vector form of the complex equations follows directly from (21), (33):

$$(a_2\tilde{\mathbf{N}}_1)_{,1}+(a_1\tilde{\mathbf{N}}_2)_{,2}+ih'(\frac{\partial}{\partial x^1}\frac{a_2}{a_1}\mathbf{n}^*\frac{\partial}{\partial x^1} + \frac{\partial}{\partial x^2}\frac{a_1}{a_2}\mathbf{n}^*\frac{\partial}{\partial x^2})(\tilde{N}_1+\tilde{N}_2) =$$

$$= -\mathbf{q}\sqrt{a}+iEhh'a_1a_2\boldsymbol{\varkappa}_1 \times \boldsymbol{\varkappa}_2+ih'(1+\nu)[(a_2q_1)_{,1}+(a_1q_2)_{,2}]\mathbf{n}^* \tag{34}$$

The $h'q_\alpha$ terms here and in (33) can apparently in most cases be dropped. The equation (34) is equivalent to two vector equations for four surface vectors $\mathbf{N}_\alpha$, $\boldsymbol{\varkappa}_\alpha$ or to six scalar equations. The system determines six stress and strain resultants N_1, N_2, S, $\varkappa_1$, $\varkappa_2$, τ constituting the $\mathbf{N}_\alpha$, $\boldsymbol{\varkappa}_\alpha$.

These six equations display certain affinity to the canonical-form intrinsic equations of Koiter and Simmonds [12, 17].

The equation (34) contains only vector parameters ($\tilde{\mathbf{N}}_\alpha$, $\boldsymbol{\varkappa}_\beta = \mathrm{Im}\tilde{\mathbf{N}}_\alpha e_{\alpha\beta}/C$) and the (first) invariant $\tilde{N}_1 + \tilde{N}_2$ of the surface tensor of stress resultants. This allows a direct extension of the equations to arbitrary *nonorthogonal* coordinates x^α.

7. Flexible Shell Equations.

The simple and concise form of the governing equations (34) makes them a good basis for a derivation of yet another form of FS equations. (The earlier version of complex FS equations [14] encompasses only linear approximation).

To derive the FS equations we need only to introduce into (34) the simplification corresponding to the basic common feature of the stress state of these shells, mentioned in Sect. 1: The stress state varies substantially less intensively with respect to one surface coordinate (x^1) than with respect to the other (x^2).

With F(x) standing for any vector or scalar function describing the significant stress and strain parameters the hypothesis is expressed by the relation

$$\frac{\partial}{\partial x^1}\frac{a_2}{a_1}\mathbf{n}*\frac{\partial F}{\partial x^1} + \frac{\partial}{\partial x^2}\frac{a_1}{a_2}\mathbf{n}*\frac{\partial F}{\partial x^2} \approx \frac{\partial}{\partial x^2}\frac{a_1}{a_2}\mathbf{n}*\frac{\partial F}{\partial x^2} \quad . \tag{35}$$

This is of course possible only for a certain class of shell-shapes and for a certain choice of coordinates. The basis unit vector $\mathbf{n}^*$ and the parameters $a_\alpha = |\mathbf{a}_\alpha|$ must vary less intensively with x^1 than with x^2. But it could hardly be otherwise when the stress state has this quality.

With the relation (35) the governing equation (34) becomes

$$(a_2\tilde{\mathbf{N}}_1)_{,1} + (a_1\tilde{\mathbf{N}}_2)_{,2} + ih' \left[\frac{a_1}{a_2} \mathbf{n}^* (\tilde{N}_1 + \tilde{N}_2)_{,2}\right]_{,2} = \tag{36}$$

$$= -\mathbf{q}\sqrt{a} + iEhh' a_1 a_2 \boldsymbol{\varkappa}_1 \times \boldsymbol{\varkappa}_2 \ .$$

The load terms with the small factor h' have been dropped.

The scalar equations resulting from (36) display the mechanical model of a flexible shell. Namely, the stressed state described by (35) is semi-membrane (or semi-momentless): In the equations of equilibrium are negligibly small all terms with the moment stress-resultants M_1, H_1 in the section x^1 = const, in the compatibility equations vanish all terms with the dual parameters ε_2, γ. The terms with $H_2 \approx H_1$ and with $Q_1 = Q_1(M_1, H_2, H_1)$, $\lambda_2(\varepsilon_2, \gamma)$ disappear also.

The conditions of the applicability of the flexible-shell theory in case of the edge loading can be deduced from the scalar equations following from (36). For orthogonal coordinates one finds the error of the FS assumption (35) to be of the order of magnitude not larger then

$$\left| \frac{R'_2}{R'_1} + \frac{ihR'_2}{3(L_2)^2} \right| \tag{37}$$

with L_2 denoting the interval of variation of the deformation with respect to the second surface coordinate: $|F_{,2}/a_2| \sim |F/L_2|$.

The system of flexible-shell equations is of order eight with respect to x^2 but merely of order four with respect to x^1. Thus, on an edge x^2 = const can and must be fulfilled all the four conditions of the general thin-shell theory while on an edge x^1 = const there are only two conditions - those of the membrane theory. The two excluded conditions determine the boundary effect (cf. Sect. 1).

The FS equations (36) and the corresponding shell model extend the previous analysis [18, 20] which has been restricted to shapes with $a,_1 = 0, R_{\alpha,1} = 0$. This applies in particular to the Wlassow-type FS equations. - Obtained with the assumption of the type (35) alone, without the assumptions of the Novozhilov complex theory, these equations [of 18, 20] differ only unsubstantially from the scalar equations following from (34) by $a,_1 = 0$, $R_{\alpha,1} = 0$.

The formulation of equations equivalent to (34) and (36) in terms of general oblique coordinates will be presented elsewhere.

References

1. Kármán, Th. v.: Über die Formänderung dünnwandiger Rohre, insbesondere federnder Ausgleichsrohre. VDI Z. 55(1911) 1889-1895.

2. Beskin, L.: Bending of curved thin-walled tubes, J.Appl. Mech.12(1945)A1 - A7.

3. Reissner, E.: On bending of curved thin-walled tubes. Proc. Nat'l. Acad. Sci. USA 35(1949)204-208.

4. Reissner, E.: On axisymmetrical deformations of thin shells of revolution, Proc. Sympos. Appl. Math.3(1950)27-52.

5. Reissner, E.: On finite bending of pressurized tubes. J. Appl. Mech.26(1959)386-392.

6. Axelrad, E.L.: Nonlinear equations of axisymmetric shells and bending of thin-walled beams. Izv. AN SSSR, OTN, Mechanika i Mash. (1960) n. 4,84-92 (In Russian), Translation in Amer. Rocket Soc. Journal Supplement 32(1962) 1147-1151.

7. Reissner, E.: On finite pure bending of cylindrical tubes. Österr. Ing.-Arch.15(1961)165-172.

8. Reissner, E.: Weinitschke H,: Finite pure bending of circular cylindrical tubes. Quart.Appl.Math.20(1963)305-319.

9. Axelrad, E.L.: Refinement of buckling-load analysis for tube flexure by way of considering precritical deformation Izv. AN SSSR, Mechanika (1965), n. 4, 133 - 139 (In Russian)

10. Koiter, W.T.: On the nonlinear theory of thin elastic shells, Koninkl. Nederl. Akad. van Wetensch. Proc., Ser. B 69(1966)1-54.

11. Simmonds, J.G.; Danielson, D.A.: Nonlinear shell theory with finite rotation and stress-function vectors, J.Appl.Mech. 39(1972)1085-1090.

12. Koiter, W.T.; Simmonds, J.G.: Foundations of shell theory, Proc. 13th ICTAM, E. Becker, G.K. Michailov eds., Springer-Verlag (1973).

13. Reissner, E.: Linear and nonlinear theory of shells, Proc. Sympos. on Thin-Shell Struct., Y.C.Fung, E.E. Sechler eds., Prentice Hall, New Jersey, 1974.

14. Axelrad, E.L.: Flexible shells, "Nauka", Moscow, 1976 (In Russian)

15. Oden, J.T.; Bathe, K.J.: A commentary on computational mechanics. Appl. Mech. Rev. 31(1978)1053-1058.

16. Pietraszkiewicz, W.: Finite rotations and Lagrangean discription in the non-linear theory of shells, Warszawa, 1979. Polish. Sci. Publ..

17. Koiter, W.T.: The intrinsic equations of shell theory with some applications, Mechanics Today, E. Reissner Anniv. Volume, S. Nemat-Nasser ed., Pergamon Press, 1980,139-154.

18. Axelrad, E.L.: Flexible shells, Proc. 15th ICTAM, Toronto 1980, F.P.J. Rimrott and B. Tabarrok eds., North-Holland, Amsterdam 1981.

19. Emmerling, F.A.: Nonlinear bending and buckling of cylinder and curved tubes under normal pressure. Ing.-Archiv 52 (1982)1-16.

20. Axelrad, E.L.: On vector description of arbitrary deformation of shells. Int. J. Solids Structures, 17 (1981) 301-304.

21. Axelrad, E.L.; Emmerling, F.A.: Finite bending and collapse of elastic pressurized tubes. Ing.-Arch. 53(1983) 41-52.

22. Libai, A.; Simmonds, J.G.: Nonlinear elastic shell theory. Advances in Appl.Mech. 23(1983)271-371.

23. Axelrad, E.L.: Schalentheorie, B.G. Teubner, Stuttgart 1983.

Seismic Behavior of Liquid Filled Shells

W. A. NASH, S. H. SHAABAN, L. WATAWALA AND S. C. LEE

University of Massachusetts, Amherst, Massachusetts USA

Summary

The elastic behavior of vertical-axis cylindrical liquid storage tanks subject to horizontal seismically induced motions of the base is considered. Tanks without as well as with a dome are examined. The effect of initial geometric imperfections is studied, as is the influence of circumferential reinforcing rings on dynamic behavior of the shell-liquid system.

Background

The initial investigations into determination of the response of ground supported liquid storage tanks to seismic motions began more than three decades ago. During the 1960's related problems of liquid filled cylindrical tanks subjected to acceleration in the direction of the geometric axis of the shell were investigated because of their utilization in the space program. These studies will not be reviewed here.

Perhaps the first study to employ modern computational techniques was due to N. W. Edwards [1] in 1969. He considered the response of a liquid-filled slab supported cylindrical tank subjected to impulsive ground motions and employed ring-shaped finite elements to determine the response of the system to ground acceleration characterized by a saw-tooth type function. In 1976 investigations by J. Y. Yang [2] and A. S. Veletsos and J. Y. Yang [3] examined the same problem by dividing the hydrodynamic effects into two parts (a) the impulsive effects which are computed by neglecting the effect of liquid surface waves, and (b) the convective forces, which are associated with the sloshing of the liquid in the tank. It was assumed that the convective effects could be obtained approximately from the solution for rigid tanks and the approach then involved evaluation of natural frequencies of the coupled liquid-tank system for lateral, beam-like modes of vibration for which there was a single sine wave of deflection in the circumferential direction. Behavior of empty as well as full tanks was described in terms of cantilever beam-type behavior together with extensional vibrations of a semi-circular arch. In 1981 A. Kumar [4] extended this approach to include the effects of the vertical component of ground motion as well as considering the effect of liquid compressibility, which was found to be negligible for problems of interest to the petrochemical industry. In this same program at Rice University, A.S. Veletsos and J. W. Turner [5] presented an approximate investigation of the seismic response of an initially out-of-round tank. This was done by computing the hydrodynamic pressure in an irregular rigid tank and then applying this loading to a flexible tank.

An extensive experimental program has been in progress at the University of California, Berkeley, utilizing the large shake table facility. This has involved tests on relatively large models of broad tanks by D. P. Clough [6] and also tall, slender tanks by A. Niwa [7]. The presence of significant radial displacement components stemming from the third and fourth circumferential harmonics was an unanticipated result of the broad tank tests [6], [8]. Another experimental investigation involving a 1/30 scale model of a 120 foot (36.58 m) diameter tank containing 30 feet (9.14 m) of water was carried out by D. D. Kana and F. T. Dodge [9] in 1975. The model was subjected to a typical seismic excitation and limits of validity of the simple analtyical model presented by the authors were determined. Mention should also be made of a recent study by T. J. Marchaj [10] concerned with the importance of vertical ground accelerations on the behavior of liquid storage tanks. The simplified analysis led the author to the conclusion that the familiar "elephant foot" phenomenon could be explained by consideration of the simultaneous vertical and horizontal motions.

In 1980 G. W. Housner and M. A. Haroun [11] presented a paper dealing with the superposition of the free lateral vibrational modes of a liquid storage tank. The accuracy of the approach was confirmed by vibration tests on full-scale tanks ranging from 48 feet (14.6 m) to 60 feet (18.3 m) in diameter. Both ambient vibration tests (the ambient forces being the result of wind currents and microseismic waves) as well as forced vibration tests were carried out on these structures. The mode shapes and natural frequencies found during test were in good agreement with those predicted by an analysis due to M. A. Haroun [12] in which the shell was modeled by finite elements and the liquid was treated as a continuum. Mention should also be made of several interesting Japanese contributions of this same nature.

A step forward in development of generalized finite element procedures for fluid-elastic solid interaction problems was taken by W. K. Liu [13] in 1980. A general mixed Lagragian-Eulerian description of a continuum was developed and a number of examples investigated, including the case of a liquid filled cylindrical tank subjected to a periodic saw-tooth ground acceleration excitation.

Behavior of Perfect, Unstiffened Cylindrical Shells

Let us consider a thin, vertical-axis right circular cylindrical shell of radius R and altitude L. The quantities x, y, and z denote axial, tangential, and inward radial coordinates respectively with the corresponding middle surface displacements being given by u, v, and w. The equation due to Flügge [14] for small elastic displacements of such a shell in the radial direction is

$$\frac{h^2}{12}\nabla^4 w + \frac{w}{R^2} + \frac{1}{R^2}\frac{\partial v}{\partial y} + \frac{\nu}{R}\frac{\partial u}{\partial x} + \frac{h^2}{12R^2}\left[\left(\frac{1-\nu}{2R}\right)\frac{\partial^3 u}{\partial x\partial y^2} - R\frac{\partial 3_u}{\partial x^2} - \left(\frac{3-\nu}{2}\right)\frac{\partial^3 v}{\partial x^2\partial y} + \frac{2}{R^2}\frac{\partial^2 w}{\partial y^2} + \frac{w}{R^2}\right] - \frac{p_z}{D_e} = 0 \qquad (1)$$

with two other coupled equations in the axial and tangential directions which are omitted for brevity but are given in [14]. Here h denotes shell thickness, E is Young's modulus, and $D = Eh/(1 - \nu^2)$ with ν being Poisson's ratio.

The equations governing motion of an inviscid, incompressible liquid are the Laplace equation

$$\nabla^2\Phi = 0 \qquad (2)$$

where

$$\nabla^2 = \frac{\partial^2}{\partial r^2} + \frac{1}{r}\cdot\frac{\partial}{\partial r} + \frac{1}{r^2}\frac{\partial^2}{\partial\theta^2} + \frac{\partial^2}{\partial x^2} \qquad (3)$$

and the Bernoulli equation

$$\frac{\partial\Phi}{\partial t} + \frac{p}{\rho_f} + gx = 0 \qquad (4)$$

where (r, θ, x) is the velocity potential, p is liquid pressure, ρ_f is liquid density, and g is the gravitational constant. Also, t denotes time. Determination of liquid motions involves simultaneous solution of the above equations subject to appropriate boundary conditions. Knowledge of Φ permits determination of the pressure distribution throughout the liquid from the Bernoulli equation.

The portion of the cylindricak tank in contact with the liquid is designated as region I and the portion of the tank above the liquid is region II. Region I (liquid depth) is of height H. The velocity vector of the liquid may be written as

$$\vec{V} = \text{grad}$$
$$= \nabla\Phi \qquad (5)$$

Consequently the boundary conditions expressing liquid-solid interaction along the elastic wall of the cylindrical tanks as well as at the rigid bottom of the tank may be written as:

$$\vec{V} \cdot \vec{n} \quad \begin{cases} \left[\frac{\partial w_r}{\partial t}\right]_{R = a} & \text{along the cylindrical wall in region I} \\ [\ 0\ ]_{x = 0} & \text{at the rigid tank bottom} \end{cases} \qquad (6)$$

Here, n is a unit vector normal to the shell boundary. Thus, along the elastic tank wall we have:

$$\left(\frac{\partial\Phi}{\partial z} - \frac{\partial w_r}{\partial t}\right)_{r = a} = 0 \qquad (7)$$

Since the liquid velocity in the x-direction is zero at the tank bottom it follows that

$$\left(\frac{\partial\Phi}{\partial x}\right)_{x = 0} = 0 \qquad (8)$$

Finally, the linearized liquid free surface condition may be expressed in the form:

$$\left(\frac{\partial^2\Phi}{\partial t^2} + g\,\frac{\partial\Phi}{\partial x}\right)_{x = H} = 0 \qquad (9)$$

It is also necessary to enforce compatibility of the elastic shell displacements, in-plane normal and shear forces, transverse shear, and bending moment at the junction of regions I and II. The computer programs developed permit consideration of a clamped base at the tank and a choice of clamped, simply supported, or free upper extremity.

The loading p_z in (1) together with p_x and p_y represent applied forces per unit area. Here, they must be taken to be the inertial loadings together with the dynamic pressure (i.e., the hydrostatic pressure has been subtracted) in the liquid which is

$$p = \rho_f \frac{\partial \Phi}{\partial t} \qquad (10)$$

We investigate solutions of (1) of the form

$$\Phi(z,o,x,t) = \sin \omega t[\Phi(x,z)\cos m\theta] \qquad (11)$$

$$w = (\cos \omega t)\sum_{n=1}^{\infty} U_{mn}(x)\cos m\theta \qquad (12)$$

$$v = (\cos \omega t)\sum_{n=1}^{\infty} V_{mn}(x)\sin m\theta \qquad (13)$$

$$u = (\cos \omega t)\sum_{n=1}^{\infty} W_{mn}(x)\cos m\theta \qquad (14)$$

where m denotes the number of circumferential waves and ω denotes the natural frequency of the coupled liquid-solid system.

To satisfy (3) and (8) we write Φ in the form

$$\Phi(r,z) = \sum_{n=1}^{\infty} A_{mn} I_m(\lambda_{mn} z_R)\cos(\lambda_{mn} x_R) \qquad (15)$$

where I_m denotes the modified Bessel functions of the first kind and m-th order. Substitution of (11) and (15) into (9) yields

$$\omega^2 = -\frac{\lambda_{mn}\, g}{R} \tan(\lambda_{mn} \frac{H}{R}) \quad \begin{matrix} m = 0,1,2,\ldots \\ n = 1,2,3,\ldots \end{matrix} \qquad (16)$$

where λ_{mn} denotes the eigenvalues.

Substitution of (11) through (14) in (1) leads to a set of three coupled ordinary differential equations in U_{mn}, V_{mn}, and W_{mn}. Solutions for each of these are assumed to be of the form

$$\begin{Bmatrix} W^I_{mn} \\ W^{II}_{mn} \end{Bmatrix} = \begin{Bmatrix} B^I_{mn} \\ B^{II}_{mn} \end{Bmatrix} e^{\Lambda z} + \begin{cases} \alpha^w_{mn} \cos(\lambda_{mn} x_R) \\ 0 \end{cases} \qquad (17)$$

where I pertains to the wetted region of the tank and II to the portion of the tank above the liquid. Substitution of these assumed solutions into the ordinary differential equations mentioned leads to a set of three algebraic equations in B_{mn} and the counterpart parameters for the u and v displacements. A nontrivial solution to this set of equations requires that an eighth order algebraic equation be satisfied, leading to the natural frequencies of the coupled liquid-elastic tank system. A computer program is available for determination of these natural frequencies and associated mode shapes for given geometric and elastic input parameters for shell and liquid [15]. An extension of this analysis to the case of a domed tank, as shown in Figure 1, is given in [16].

Example

Let us consider an empty steel liquid storage tank 40 feet (12.12m) tall, of radius 60 feet (18.29m), and wall thickness 1 inch (2.54cm). Based upon the above analysis, the computer program indicates the final six axial modes (considering only m = 1 circumferential mode) with the tank clamped at the base slab and free at the top to be: 33.89, 43.54, 44.16, 45.10, and 46.07 Hz. When this same tank is considered to be filled to a depth of 20 feet (6.09m) with water, the first three axial frequencies are 9.40, 15.90, and 20.38 Hz. For the tank filled with 10 feet (3.048m) of water, these first three axial frequencies become 13.30, 22.60, and 35.30 Hz.

Behavior of Imperfect, Unstiffened Cylindrical Shells

Many previous investigations have pointed to the importance of consideration of finite amplitude displacements when initial geometric imperfections are under consideration. Thus, we employ the equations of motion and compatibility in the form:

$$\frac{D}{h}\nabla^4 w_1 = \left(\frac{\partial^2 w_0}{\partial x^2} + \frac{\partial^2 w_1}{\partial x^2}\right)\frac{\partial^2 \Phi}{\partial y^2} - 2\left(\frac{\partial^2 w_0}{\partial x \partial y} + \frac{\partial^2 w_1}{\partial x \partial y}\right)\frac{\partial^2 \Phi}{\partial x \partial y}$$

$$+ \left(\frac{\partial^2 w_0}{\partial y^2} + \frac{\partial^2 w_1}{\partial y^2}\right)\frac{\partial^2 \Phi}{\partial x^2} + \frac{1}{R}\frac{\partial^2 \Phi}{\partial x^2} - \frac{\partial^2 w_1}{\partial t^2} + \frac{q_e}{h} \tag{18}$$

$$\frac{1}{E}\nabla^4 \Phi = \left(\frac{\partial^2 w_1}{\partial x \partial y}\right)^2 - \frac{\partial^2 w_1}{\partial x^2}\frac{\partial^2 w_1}{\partial y^2} - \frac{1}{R}\frac{\partial^2 w_1}{\partial x^2}$$

$$- \left(\frac{\partial^2 w_1}{\partial x^2}\frac{\partial^2 w_0}{\partial y^2} + 2\frac{\partial^2 w_1}{\partial x \partial y}\frac{\partial^2 w_0}{\partial x \partial y} + \frac{\partial^2 w_0}{\partial x^2}\frac{\partial^2 w_1}{\partial y^2}\right)$$

where ρ represents shell density, D is shell flexural rigidity, and Φ denotes a stress function of the middle surface forces.

To gain an understanding of the influence of initial geometric imperfections on the dynamic behavior of the shell a simple model of the imperfection is considered, viz:

$$w_o(x,y) = W_o \sin\frac{jy}{R}\sin\frac{m\pi x}{L} \tag{20}$$

where W_0 is the amplitude of the imperfection. It is recognized that realistic imperfection configurations may, at times, require some, if not many, additional terms. However, the analysis and accompanying computer effort would become of almost prohibitive length were more terms employed.

The characterization of the dynamic response is taken to be

$$w_1(x,y,t) = f_1(t)\cos y \sin Rx + f_2(t)\sin sy \sin Rx + f_3(t) \sin^2 Rx \tag{21}$$

where the $f_1(t)$ are time-dependent amplitude functions and $s = n|R$, $r = m\pi|L$. The first two terms of (20) represent a harmonic distribution of response around the shell circumference and the third term corresponds to an axisymmetric response. The third term is necessary to permit satisfaction of periodicity of circumferential displacement. The vibratory configuration (21) corresponds to zero radial displacement at the ends of the shell, but does not represent a moment-free condition at those ends. Accordingly the boundary condition represented lies intermediate between simply supported and clamped ends.

Let us assume a circumferential wave pattern of the form:

$$w_0(s,y) = W_0 \sin sy \sin Rx$$
$$w_1(x,y,t) = f_1(t)\cos sy \sin Rx + f_2(t)\sin sy \sin Rx + f_3(t) \sin^2 Rx. \tag{22}$$

Substitution of these relations into (19) yields an exact solution for the stress function in terms of harmonic functions. Coefficients of these terms are functions of the time-dependent amplitude functions. This value of the stress function together with (20) and (21) makes it possible to enforce the condition of periodicity of circumferential displacement which leads to:

$$f_3(t) = \frac{s^2 R}{4} [f_1^2(t) + f_2^2(t) W_0] \tag{23}$$

Consideration must now be given to the equation of motion. Substitution of the above values of w_0, w_1, and Φ yields a relation containing the two unknown functions $f_1(t)$ and $f_2(t)$. Direct solution is impossible, so the Galerkin technique was employed to obtain an approximate solution. The Galerkin weighing functions employed were

$$G_1 = \frac{w_1}{f_1} = \cos sy \sin Rx + \frac{s^2 R}{2} f_1 \sin^2 Rx$$
$$G_2 = \frac{w_1}{f_2} = \sin sy \sin Rx + \frac{s^2 R}{2} (f_2 + W_0) \sin^2 Rx \tag{24}$$

These functions when employed in the Galerkin approach yield two coupled equations involving f_1, f_2, their second derivatives with respect to time, as well as R, s, sin Rx, cos sy, and W_0, together with functions representing the generalized forces acting on the shell. These forces are given by integral equations involving $q_e(x,y,t)$, G_1 and G_2. The external radial force $q_e(x,y,t)$ is assumed to be fixed in space and harmonic in time in the form,

$$q_e(x,y,t) = Q(x,y)\cos\omega t \tag{25}$$

Therefore the generalized forcing functions c(t) and s(t) are given by

$$\begin{aligned} c(t) &= C_{mn}\cos\omega t \\ s(t) &= S_{mn}\cos\omega t \end{aligned} \tag{26}$$

The two coupled ordinary non-linear differential equations cannot yet be solved exactly. But an approximate solution to them can be obtained by the procedure known as the method of averaging. The unknown functions $f_1(t)$ and $f_2(t)$ are taken to be in the form

$$\begin{aligned} f_1(t) &= B_1(t)\cos\omega t \\ f_2(t) &= B_1(t)\cos\omega t \end{aligned} \tag{27}$$

and then applying the method of averaging gives the approximate solution as

$$\begin{aligned} f_1(t) &= \bar{B}_1 \cos\omega t \\ f_2(t) &= \bar{B}_2 \cos\omega t \end{aligned} \tag{28}$$

Free vibrations of the imperfect shell may be investigated in this manner, Forced excitation of the shell having its geometric axis oriented vertically and subject to harmonic lateral excitation at its lower extremity is of importance in the case of seismically excited liquid storage tanks. For this case the support motion is:

$$\ddot{U}_s = A_s \cos\omega_s t \tag{29}$$

It is relatively direct to consider the influence of the perfect liquid when it completely fills the tank. The constitutive equation of the liquid, together with boundary conditions mentioned previously for the perfect shell are available and from these the velocity potential may be written in terms of products of harmonic and Bessel functions. Satisfaction of boundary as well as free surface conditions lead to coupled equations through the liquid-shell behavior which may be solved through the Galerkin method and then application of averaging.

As a necessary check on the numerical accuracy of the present approach, the case of free vibrations of an initially perfect cylindrical shell was considered. The first four circumferential modes as well as the first four axial modes of a cylinder having a thickness/radius ratio of 0.01 and a length/radius ratio of 1.0 were determined by the present approach and compared with the tabulated values given in [17] which were obtained using a three-dimensional wave propagation theory. In all cases the natural frequencies differed by no more than two percent, thus leading confidence to the current approach.

Next, free vibrations of an imperfect cylindrical shell were considered. The case of a shell having a thickness/radius ratio of 1/720, a length/radius ratio of 2/3 and Poisson's ratio of 0.272 was examined in detail. These parameters were selected because they are representative of contemporary prototype liquid storage tanks. It was found that in the case of a cylindrical shell having a maximum initial imperfection equal in magnitude to the shell thickness, the first natural frequency of radial vibration decreased approximately fifteen percent from that found for a perfect shell.

Response of an empty cylindrical shell with axis vertical and lower extremity clamped to a rigid slab undergoing horizontal seismic motion represented by the standard artificial earthquake [18] with peak acceleration of 0.5 g is represented in Fig. 2. In this case, the upper extremity of the shell was considered to be unsupported and the forced response indicated represents peak radial displacement of a generator lying in a diametral plane about which the base excitation is symmetric.

Natural frequencies of the same cylindrical shell completely filled with water were determined for a range of values of amplitude of initial imperfection up to and including a value equal to the shell thickness. For very small dimensionless amplitudes of vibration (0.001) it was found that the decrease in natural frequency is very nearly the same as for the empty shell, namely slightly more than ten percent. Maximum decrease was found to occur at an imperfection amplitude of approximately half the shell thickness, with less severe decreases with increasing values of imperfection.

Behavior of Perfect, Ring Stiffened Cylindrical Shells

Here we consider the behavior of a vertical axis cylindrical shell subject to seismic motion of its base. The shell is filled to an arbitrary depth with a perfect liquid. The shell strain-displacement relations corresponding to the Flügge equations (1) are employed to represent bending and membrane energies of the shell. Likewise shell kinetic energy as well as work due to applied forces is formulated. Finally, bending as well as torsional energies of the ring stiffeners are determined and variational methods employed to formulate the dynamic behavior of the system. The case considered previously, i.e. a cylindrical shell of height 40 feet (11.1 m), radius 60 feet (18.29 m), and wall thickness 1 inch (2.54 cm) having top free and base clamped was considered in detail. For the case of three ring stiffeners equally spaced along the tank height, the radial displacements corresponding to an artificial earthquake accelerogram [18] are shown in Fig. 3 for a tank completely filled with water. In general, it was found that the use of ring stiffeners increases the load-carrying capacity of a liquid storage tank from a static point of view but is usually not effective in reducing the dynamic response of the system when subject to horizontal base excitation. In fact, intra-ring deformations may add unfavorably to the dynamic response.

Acknowledgement

The authors would like to express their gratitude to the National Science Foundation for their support of this study under grant CEE-76-14833.

References

1. Edwards, N.W., A Procedure for Dynamic Analysis of Thin-Walled Cylindrical Liquid Storage Tanks Subjected to Lateral Ground Motions, Ph.D. Dissertation, University of Michigan, Ann Arbor, 1969.

2. Yang, J.W., Dynamic Behavior of Fluid-Tank Systems, Ph.D. Dissertation, Rice University, Houston, Texas, 1976.

3. Veletsos, A.S., and Yang, J.Y., Earthquake Response of Liquid Storage Tanks, Advances in Civil Engineering Through Engineering Mechanics, ASCE, 1977, pp. 1-24.

4. Kumar, A., Studies of Dynamic and Static Response of Cylindrical Liquid Storage Tanks, Ph.D. Dissertation, Rice University, Houston, Texas, 1981.

5. Veletsos, A.S., and Turner, J.W., Dynamics of Out-of-Round Liquid Storage Tanks, Proceedings of the Third EMD Specialty Conference, ASCE, Austin, Texas,]979, pp. 471-474.

6. Clough, D.P., Experimental Evaluation of Seismic Design Methods for Broad Cylindrical Tanks, Dept. of Civil Engineering, University of California, Berkeley, Report No. EERC/77-10, 1977.

7. Niwa, A., Seismic Behavior of Tall Liquid Storage Tanks, Dept. of Civil Engineering, University of California, Berkeley, Report No. EERC/78-04, 1978.

8. Clough, R.W., Niwa, A., and Clough, D.P., Experimental Seismic Study of Cylindrical Tanks, Journal of the Structural Division, Proc. of the ASCE, Vol. 105, No. ST12, 1979, pp. 2565-2590.

9. Kana, D.D., and Dodge, F.T., Design Support Modeling of Liquid Slosh in Storage Tanks Subject to Seismic Excitation, ASCE Specialty Conference on Structural Design of Nuclear Plant Facilities, New Orleans, 1975.

10. Marchaj, T.J., Importance of Vertical Acceleration in the Design of Liquid Containing Tanks, Proceedings of the Second U.S. National Conference on Earthquake Engineering, EERI, 1979, pp. 146-155.

11. Housner, G.W., and Haroun, M.A., Earthquake Response of Deformable Liquid Storage Tanks, ASME Paper No. 80-C2/PVP-79, presented in San Francisco, 1980.

12. Haroun, M.A., Dynamic Analyses of Liquid Storage Tanks, Report EERL 80-04, California Institute of Technology, Pasadena, 1980.

13. Liu, W.K., Development of Finite Element Procedures for Fluid-Structures Interaction, Ph.D. Dissertation, California Institute of Technology, Pasadena, 1980.

14. Flügge, W., Stresses in Shells, Springer-Verlag, Berlin, 1960.

15. Shaaban, S.H., and Nash. W.A., Finite element analysis of a seismically excited cylindrical storage tank, ground supported and partially filled with liquid, Program available as EXLITANK from the National Information Service, Earthquake Engineering, University of California, Berkeley, 1977.

16. Balendra, T., and Nash, W.A., Earthquake Analysis of a Cylindrical Liquid Storage Tank with a Dome by the Finite Element Method, Program available as EXDOMTANK from the National Information Service, Earthquake Engineering, University of California, Berkeley, 1980.

17. Armenakas, A.E., Gazis, D.C., and Herrmann, T., Free Vibrations of Circular Cylindrical Shells, Pergamon Press, 1969.

18. Ruiz, P., and Penzien, J., Artificial Generation of Earthquake Accelerograms, Program available as PSEQGN from the National Information Service, Earthquake Engineering, University of California, Berkeley, California.

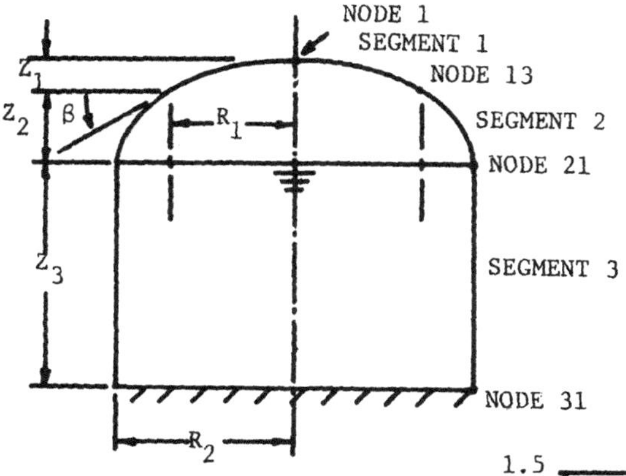

FIGURE 1

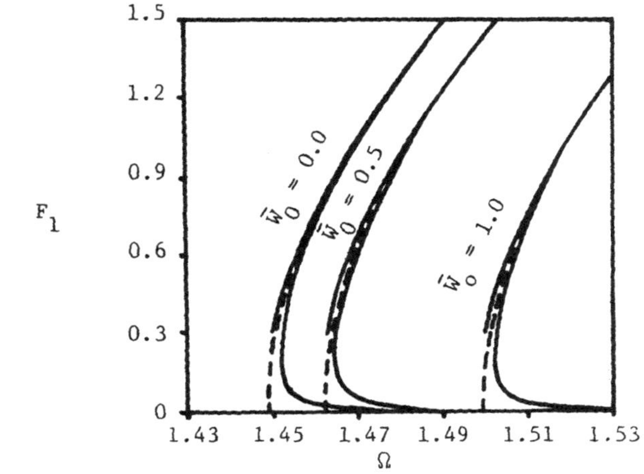

Figure 2 Amplitude of forced vibration vs frequency of excitation

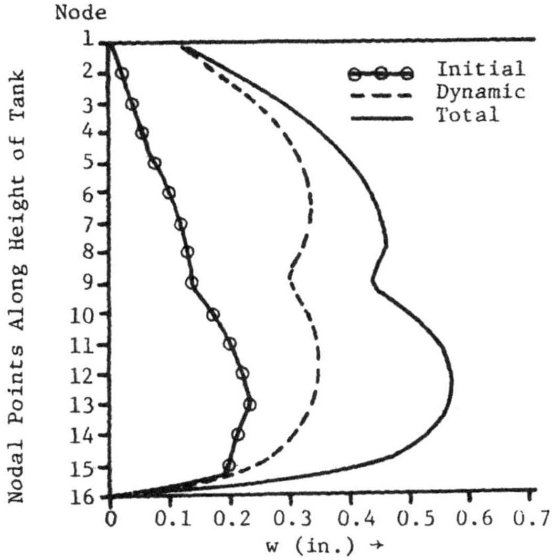

Figure 3 Distribution of Radial Displacements Along Height of Liquid-Filled Tank at Time t = 3.8 sec. (With 3 Ring Stiffeners)

On Geometrically Non-Linear Theories for Thin Elastic Shells

R. SCHMIDT

Lehrstuhl für Mechanik II
Ruhr-Universität Bochum
4630 Bochum 1, West-Germany

Summary

The present paper deals with geometrically non-linear first approximation Kirchhoff-Love type theories for thin elastic shells undergoing small strains accompanied by moderate, large or unrestricted rotations. All theories will be given in an entirely Lagrangian description. We shall start our considerations with a general theory valid for small strains and arbitrary, unrestricted rotations. Then, this general theory will be simplified for shell problems in which the shell material elements undergo large rotations according to the classification scheme given below. Three variants will be derived which admit large rotations about tangents to the shell middle surface and either large, moderate or small rotations about the normal. Finally, the general shell equations will be simplified for shells undergoing moderate rotations about tangents to the shell middle surface and either moderate or small rotations about the normal. All theories presented here are derivable from variational principles.

Introduction

The search for appropriate geometrically non-linear small strain Kirchhoff-Love type theories which are able to describe the large deflection and stability behaviour of thin elastic shells has already led to the formulation of many successful variants. We only mention here the classic theories of Marguerre [1], Donnell, Mushtari, Vlasov [2,3,4], Mushtari and Galimov [5], Leonard [6], Sanders [7], Koiter [8], and Reissner [9] which have been derived under certain simplifying assumptions with respect to the magnitude of displacements, displacement gradients, linearized rotations [1-8] or rotation angles [9], and the fully non-linear theories of Budiansky [10] and Reissner [11] which are valid for arbitrary, unrestricted rotations. These theories have already led to a vast number of numerical applications which have proved their validity within the bounds given by the respectively underlying assumptions.

A new approach to the derivation of geometrically non-linear small strain shell theories, which provides a better understanding of their range of applicability and gives a deeper insight into questions of consistency, is given in the works of Pietraszkiewicz [12 - 15]. In [15] it is shown how a consistent, entirely Lagrangian description of geometric and static boundary conditions can be achieved in the frame of a geometrically non-linear first approximation Kirchhoff-Love type shell theory for small strains and unrestricted rotations. In [12 - 14] the polar decomposition theorem is applied to shell structures according to which the deformation in each point of the shell may be decomposed exactly, apart from a rigid-body translation, into a pure stretch along the principal directions of strain and a rigid body rotation. This makes it possible to impose restrictions on strains and rotations, independently. As a result it is suggested to classify shell theories which assume small strains of $O(\eta)$, $\eta << 1$, according to the magnitude of the rotation angle ω of the shell material elements as follows:

small rotations:	$\omega = O(\eta)$,	i. e. $\omega << 1$;	(I)
moderate rotations:	$\omega = O(\eta^{1/2})$,	i. e. $\omega^2 << 1$;	(II)
large rotations:	$\omega = O(\eta^{1/4})$,	i. e. $\omega^4 << 1$;	(III)
unrestricted rotations:	$\omega \geq O(1)$.		(IV)

In literature consistent with this classification scheme several entirely Lagrangian geometrically non-linear first approximation shell theories, associated variational principles, stability and post-buckling equations, and numerical studies have been already given in the range of unrestricted rotations in [15 - 20] , for large rotations in [12,13,16,21 - 23] and for moderate rotations in [12,13,24 - 33].

The present paper deals with the derivation and justification of first approximation Kirchhoff-Love type theories for different levels of geometrically non-linear shell deformation described in the summary. The hierarchical system of shell equations to be derived here covers all types of small strain shell deformations which may be of practical or theoretical interest. Special attention is focussed on the following items:

(i) Each theory is given in an entirely Lagrangian description and includes appropriate static boundary conditions for the four resultant Kirchhoff boundary loads as well as static corner conditions for the resultant concentrated corner forces. This requires a generalization of Kirchhoff's concept for different levels of non-linear boundary deformation.

(ii) All theories are consistent in the sense that the equilibrium equations, static boundary conditions, and static corner conditions are obtainable as Euler-Lagrange equations from the associated principle of stationary total potential energy, and that also variational principles of the complementary type may be constructed.

The theories derived here generalize or improve others given previously in [12,13,15,16,21 - 23] for the range of large and arbitrary unrestricted rotations and justify the consistency of the moderate rotation shell theory treated in [12,13,24 - 33].

Notations and Basic Shell Relations

Let M denote the shell middle surface in the undeformed reference configuration with convected curvilinear Gaussian surface coordinates θ^α, $\alpha = 1,2$. In each point $M \in M$ we define a position vector $\underset{\sim}{r}(\theta^\alpha)$ with respect to a fixed Cartesian frame, covariant surface base vectors $\underset{\sim}{a}_\alpha = \underset{\sim}{r}_{,\alpha}$, a unit normal vector $\underset{\sim}{n} = 1/2\ \epsilon^{\alpha\beta}\underset{\sim}{a}_\alpha \times \underset{\sim}{a}_\beta$, covariant components of the metric tensor $a_{\alpha\beta} = \underset{\sim}{a}_\alpha \cdot \underset{\sim}{a}_\beta$ (with $\det(a_{\alpha\beta}) = a$) and of the curvature tensor $b_{\alpha\beta} = \underset{\sim}{a}_{\alpha,\beta} \cdot \underset{\sim}{n}$. The notations $(\ldots)_{,\alpha}$ and $(\ldots)|_\alpha$ stand for partial and covariant differentiation on M, respectively. $\epsilon^{\alpha\beta}$ denotes the skew-symmetric surface permutation tensor and δ^α_β is the Kronecker symbol. Indices of surface tensors will be raised by means of contravariant components of the metric tensor $a^{\alpha\beta}$ satisfying the relations $a^{\alpha\lambda}a_{\beta\lambda} = \delta^\alpha_\beta$. Einstein's summation convention will be used with Greek indices assuming the values 1,2. By $\underset{\sim}{u} = u^\alpha \underset{\sim}{a}_\alpha + w\underset{\sim}{n}$ we denote the displacement field mapping each point $M \in M$ into a point $\bar{M}$ on the deformed shell middle surface $\bar{M}$ with position vector $\bar{\underset{\sim}{r}} = \underset{\sim}{r} + \underset{\sim}{u}$. All quantities given above for M are defined analogously for $\bar{M}$ and marked with a bar. Then we find

$$\bar{\underset{\sim}{a}}_\alpha = \bar{\underset{\sim}{r}}_{,\alpha} = l^\lambda_{\cdot\alpha}\underset{\sim}{a}_\lambda + \varphi_\alpha \underset{\sim}{n} \quad , \quad \bar{\underset{\sim}{n}} = n^\lambda \underset{\sim}{a}_\lambda + n\underset{\sim}{n} \ ;$$
$$\bar{\underset{\sim}{a}}_\alpha = \underset{\sim}{G} \cdot \underset{\sim}{a}_\alpha \quad , \quad \bar{\underset{\sim}{n}} = \underset{\sim}{G} \cdot \underset{\sim}{n} \quad , \quad \underset{\sim}{G} = \bar{\underset{\sim}{a}}_\alpha \underset{\sim}{a}^\alpha + \bar{\underset{\sim}{n}}\underset{\sim}{n} \ . \tag{1.1}$$

Here $\underset{\sim}{G}$ denotes the spatial deformation gradient tensor at M and we have introduced the abbreviations

$$\sqrt{\frac{\bar{a}}{a}}\, n^\lambda = m^\lambda = -\epsilon^{\kappa\rho}\epsilon^{\lambda\mu}\varphi_\kappa l_{\mu\rho} = -(1+\theta^\mu_\mu)\varphi^\lambda + \varphi_\mu(\theta^{\mu\lambda} - \omega^{\mu\lambda}) \ , \tag{1.2}$$

$$\sqrt{\frac{\bar{a}}{a}}\, n = m = \frac{1}{2}\,\epsilon^{\kappa\rho}\epsilon^{\lambda\mu} l_{\lambda\kappa} l_{\mu\rho} = 1 + \theta^\lambda_\lambda + \frac{1}{2}\,\theta^\lambda_\lambda\theta^\mu_\mu - \frac{1}{2}\,\theta^\lambda_\mu\theta^\mu_\lambda + \varphi^2 \ , \tag{1.3}$$

$$l_{\alpha\beta} = a_{\alpha\beta} + \varphi_{\alpha\beta} \ , \quad \varphi_{\alpha\beta} = u_{\alpha|\beta} - b_{\alpha\beta}w \quad , \quad \omega_{\alpha\beta} = \frac{1}{2}\,(u_{\beta|\alpha} - u_{\alpha|\beta}) \ , \tag{1.4}$$

$$\theta_{\alpha\beta}=\frac{1}{2}(u_{\alpha|\beta}+u_{\beta|\alpha})-b_{\alpha\beta}w\ ,\ \varphi_\alpha=w_{,\alpha}+b^\beta_\alpha u_\beta\ ,\ \varphi=\frac{1}{2}\epsilon^{\alpha\beta}u_{\beta|\alpha}\ . \tag{1.5}$$

As usual we define covariant components of the symmetric membrane and bending strain tensors $\gamma_{\alpha\beta}$ and $\kappa_{\alpha\beta}$, respectively, by

$$\gamma_{\alpha\beta}=\frac{1}{2}(\bar{a}_{\alpha\beta}-a_{\alpha\beta})\quad ,\quad \kappa_{\alpha\beta}=-(\bar{b}_{\alpha\beta}-b_{\alpha\beta})\quad . \tag{1.6}$$

Let C denote the boundary curve of M with length parameter s, unit tangent vectors $\underset{\sim}{t}=\underset{\sim}{r}_{,s}=t^\alpha\underset{\sim}{a}_\alpha$, unit outward normal vectors $\underset{\sim}{\nu}=\underset{\sim}{t}\times\underset{\sim}{n}=\nu^\alpha\underset{\sim}{a}_\alpha$, normal curvature $\sigma_t=b_{\alpha\beta}t^\alpha t^\beta$, geodesic torsion $\tau_t=-b_{\alpha\beta}\nu^\alpha t^\beta$ and geodesic curvature $\kappa_t=t^\alpha t^\beta\nu_{\alpha|\beta}$. Here $(\ldots)_{,s}=\partial(\ldots)/\partial s$. At C we may use the decomposition $\underset{\sim}{u}=u_\nu\underset{\sim}{\nu}+u_t\underset{\sim}{t}+w\underset{\sim}{n}$ and $\underset{\sim}{\bar{n}}=n_\nu\underset{\sim}{\nu}+n_t\underset{\sim}{t}+n\underset{\sim}{n}$ with $u_\nu=u_\alpha\nu^\alpha$, $u_t=u_\alpha t^\alpha$, $n_\nu=n_\alpha\nu^\alpha$, $n_t=n_\alpha t^\alpha$. Tangent vectors of the deformed boundary $\bar{C}$ are defined by $\underset{\sim}{\bar{a}}_t=\underset{\sim}{\bar{r}}_{,s}$:

$$\begin{aligned}
\underset{\sim}{\bar{a}}_t &= c_\nu\underset{\sim}{\nu}+c_t\underset{\sim}{t}+c_n\underset{\sim}{n}\ ,\ |\underset{\sim}{\bar{a}}_t|^2=c_\nu^2+c_t^2+c_n^2=1+2\gamma_{tt}\ ,\\
c_\nu &= u_{\nu,s}+\tau_t w-\kappa_t u_t=\theta_{\nu t}-\varphi\ ,\\
c_t &= 1+u_{t,s}+\kappa_t u_\nu-\sigma_t w=1+\theta_{tt}\ ,\\
c_n &= w_{,s}+\sigma_t u_t-\tau_t u_\nu=\varphi_t\ ,
\end{aligned} \tag{1.7}$$

with $\gamma_{tt}=\gamma_{\alpha\beta}t^\alpha t^\beta$, $\theta_{tt}=\theta_{\alpha\beta}t^\alpha t^\beta$, $\theta_{\nu t}=\theta_{\alpha\beta}\nu^\alpha t^\beta$, $\varphi_t=\varphi_\alpha t^\alpha$ and so on. Under Kirchhoff-Love constraints we have $\underset{\sim}{\bar{a}}_t\cdot\underset{\sim}{\bar{n}}=0$ and $\underset{\sim}{\bar{n}}\cdot\underset{\sim}{\bar{n}}=1$ yielding [15]

$$\begin{pmatrix}n_t\\ n\end{pmatrix}=\frac{-1}{|\underset{\sim}{\bar{a}}_t|^2-c_\nu^2}\left\{c_\nu n_\nu\begin{pmatrix}c_t\\ c_n\end{pmatrix}+\sqrt{|\underset{\sim}{\bar{a}}_t|^2(1-n_\nu^2)-c_\nu^2}\begin{pmatrix}c_n\\ -c_t\end{pmatrix}\right\} \tag{1.8}$$

$$\begin{pmatrix}\delta n_t\\ \delta n\end{pmatrix}=\begin{pmatrix}n\\ -n_t\end{pmatrix}\frac{1}{D}(k_\nu\delta u_\nu+k_t\delta u_t+k_n\delta w+ + n_\nu\delta u_{\nu,s}+n_t\delta u_{t,s}+n\delta w_{,s})+\begin{pmatrix}F\\ G\end{pmatrix}\delta n_\nu\ , \tag{1.9}$$

which show that the rotational boundary variables n_t and n are dependent quantities which may be calculated if $\underset{\sim}{u}(s)$ and $n_\nu(s)$ are given. In (1.9) we have introduced the first variation δ and the notations

$$\begin{aligned}
&k_\nu=\kappa_t n_t-\tau_t n\ ,\ k_t=\sigma_t n-\kappa_t n_\nu\ ,\ k_n=\tau_t n_\nu-\sigma_t n_t\ ,\\
&D=c_n n_t-c_t n\ ,\ F=(c_\nu n-c_n n_\nu)D^{-1}\ ,\ G=(c_t n_\nu-c_\nu n_t)D^{-1}\ .
\end{aligned} \tag{1.10}$$

Let C_u denote that part of C where at least one of the four independent geometric boundary variables u_ν, u_t, w, and n_ν are prescribed as u_ν^*, u_t^*, w^*, and n_ν^*.

By C_f we denote that part of C where an external stress vector distribution $\underset{\sim}{t}^*(s,\theta^3)$ is applied per unit area of the undeformed lateral boundary surface with infinitesimal area elements dF. After integration through the shell thickness h, $-\frac{h}{2} \leq \theta^3 \leq \frac{h}{2}$, this yields a prescribed external boundary force $\underset{\sim\nu}{T}^* = T^*_{\nu\nu}\underset{\sim}{\nu} + T^*_{t\nu}\underset{\sim}{t} + T^*_{n\nu}\underset{\sim}{n}$ and a "static moment" $\underset{\sim\nu}{H}^* = H^*_{\nu\nu}\underset{\sim}{\nu} + H^*_{t\nu}\underset{\sim}{t} + H^*_{n\nu}\underset{\sim}{n}$ defined by

$$\underset{\sim\nu}{T}^* ds = \int_{-h/2}^{h/2} \underset{\sim}{t}^* dF \quad , \quad \underset{\sim\nu}{H}^* ds = \int_{-h/2}^{h/2} \underset{\sim}{t}^* \theta^3 dF \; . \tag{1.11}$$

Since under Kirchhoff-Love constraints the displacement vector $\underset{\sim}{v}(\theta^\alpha,\theta^3) = \underset{\sim}{u}(\theta^\alpha) + \theta^3(\bar{\underset{\sim}{n}} - \underset{\sim}{n})$ the work of a dead-load type stress vector distribution $\underset{\sim}{t}^*$ is calculated straightforward as [15]

$$\int_{C_f} \int_{-h/2}^{h/2} \underset{\sim}{v} \cdot \underset{\sim}{t}^* dF = \int_{C_f} [\underset{\sim\nu}{T}^* \cdot \underset{\sim}{u} + \underset{\sim\nu}{H}^* \cdot (\bar{\underset{\sim}{n}} - \underset{\sim}{n})] ds \; . \tag{1.12}$$

Finally, C_u and C_f may have corner points located at $s = s_{uk}$ and $s = s_{fk}$, $k = 1,2,\ldots$, respectively. In general C_u and C_f may coincide. In this case the static and geometric boundary or corner conditions, respectively, must be mutually complementary.

For an elastic, isotropic shell the strain energy Σ, per unit area of M, may be approximated as [34 - 37,12]

$$\begin{aligned} \Sigma(\gamma_{\alpha\beta},\kappa_{\alpha\beta}) &= \frac{h}{2} H^{\alpha\beta\lambda\mu}(\gamma_{\alpha\beta}\gamma_{\lambda\mu} + \frac{h^2}{12}\kappa_{\alpha\beta}\kappa_{\lambda\mu}) + O(Eh\eta^2\theta^2) \; , \\ H^{\alpha\beta\lambda\mu} &= \frac{E}{2(1+\nu)} (a^{\alpha\lambda}a^{\beta\mu} + a^{\alpha\mu}a^{\beta\lambda} + \frac{2\nu}{1-\nu} a^{\alpha\beta}a^{\lambda\mu}) \quad , \end{aligned} \tag{1.13}$$

where E and ν denote Young's modulus and Poisson's ratio, respectively. This first approximation of Σ holds to within a relative error of $O(\theta^2)$, where $\theta = \max(\sqrt{h/R}, \sqrt{\eta}, h/L, h/L^*, h/d)$, $\theta^2 \ll 1$, is a parameter which expresses the smallness of various quantities. Here R is the smallest radius of curvature of M, η is the largest principal strain in the shell, L and L* denote the smallest wave lengths of deformation and curvature patterns, respectively, on M, and d is the distance of the point under consideration from the shell boundary. A second estimation parameter is $\lambda = h/\theta = \min(\sqrt{hR}, h/\sqrt{\eta}, L, L^*, d)$ which allows to estimate derivatives of displacements, strains, stresses and curvatures as $(\;)|_\alpha = O((\;)/\lambda)$. For a detailed discussion we refer to Koiter [36, pp. 158 - 161], [37, p. 140]. With Σ given above we obtain the linear constitutive equations

$$N^{\alpha\beta} = \frac{\partial \Sigma}{\partial \gamma_{\alpha\beta}} = hH^{\alpha\beta\lambda\mu}\gamma_{\lambda\mu} \quad , \quad M^{\alpha\beta} = \frac{\partial \Sigma}{\partial \kappa_{\alpha\beta}} = \frac{h^3}{12} H^{\alpha\beta\lambda\mu}\kappa_{\lambda\mu} \quad . \tag{1.14}$$

Let $\underset{\sim}{p} = p^{\alpha}\underset{\sim}{a}_{\alpha} + p\underset{\sim}{n}$ denote the distributed surface load per unit area of M. With (1.12) and (1.13) the principle of stationary total potential energy reads $\delta I = 0$ for the functional

$$I = \iint_M [\Sigma(\underset{\sim}{u}) - \underset{\sim}{p} \cdot \underset{\sim}{u}]dA - \int_{C_f} [\underset{\sim}{T}^*_{\nu} \cdot \underset{\sim}{u} + H^*_{\nu\nu} n_{\nu} + H^*_{t\nu} n_t(\underset{\sim}{u}, n_{\nu}) + H^*_{n\nu}(n(\underset{\sim}{u}, n_{\nu}) - 1)]ds \; , \tag{1.15}$$

where the geometric boundary and corner conditions

$$\begin{aligned} &\underset{\sim}{u} = \underset{\sim}{u}^* \;, \quad n_{\nu} = n^*_{\nu} \quad \text{on } C_u \;, \\ &\underset{\sim}{u} = \underset{\sim}{u}^* \quad \text{at } s = s_{uk} + 0 \text{ and } s = s_{uk} - 0 \end{aligned} \tag{1.16}$$

are imposed as subsidiary conditions.

According to the polar decomposition theorem the deformation gradient tensor $\underset{\sim}{G}$ $(1.1)_2$ at M may be decomposed exactly into a symmetric right stretch tensor $\underset{\sim}{U}$ and a rigid-body rotation tensor $\underset{\sim}{R}$:

$$\underset{\sim}{G} = \underset{\sim}{R} \cdot \underset{\sim}{U}, \quad \underset{\sim}{U} = \check{\underset{\sim}{a}}_{\alpha}\underset{\sim}{a}^{\alpha} + \check{\underset{\sim}{n}}\underset{\sim}{n}, \quad \underset{\sim}{R} = \bar{\underset{\sim}{a}}_{\alpha}\check{\underset{\sim}{a}}^{\alpha} + \bar{\underset{\sim}{n}}\check{\underset{\sim}{n}} \quad . \tag{1.17}$$

Here, $\check{\underset{\sim}{a}}_{\alpha}$ and $\check{\underset{\sim}{n}} = \underset{\sim}{n}$ denote an intermediate basis after pure stretch along the principal directions of strain with $\check{\underset{\sim}{a}}_{\alpha} \cdot \check{\underset{\sim}{a}}_{\beta} = \bar{a}_{\alpha\beta}$. Alternatively the decomposition (1.17) may be described by means of the engineering strain tensor $\check{\underset{\sim}{\gamma}} = \check{\gamma}_{\alpha\beta}\underset{\sim}{a}^{\alpha}\underset{\sim}{a}^{\beta}$ and a finite rotation vector $\underset{\sim}{\Omega}$ defined by [12 - 14]

$$\check{\underset{\sim}{\gamma}} = \underset{\sim}{U} - \underset{\sim}{1} \quad , \quad \underset{\sim}{\Omega} = \sin\omega\,\underset{\sim}{e} = \frac{1}{2}(\check{\underset{\sim}{a}}_{\alpha} \times \bar{\underset{\sim}{a}}^{\alpha} + \underset{\sim}{n} \times \bar{\underset{\sim}{n}}). \tag{1.18}$$

Here $\underset{\sim}{1}$ is the identity tensor, $\underset{\sim}{e}$ is a unit vector along the axis of rigid-body rotation and ω denotes the rotation angle. For small strains $(1.18)_2$ yields [12 - 14]

$$\underset{\sim}{\Omega} = \frac{1}{2}\epsilon^{\beta\alpha}[(2 + \theta^{\lambda}_{\lambda})\varphi_{\alpha} - \varphi^{\lambda}(\theta_{\lambda\alpha} - \omega_{\lambda\alpha})]\underset{\sim}{a}_{\beta} + \varphi\underset{\sim}{n} \quad . \tag{1.19}$$

First Approximation Theory for Small Strains and Unrestricted Rotations

First, let us recall the equations of the entirely Lagrangian first approximation shell theory which we have given previously in [17] based on the complete non-linear strain-displacement relations for the membrane and bending strain tensors $\gamma_{\alpha\beta}$ and $\kappa_{\alpha\beta}$, respectively,

$$\gamma_{\alpha\beta} = \frac{1}{2}(\bar{a}_{\alpha\beta} - a_{\alpha\beta}) = \theta_{\alpha\beta} + \frac{1}{2}\varphi_\alpha\varphi_\beta + \frac{1}{2}\varphi^\lambda_{\cdot\alpha}\varphi_{\lambda\beta} \,, \tag{2.1}$$

$$\kappa_{\alpha\beta} = -(\bar{b}_{\alpha\beta} - b_{\alpha\beta}) = (n^\lambda|_\beta - b^\lambda_\beta n)\,1_{\lambda\alpha} + (b_{\lambda\beta}n^\lambda + n_{,\beta})\varphi_\alpha + b_{\alpha\beta}. \tag{2.2}$$

If (2.1) and (2.2) are introduced into the principle of stationary total potential energy (1.15) and (1.16) one obtains after rather involved transformations by means of partial integration and Gauss' divergence theorem with the help of (1.9) the exact equilibrium equations in M

$$T^{\alpha\beta}|_\beta - b^\alpha_\beta T^\beta + p^\alpha = 0 \quad , \quad T^\beta|_\beta + b_{\alpha\beta}T^{\alpha\beta} + p = 0 \quad , \tag{2.3}$$

where the components of the internal stress resultants are

$$\begin{aligned} T^{\lambda\beta} &= 1^\lambda_{\cdot\alpha}N^{\alpha\beta} + (n^\lambda|_\alpha - b^\lambda_\alpha n)M^{\alpha\beta} + \sqrt{\frac{a}{\bar{a}}}\,\epsilon^{\alpha\beta}\epsilon^{\lambda\mu}(1_{\mu\alpha}A - \varphi_\alpha B_\mu) - C^{\lambda\beta}H, \\ T^\beta &= \varphi_\alpha N^{\alpha\beta} + (n_{,\alpha} + b^\lambda_\alpha n_\lambda)M^{\alpha\beta} + \sqrt{\frac{a}{\bar{a}}}\,\epsilon^{\alpha\beta}\epsilon^{\lambda\mu}1_{\lambda\alpha}B_\mu - D^\beta H \,, \end{aligned} \tag{2.4}$$

with the abbreviations

$$\begin{aligned} A &= (\varphi_\kappa M^{\kappa\rho})|_\rho + 1_{\gamma\kappa}b^\gamma_\rho M^{\kappa\rho} \,, \qquad B_\mu = (1_{\mu\kappa}M^{\kappa\rho})|_\rho - \varphi_\kappa b_{\mu\rho}M^{\kappa\rho} \,, \\ C^{\lambda\beta} &= (1_{\mu\alpha}1^{\mu\alpha} + \varphi_\alpha\varphi^\alpha)1^{\lambda\beta} - (1_{\mu\alpha}1^{\mu\beta} + \varphi_\alpha\varphi^\beta)1^{\lambda\alpha}, \\ D^\beta &= 1_{\mu\alpha}1^{\mu\alpha}\varphi^\beta - 1_{\mu\alpha}1^{\mu\beta}\varphi^\alpha \,, \\ H &= \frac{a}{\bar{a}}\sqrt{\frac{a}{\bar{a}}}\,\frac{1}{2}\epsilon^{\alpha\rho}\epsilon^{\kappa\mu}[(1_{\mu\alpha}A - \varphi_\alpha B_\mu)1_{\kappa\rho} + 1_{\kappa\alpha}B_\mu\varphi_\rho] \,. \end{aligned} \tag{2.5}$$

The static boundary conditions on C_f and static corner conditions at each corner point of C_f following from (1.15) are

$$\underset{\sim}{P} = \underset{\sim}{P}^* \quad , \qquad M = M^* \qquad \text{on } C_f \quad , \tag{2.6}$$

$$\underset{\sim}{F} = \underset{\sim}{F}^* \quad \text{at} \quad s = s_{fk} + 0 \quad \text{and} \quad s = s_{fk} - 0 \quad , \tag{2.7}$$

where the components of the internal and external resultant Kirchhoff boundary loads and concentrated corner forces, respectively, are

$$\begin{pmatrix} P_\nu \\ P_t \\ P_n \end{pmatrix} = \begin{pmatrix} T^{\alpha\beta}\nu_\alpha\nu_\beta \\ T^{\alpha\beta}t_\alpha\nu_\beta \\ T^\beta\nu_\beta \end{pmatrix} - C\begin{pmatrix} k_\nu \\ k_t \\ k_n \end{pmatrix} + \frac{d}{ds}\begin{pmatrix} F_\nu \\ F_t \\ F_n \end{pmatrix} \,, \quad \begin{pmatrix} F_\nu \\ F_t \\ F_n \end{pmatrix} = C\begin{pmatrix} n_\nu \\ n_t \\ n \end{pmatrix} \;; \tag{2.8}$$

$$\begin{Bmatrix} P^*_\nu \\ P^*_t \\ P^*_n \end{Bmatrix} = \begin{Bmatrix} T^*_{\nu\nu} \\ T^*_{t\nu} \\ T^*_{n\nu} \end{Bmatrix} - C^* \begin{Bmatrix} k_\nu \\ k_t \\ k_n \end{Bmatrix} + \frac{d}{ds} \begin{Bmatrix} F^*_\nu \\ F^*_t \\ F^*_n \end{Bmatrix} , \quad \begin{Bmatrix} F^*_\nu \\ F^*_t \\ F^*_n \end{Bmatrix} = C^* \begin{Bmatrix} n_\nu \\ n_t \\ n \end{Bmatrix} ; \tag{2.9}$$

$$M = H_{\nu\nu} + FH_{t\nu} + GH_{n\nu} , \quad M^* = H^*_{\nu\nu} + FH^*_{t\nu} + GH^*_{n\nu} ; \tag{2.10}$$

with the abbreviations

$$\begin{aligned} &H_{\nu\nu} = M^{\alpha\beta} 1^{\lambda}_{\cdot\alpha} \nu_\lambda \nu_\beta ; \quad H_{t\nu} = M^{\alpha\beta} 1^{\lambda}_{\cdot\alpha} t_\lambda \nu_\beta , \quad H_{n\nu} = M^{\alpha\beta} \varphi_\alpha \nu_\beta , \\ &C = -\frac{1}{D} (H_{t\nu} n - H_{n\nu} n_t) , \quad C^* = -\frac{1}{D} (H^*_{t\nu} n - H^*_{n\nu} n_t) . \end{aligned} \tag{2.11}$$

Here in agreement with [17,19] in the definition of the abbreviations C and C*, and of the concentrated corner forces $\underset{\sim}{F}$ and $\underset{\sim}{F}^*$, a sign convention has been chosen which is opposite to that one used in [15 - 16]. The theory given above is general in the sense that no restrictions have been imposed so far on the rotations. In what follows we shall simplify this theory consistently for shell problems in which large or moderate rotations according to assumptions (III) and (II), respectively, appear. In [17] we had shown that the bending strain tensor $\kappa_{\alpha\beta}$ can be already simplified, if we introduce the small strain assumption explicitly in (2.2) and drop all terms whose contribution to the strain energy function (1.13) is of $O(Eh\eta^2\theta^2)$. Then (2.2) reads

$$\kappa_{\alpha\beta} = (m^\lambda|_\beta - b^\lambda_\beta m) 1_{\lambda\alpha} + (b_{\lambda\beta} m^\lambda + m_{,\beta}) \varphi_\alpha + b_{\alpha\beta}(1 + \gamma^\lambda_\lambda) . \tag{2.12}$$

This third-order polynomial is identical to the modified bending strain tensor $\chi_{\alpha\beta} = -(\sqrt{\frac{\bar{a}}{a}}\, \bar{b}_{\alpha\beta} - b_{\alpha\beta}) + b_{\alpha\beta}\gamma^\lambda_\lambda$ introduced in [8,10]. If (2.12) together with (1.8 - 9) is used in the principle of stationary total potential energy, the equilibrium equations (2.3) and static boundary and corner conditions (2.6) and (2.7) appear in a simpler form where now [15]

$$\begin{aligned} &T^{\lambda\beta} = 1^\lambda_{\cdot\alpha}(N^{\alpha\beta} + a^{\alpha\beta} b_{\kappa\rho} M^{\kappa\rho}) + (m^\lambda|_\alpha - b^\lambda_\alpha m) M^{\alpha\beta} + \epsilon^{\alpha\beta}\epsilon^{\lambda\mu}(1_{\mu\alpha} A - \varphi_\alpha B_\mu) , \\ &T^\beta = \varphi_\alpha(N^{\alpha\beta} + a^{\alpha\beta} b_{\kappa\rho} M^{\kappa\rho}) + (m_{,\alpha} + b^\lambda_\alpha m_\lambda) M^{\alpha\beta} + \epsilon^{\alpha\beta}\epsilon^{\lambda\mu} 1_{\lambda\alpha} B_\mu , \\ &H_{\nu\nu} = \sqrt{\frac{\bar{a}}{a}}\, M^{\alpha\beta} 1^\lambda_{\cdot\alpha} \nu_\lambda \nu_\beta , \quad H_{t\nu} = \sqrt{\frac{\bar{a}}{a}}\, M^{\alpha\beta} 1^\lambda_{\cdot\alpha} t_\lambda \nu_\beta , \quad H_{n\nu} = \sqrt{\frac{\bar{a}}{a}}\, M^{\alpha\beta} \varphi_\alpha \nu_\beta . \end{aligned} \tag{2.13}$$

It is, however, possible to simplify also the relations for the dependent rotational boundary variables (1.8) to within the relative error margin of $O(\eta)$ compatible with the relative error of the strain energy function by expanding (1.8) in a series and introducing the small strain assumption. Then the static boundary and corner conditions assume even a more reduced form given in [17].

<u>Small Strains Accompanied by Large Rotations</u>

If the shell material elements undergo small strains of $O(\eta)$ and large rotations of $O(\eta^{1/4})$, we obtain from (1.19) and (2.1) the estimates [12 - 14]

$$\varphi_\alpha = O\,(\eta^{1/4}) \quad , \quad \varphi = O(\eta^{1/4}) \quad , \quad \theta_{\alpha\beta} = O(\eta^{1/2}) \quad . \tag{3.1}$$

If it is assumed that only the rotations about tangents to the shell middle surface are large of $O\,(\eta^{1/4})$ while those about the normal remain moderate of $O(\eta^{1/2})$ we find [12 - 14]

$$\varphi_\alpha = O(\eta^{1/4}) \quad , \quad \varphi = O(\eta^{1/2}) \quad , \quad \theta_{\alpha\beta} = O(\eta^{1/2}) \quad . \tag{3.2}$$

Now all terms of the strain-displacement relations (2.1) and (2.12) can be estimated with the help of (3.1) or (3.2), and those terms whose contribution to the strain energy function is of $O(Eh\eta^2\theta^2)$ can be neglected. On the basis of (3.1) it is permissible to approximate the tensor of changes of curvature as

$$\begin{aligned}\kappa_{\alpha\beta} = [&-(\varphi_{\alpha|\beta} + b^\lambda_\beta\varphi_{\lambda\alpha}) + \varphi^\lambda_\lambda|_\beta\varphi_\alpha - \varphi_{\lambda|\beta}\varphi^\lambda_{\cdot\alpha} - \epsilon^{\kappa\rho}\epsilon^{\lambda\mu}(\varphi_\kappa\varphi_{\mu\rho})|_\beta a_{\lambda\alpha} - \\ &- b^\lambda_\beta(\varphi_\lambda\varphi_\alpha - \underline{\varphi^\kappa_\kappa\omega_{\lambda\alpha}}) + \frac{1}{2}\, b_{\alpha\beta}\varphi^\lambda\varphi_\lambda + \epsilon^{\kappa\rho}\epsilon^{\lambda\mu}\,[\frac{1}{2}\,(\varphi_{\lambda\kappa}\varphi_{\mu\rho})|_\beta\varphi_\alpha \\ &- (\varphi_\kappa\varphi_{\mu\rho})|_\beta\varphi_{\lambda\alpha}] + \underline{b^\lambda_\beta(\omega_{\lambda\alpha}\varphi^2 + \omega_{\lambda\kappa}\varphi^\kappa\varphi_\alpha)}][1 + O(\theta^2)] \quad .\end{aligned} \tag{3.3}$$

Here and in all following relations of this chapter terms marked by a solid line may additionally be dropped, if the rotations about the normal remain moderate, that means, if the estimates (3.2) are used. For lack of space we have omitted her to write (3.3) explicitly in symmetric form. The membrane strain tensor (2.1) cannot be simplified for the two theories treated simultaneously in this chapter based on the estimates (3.1) and (3.2), respectively.

Next, appropriate approximations for the dependent rotational variables n_t and $n - 1$ have to be derived whose variations can be expressed in a proper form as a function of the four independent variations δu_ν, δu_t, δw, and δn_ν. They are obtained here by a series expansion of the exact relations (1.8), truncation according to the relative error margin $O(\eta)$ compatible with the relative error of the strain energy function (1.13), and additional transformations with the help of the identities (1.7) and (2.1) as

$$
\begin{aligned}
n_t &= -(1 + \underline{\tfrac{1}{2} c_\nu^2})\varphi_t - [c_\nu(1 + \underline{\theta_{tt}}) + \underline{c_\nu^3}]n_\nu + \tfrac{1}{2}\varphi_t n_\nu^2 \ , \\
n - 1 &= -\tfrac{1}{2}\varphi_t^2 - \tfrac{1}{2}\theta_{tt}^2 + \underline{\tfrac{1}{2} c_\nu^2 \theta_{tt} + \tfrac{3}{8} c_\nu^4} - \\
&\quad - (c_\nu + \underline{c_\nu^3})\varphi_t n_\nu - (\tfrac{1}{2} c_t + \underline{\tfrac{3}{4} c_\nu^2}) n_\nu^2 - \tfrac{1}{8} n_\nu^4 \ .
\end{aligned}
\tag{3.4}
$$

Furthermore, within the same error margin we obtain from (1.2)

$$
n_\nu = -\varphi_\nu - \varphi_\nu \theta_{tt} + \varphi_t \theta_{t\nu} + \varphi_t \varphi \ . \tag{3.5}
$$

If (2.1), (3.3) and (3.4) are introduced into the principle of stationary total potential energy (1.15), one obtains under the subsidiary conditions (1.16) the following equilibrium equations in M:

$$
T^{\alpha\beta}|_\beta - b^\alpha_\beta T^\beta + p^\alpha = 0 \ , \quad T^\beta|_\beta + b_{\alpha\beta} T^{\alpha\beta} + p = 0 \ , \tag{3.6}
$$

where the components of the internal stress resultants are

$$
\begin{aligned}
T^{\lambda\beta} &= 1^\lambda_{\cdot\alpha} N^{\alpha\beta} + (m^\lambda|_\alpha - b^\lambda_\alpha) M^{\alpha\beta} - \tfrac{1}{2}\,\underline{(b^\lambda_\alpha M^{\alpha\beta} - b^\beta_\alpha M^{\lambda\alpha})(\theta^\kappa_\kappa + \varphi^2)} + \\
&\quad + [\,\underline{b^\alpha_\rho (a^{\beta\lambda} + \omega^{\beta\lambda})\omega_{\alpha\kappa}} + \underline{\tfrac{1}{2}\,(b^\beta_\rho \varphi^\lambda - b^\lambda_\rho \varphi^\beta)}\varphi_\kappa] M^{\kappa\rho} + \\
&\quad + \epsilon^{\alpha\beta}\epsilon^{\lambda\mu}[1_{\mu\alpha}(\varphi_\kappa M^{\kappa\rho})|_\rho - \varphi_\alpha (1_{\mu\kappa} M^{\kappa\rho})|_\rho] \ , \\
T^\beta &= \varphi_\alpha (N^{\alpha\beta} + a^{\alpha\beta} b_{\kappa\rho} M^{\kappa\rho}) + [m_{,\alpha} - b^\lambda_\alpha(\varphi_\lambda + \underline{\varphi^\mu \omega_{\mu\lambda}})] M^{\alpha\beta} - \\
&\quad - b_{\mu\rho}(a^{\beta\mu} + \underline{\omega^{\beta\mu}})\varphi_\kappa M^{\kappa\rho} + \epsilon^{\alpha\beta}\epsilon^{\lambda\mu} 1_{\lambda\alpha}(1_{\mu\kappa} M^{\kappa\rho})|_\rho \ .
\end{aligned}
\tag{3.7}
$$

The static boundary conditions on C_f and static corner conditions at each corner point of C_f following as natural boundary conditions from (1.5) are

$$
\underset{\sim}{P} = \underset{\sim}{P}^* \ , \quad M = M^* \qquad \text{on } C_f \ , \tag{3.8}
$$

$$
\underset{\sim}{F} = \underset{\sim}{F}^* \qquad \text{at } s = s_{fk} + 0 \ \text{ and } \ s = s_{fk} - 0 \ , \tag{3.9}
$$

where the components of the Kirchhoff boundary loads and concentrated corner forces now take the reduced forms

$$
\begin{pmatrix} P_\nu \\ P_t \\ P_n \end{pmatrix} = \begin{pmatrix} T^{\alpha\beta}\nu_\alpha \nu_\beta \\ T^{\alpha\beta} t_\alpha \nu_\beta \\ T^\beta \nu_\beta \end{pmatrix} + \begin{pmatrix} h_\nu \\ h_t \\ h_n \end{pmatrix} + \frac{d}{ds}\begin{pmatrix} F_\nu \\ F_t \\ F_n \end{pmatrix} \ , \quad \begin{pmatrix} F_\nu \\ F_t \\ F_n \end{pmatrix} = -\begin{pmatrix} f_\nu \\ \underline{f_t} \\ f_n \end{pmatrix} H_{t\nu} - \begin{pmatrix} g_\nu \\ g_t \\ g_n \end{pmatrix} H_{n\nu} \ ;
$$

$$
\begin{pmatrix} P^*_\nu \\ P^*_t \\ P^*_n \end{pmatrix} = \begin{pmatrix} T^*_{\nu\nu} \\ T^*_{t\nu} \\ T^*_{n\nu} \end{pmatrix} + \begin{pmatrix} h^*_\nu \\ h^*_t \\ h^*_n \end{pmatrix} + \frac{d}{ds}\begin{pmatrix} F^*_\nu \\ F^*_t \\ F^*_n \end{pmatrix} \ , \quad \begin{pmatrix} F^*_\nu \\ F^*_t \\ F^*_n \end{pmatrix} = -\begin{pmatrix} f_\nu \\ \underline{f_t} \\ f_n \end{pmatrix} H^*_{t\nu} - \begin{pmatrix} g_\nu \\ g_t \\ g_n \end{pmatrix} H^*_{n\nu} \ ; \tag{3.10}
$$

$$
M = H_{\nu\nu} + F H_{t\nu} + G H_{n\nu} \ , \qquad M^* = H^*_{\nu\nu} + F H^*_{t\nu} + G H^*_{n\nu} \ ;
$$

with $H_{\nu\nu}$, $H_{t\nu}$, and $H_{n\nu}$ according to (2.13) and with the abbreviations

$$\begin{Bmatrix} h_\nu \\ h_t \\ h_n \end{Bmatrix} = \begin{Bmatrix} f_t\kappa_t - f_n\tau_t \\ f_n\sigma_t - f_\nu\kappa_t \\ f_\nu\tau_t - f_t\sigma_t \end{Bmatrix} H_{t\nu} + \begin{Bmatrix} g_t\kappa_t - g_n\tau_t \\ g_n\sigma_t - g_\nu\kappa_t \\ g_\nu\tau_t - g_t\sigma_t \end{Bmatrix} H_{n\nu} ;$$

$$\begin{aligned}
f_\nu &= - [(1 + \underline{\theta_{tt}})n_\nu + \underline{\varphi_t c_\nu} + \underline{3c_\nu^2 n_\nu}], \\
f_t &= - \underline{c_\nu n_\nu} , \quad f_n = - (1 + \tfrac{1}{2} \underline{c_\nu^2}) + \tfrac{1}{2} n_\nu^2 ; \\
g_\nu &= - (1 + \underline{3c_\nu^2})\varphi_t n_\nu + \underline{c_\nu \theta_{tt}} + \underline{\tfrac{3}{2} c_\nu^3} - \underline{\tfrac{3}{2} c_\nu n_\nu^2} , \qquad (3.11) \\
g_t &= - \theta_{tt} + \underline{\tfrac{1}{2} c_\nu^2} - \tfrac{1}{2} n_\nu^2 , \quad g_n = -\varphi_t - (c_\nu + \underline{c_\nu^3})n_\nu ; \\
F &= -c_\nu(1 + \underline{\theta_{tt}}) - \underline{c_\nu^3} + \varphi_t n_\nu , \\
G &= -(c_\nu + \underline{c_\nu^3})\varphi_t - (c_t + \underline{\tfrac{3}{2} c_\nu^2})n_\nu - \tfrac{1}{2} n_\nu^3 .
\end{aligned}$$

The quantities h_ν^*, h_t^*, and h_n^* are defined analogously to $(3.11)_1$, where $H_{t\nu}$ and $H_{n\nu}$ have to be replaced by $H_{t\nu}^*$ and $H_{n\nu}^*$, respectively.

The two theories derived above for large rotations about tangents to the shell middle surface and either large or moderate rotations about the normal improve those given in [21] for which no variational principles can be constructed, and those given in [16] which are completely consistent with respect to associated variational principles but admit a greater relative error of $O(\eta^{1/2})$ in the relations for the rotational boundary variables (3.4) and (3.5).

A Simple Small Strain Large Rotation Shell Theory Assuming Small Rotations About the Normal

The assumption of large rotations about tangents to M and small rotations about the normal is reasonable for many large rotation shell problems. For this case (1.19) and (2.1) yield the estimates [12 - 14]

$$\varphi_\alpha = O(\eta^{1/4}) \quad , \quad \varphi = (\eta) \quad , \quad \theta_{\alpha\beta} = O(\eta^{1/2}) \quad . \qquad (4.1)$$

It has been shown in [21,22] that strain-displacement relations simplified with the estimates (4.1) to within the error margin $O(Eh\eta^2\theta^2)$ of (1.13) do not allow for a variational formulation of the resulting shell theory. In [23] a variationally derivable and numerically very efficient theory was given by admitting a slightly greater error $O(Eh\eta^2\theta\sqrt{\theta})$ in (1.13) and by modifying the associated simplified bending strain tensor additionally with the help of the relation $\theta_{\alpha\beta} = -\frac{1}{2}\varphi_\alpha\varphi_\beta + O(\theta^2)$ following from (2.1). Here, we do this also but propose a modified bending strain tensor which assures the consistency of our theory with respect to complementary variational principles as well :

$$\gamma_{\alpha\beta} = [\theta_{\alpha\beta} + \frac{1}{2}\varphi_\alpha\varphi_\beta + \frac{1}{2}\theta^\lambda_\alpha\theta_{\lambda\beta} - \frac{1}{2}(\theta^\lambda_\alpha\omega_{\lambda\beta} + \theta^\lambda_\beta\omega_{\lambda\alpha})][1 + O(\theta\sqrt{\theta})] ,$$

$$\kappa_{\alpha\beta} = -\frac{1}{2}[\varphi_{\alpha|\beta} + \varphi_{\beta|\alpha} + b^\lambda_\alpha(\theta_{\lambda\beta} - \underline{\omega_{\lambda\beta}}) + b^\lambda_\beta(\theta_{\lambda\alpha} - \underline{\omega_{\lambda\alpha}}) + \frac{1}{2}(\varphi_\alpha\varphi^\lambda\varphi_{\lambda|\beta} +$$
$$+ \varphi_\beta\varphi^\lambda\varphi_{\lambda|\alpha}) + b^\lambda_\beta\varphi_\lambda\varphi_\alpha + b^\lambda_\alpha\varphi_\lambda\varphi_\beta - b_{\alpha\beta}\varphi^\lambda\varphi_\lambda][1 + O(\theta\sqrt{\theta})] . \qquad (4.2)$$

Since we want to construct here a theory which is also applicable at such stages of an (incremental) loading process or in such local zones of the finitely deformed shell where the rotations remain small, we have retained in (4.2) the complete linear part of $\kappa_{\alpha\beta}$. For shell problems with small rotations about the normal considered here the terms underlined in (4.2) may be dropped if the rotations about tangents to M exceed $O(\eta)$.

If $(4.2)_1$ and $(4.2)_2$ are used in the principle of stationary total potential energy (1.15), it turns out that in order to obtain a consistent theory the rotational boundary variables should be approximated as

$$n_\nu = -\varphi_\nu , \quad n_t = -\varphi_t , \quad n - 1 = -\frac{1}{2}\varphi_\nu^2 - \frac{1}{2}\varphi_t^2 \qquad (4.3)$$

which holds to within a relative error of $O(\eta^{1/2})$. The equilibrium equations obtained as Euler-Lagrange equations from (1.15) are

$$T^{\alpha\beta}|_\beta - b^\alpha_\beta T^\beta + p^\alpha = 0 , \quad T^\beta|_\beta + b_{\alpha\beta}T^{\alpha\beta} + p = 0 , \qquad (4.4)$$

where now

$$T^{\lambda\beta} = N^{\lambda\beta} - \frac{1}{2}(b^\lambda_\alpha M^{\alpha\beta} + b^\beta_\alpha M^{\alpha\lambda}) - \underline{\frac{1}{2}(b^\lambda_\alpha M^{\alpha\beta} - b^\beta_\alpha M^{\alpha\lambda})} +$$
$$+ \frac{1}{2}(\varphi^{\lambda\alpha} + \theta^{\lambda\alpha})N^\beta_\alpha - \frac{1}{2}(\varphi^{\alpha\beta} - \theta^{\alpha\beta})N^\lambda_\alpha ,$$
$$T^\beta = \varphi_\alpha(N^{\alpha\beta} + a^{\alpha\beta}b_{\kappa\rho}M^{\kappa\rho}) + [(\delta^\beta_\lambda + \frac{1}{2}\varphi_\lambda\varphi^\beta)M^{\rho\lambda}]|_\rho -$$
$$- \varphi_\alpha[(\frac{1}{2}\varphi^\alpha|_\lambda + b^\alpha_\lambda)M^{\beta\lambda} + (\frac{1}{2}\varphi^\beta|_\lambda + b^\beta_\lambda)M^{\alpha\lambda}] . \qquad (4.5)$$

The static boundary conditions on C_f and static corner conditions at each corner point of C_f obtained as natural boundary conditions from (1.15) are

$$\underset{\sim}{P} = \underset{\sim}{P}^* , \quad M = M^* \quad \text{on } C_f , \qquad (4.6)$$

$$F_n(s_{fk}+0) - F_n(s_{fk}-0) = F^*_n(s_{fk}+0) - F^*_n(s_{fk}-0) , \qquad (4.7)$$

where the components of the internal and external resultant Kirchhoff boundary loads and concentrated corner forces, respectively, are

$$
\begin{aligned}
P_\nu &= T^{\alpha\beta}\nu_\alpha\nu_\beta + \tau_t F_n \quad , & P^*_\nu &= T^*_{\nu\nu} + \tau_t F^*_n \quad , \\
P_t &= T^{\alpha\beta}t_\alpha\nu_\beta - \sigma_t F_n \quad , & P^*_t &= T^*_{t\nu} - \sigma_t F^*_n \quad , \\
P_n &= T^\beta\nu_\beta + \frac{d}{ds} F_n \quad , & P^*_n &= T^*_{n\nu} + \frac{d}{ds} F^*_n \quad ;
\end{aligned}
\tag{4.8}
$$

$$
M = H_{\nu\nu} + \varphi_\nu H_{n\nu} \quad , \quad M^* = H^*_{\nu\nu} + \varphi_\nu H^*_{n\nu} \quad , \quad F_n = H_{t\nu} + \varphi_t H_{n\nu} \quad ,
$$

$$
F^*_n = H^*_{t\nu} + \varphi_t H^*_{n\nu} \quad ;
$$

with the abbreviations

$$
\begin{aligned}
H_{\nu\nu} &= (1 - \tfrac{1}{2}\varphi_\nu^2)M_{\nu\nu} - \tfrac{1}{2}\varphi_\nu\varphi_t M_{t\nu} \quad , \\
H_{t\nu} &= (1 - \tfrac{1}{2}\varphi_t^2)M_{t\nu} - \tfrac{1}{2}\varphi_\nu\varphi_t M_{\nu\nu} \quad , \qquad H_{n\nu} = \varphi_\nu M_{\nu\nu} + \varphi_t M_{t\nu} \ .
\end{aligned}
\tag{4.9}
$$

Small Strains Accompanied by Moderate Rotations

If the shell material elements undergo small strains of $O(\eta)$ and moderate rotations of $O(\eta^{1/2})$, (1.19) and (2.1) yield the estimates [12 - 14]

$$
\varphi_\alpha = O(\eta^{1/2}) \quad , \quad \varphi = O(\eta^{1/2}) \quad , \quad \theta_{\alpha\beta} = O(\eta) \quad . \tag{5.1}
$$

If only the rotations about tangents to the shell middle surface are moderate of $O(\eta^{1/2})$, while those about the normal are assumed small, we find [12 - 14]

$$
\varphi_\alpha = O(\eta^{1/2}) \quad , \quad \varphi = O(\eta) \quad , \quad \theta_{\alpha\beta} = O(\eta) \quad . \tag{5.2}
$$

On the basis of (5.1) it is possible to simplify the strain-displacement relations consistently with the error margin $O(Eh\eta^2\theta^2)$ of the strain energy function Σ (1.13) as

$$
\begin{aligned}
\gamma_{\alpha\beta} &= [\theta_{\alpha\beta} + \tfrac{1}{2}\varphi_\alpha\varphi_\beta + \underline{\tfrac{1}{2}a_{\alpha\beta}\varphi^2} - \underline{\tfrac{1}{2}(\theta^\lambda_\alpha\omega_{\lambda\beta} + \theta^\lambda_\beta\omega_{\lambda\alpha})}][1 + O(\theta^2)] \ , \\
\kappa_{\alpha\beta} &= -\frac{1}{2}[\varphi_{\alpha|\beta} + \varphi_{\beta|\alpha} + b^\lambda_\alpha(\underset{\cdots}{\theta_{\lambda\beta}} - \underset{-\cdot-}{\omega_{\lambda\beta}}) + b^\lambda_\beta(\underset{\cdots}{\theta_{\lambda\alpha}} - \underset{-\cdot-}{\omega_{\lambda\alpha}})][1 + O(\theta^2)]
\end{aligned}
\tag{5.3}
$$

where terms marked by a solid line may be omitted additionally,if the rotations about the normal remain small, that means if (5.2) is valid. On the basis of (5.1) the terms marked by dots, and on the basis of (5.2) additionally the terms marked by a dot and dash line may be neglected as well, if the rotations about tangents to M really exceed $O(\eta)$. With (5.3) the principle of stationary total potential energy yields that for reasons of consistency the approximations $n_\nu = -\varphi_\nu$, $n_t = -\varphi_t$, $n = 1$ have to be used which hold to within a relative error margin $O(\eta^{1/2})$. The equilibrium equations are obtained in the form (4.4) where now [12,13,24,26]

$$T^{\lambda\beta} = N^{\lambda\beta} - \frac{1}{2}(b^{\lambda}_{\alpha}M^{\alpha\beta} + b^{\beta}_{\alpha}M^{\alpha\lambda}) - \frac{1}{2}(b^{\lambda}_{\alpha}M^{\alpha\beta} - b^{\beta}_{\alpha}M^{\alpha\lambda}) - \frac{1}{2}\omega^{\lambda\beta}N^{\alpha}_{\alpha} +$$
$$+ \frac{1}{2}\varphi^{\lambda\alpha}N^{\beta}_{\alpha} - \frac{1}{2}\varphi^{\alpha\beta}N^{\lambda}_{\alpha} \quad , \qquad T^{\beta} = M^{\alpha\beta}|_{\alpha} + \varphi_{\alpha}N^{\alpha\beta} \; . \tag{5.4}$$

The static boundary conditions on C_f and static corner conditions at each corner point of C_f take the form (4.6) - (4.8) where now

$$M = M_{\nu\nu} \quad , \quad M^* = H^*_{\nu\nu} \quad ; \quad F_n = M_{t\nu} \quad , \quad F^*_n = H^*_{t\nu} \; . \tag{5.5}$$

In [24,26] we have indicated explicitly the simplifications which have to be introduced into the moderate rotation shell theory given here in order to reduce it to those derived in [1 - 8]. Finally, if all rotations are assumed small, the shell equations given here yield the classic linear shell theory.

References

1. Marguerre, K.: Proc. 5th Int. Congr. of Appl. Mech. Cambridge/Mass. 1938, 93-101, Wiley and Sons 1939.
2. Donnell, L. H.: NACA TR 479, 1933.
3. Mushtari, K. M.: Izv. fiz.-mat. ob-va pri kazan' un-te 11 (1938), 71 - 150 .
4. Vlasov, V. Z.: NASA TT F-99, 1964.
5. Mushtari, K. M.; Galimov, K. Z.: Non-Linear Theory of Thin Elastic Shells, Jerusalem: The Israel Program for Sci. Transl. 1961.
6. Leonard, R. W.: Nonlinear First Approximation Thin Shell and Membrane Theory, Diss., Virginia Polytechnic Institute 1961.
7. Sanders, J. L.: Quart. Appl. Math. 21 (1963), 21 - 36.
8. Koiter, W. T.: Proc. Kon. Ned. Ak. Wet., Ser. B, 69 (1966), 1 - 54.
9. Reissner, E.: Proc. Symp. in Appl. Math., vol. 3, 27 - 52, McGraw-Hill 1950.
10. Budiansky, B.: J. Appl. Mech. 35 (1968), 393 - 401.
11. Reissner, E.: H. Reissner Anniv. Vol., 231 - 247, Ann Arbor 1949.
12. Pietraszkiewicz, W.: Mitt. Inst. f. Mech., Ruhr-Universität Bochum, 10, 1977.
13. Pietraszkiewicz , W.: In: Thin Shell Theory, New Trends and Applications, 153 - 208, Wien-New York: Springer-Verlag 1980.
14. Pietraszkiewicz, W.: In: Theory of Shells, 445 - 471, Amsterdam-New York-Oxford : North-Holland Publ. Co. 1980.
15. Pietraszkiewicz, W.; Szwabowicz, M. L.: Arch. of Mech. 33 (1981), 273 -288.
16. Schmidt, R.: Proc. Int. Conf. on Finite Element Methods, Shanghai (China), 621 - 626, New York: Gordon and Breach 1982.

17. Schmidt R.: On the Entirely Lagrangian First Approximation Theory of Thin Elastic Shells at Small Strains and Arbitrary Rotations. Lecture GAMM Annual Sci. Conf., Hamburg, March 1983, accepted for publication in ZAMM 64 (1984), presumably Nr. 6 .

18. Schmidt, R.: ZAMM 62 (1982), T165 - T167.

19. Schmidt, R.; Stumpf, H.: On the Stability and Post-Buckling of Thin Elastic Shells with Unrestricted Rotations, submitted for publication in Mechanics Research Communications.

20. Nolte, L. P.: ZAMM 63 (1983), T79 - T82.

21. Pietraszkiewicz, W.: Mitt. Inst. f. Mech., Ruhr-Univ. Bochum, 26, 1981.

22. Pietraszkiewicz, W.: ZAMM 63 (1983), T200 - T202.

23. Nolte, L.-P.; Stumpf, H.: Mech. Res. Comm. 10 (1983), 213 - 221.

24. Schmidt, R.: Variational Principles for Geometrically Non-Linear Theories of Shells Undergoing Moderate Rotations (in German), Diss., Ruhr-Univ. Bochum 1980.

25. Schmidt, R.: Proc. 4th Seminar about Finite Element Method and Variational Method, Plzen (CSSR), vol. II, 363 - 365, 1981.

26. Schmidt, R.; Pietraszkiewicz, W.: Ing.-Arch. 50 (1981), 187 - 201.

27. Stumpf, H.: Ing.-Arch. 48 (1979), 221 - 237.

28. Stein, E.: In: Theory of Shells, 509 - 535, Amsterdam-New York-Oxford: North-Holland Publ. Co. 1980.

29. Stumpf, H.: Ing.-Arch. 51 (1981), 195 - 213.

30. Stumpf, H.: In: Stability in the Mechanics of Continua, 89 - 100, Berlin-Heidelberg-New York: Springer-Verlag 1982.

31. Stumpf, H.: Mitt. Inst. f. Mech., Ruhr-Univ. Bochum, 34, 1982.

32. Stein, E.; Berg, A.; Wagner, W.: In: Buckling of Shells, 91 - 136, Berlin-Heidelberg-New York: Springer-Verlag 1982.

33. Makowski, J.: Linear and Non-Linear Analysis of Elastic Stability of Thin Shells (in Polish), Diss., Gdańsk 1981.

34. Koiter, W. T.: In: Theory of Thin Elastic Shells, 12 - 33, Amsterdam: North-Holland Publ. Co. 1960.

35. John, F.: Comm. Pure Appl. Math. 18 (1965), 235 - 267.

36. Koiter, W. T.; Simmonds, J. G.: In: Theoretical and Applied Mechanics, 150 - 176, Berlin: Springer-Verlag 1973.

37. Koiter, W.T.: Mechanics Today, vol. 5, 139 - 154, 1980.

Buckling and Post-Buckling of Shells for Unrestricted and Moderate Rotations

H. STUMPF

Lehrstuhl für Mechanik II
Ruhr-Universität Bochum
4630 Bochum 1, West-Germany

Summary

In the first part of this paper a Lagrangean nonlinear theory of thin elastic shells for unrestricted rotations is considered, where the boundary conditions are non-rational functions of the shell deformations. Using energy considerations the equations of critical equilibrium are derived, which define the general eigenvalue problem of shell buckling. Furthermore post-buckling equations are obtained by application of the static perturbation technique.
If the rotations of the shell elements can be restricted to be moderate, essential simplifications in the prebuckling, buckling and post-buckling equations are achieved, which is shown in the second part.

Zusammenfassung

Im ersten Teil dieser Arbeit wird eine Lagrangesche nichtlineare Theorie dünner elastischer Schalen für unbeschränkte Rotationen betrachtet, bei der die Randbedingungen nicht-rationale Funktionen der Schalendeformationen sind. Unter Benutzung energetischer Überlegungen werden die Gleichungen für kritisches Gleichgewicht hergeleitet, die das allgemeine Eigenwertproblem des Schalenbeulens definieren. Darüber hinaus lassen sich mit Hilfe der statischen Perturbationsmethode Gleichungen zur Beschreibung des Nachbeulverhaltens bestimmen.
Können die Rotationen der Schalenelemente auf eine mittlere Größenordnung beschränkt werden, so ergeben sich erhebliche Vereinfachungen in den Vorbeul-, Beul- und Nachbeulgleichungen, was im zweiten Teil gezeigt wird.

Introduction

In the frame of a geometrically nonlinear first-approximation shell theory, where strains are assumed to be small, energy-consistent shell equations in Lagrangean description or their variational formulation are of special interest for numerical applications [1 - 23]. Consistently simplified shell theories can be obtained by introducing restrictions on the rotations of shell elements, which can be small, moderate, large or, in the general case, unrestricted [5].
The moderate rotation shell theory yields an adequate description for a wide area of shell problems of engineering interest. Associated variational

principles are given in [6 - 9], stability equations in [10 - 13] and energy-consistent buckling and post-buckling equations including the geometric and nonlinear static boundary conditions in [14 - 16].
A Lagrangean shell theory for unrestricted rotations had been presented in [17] using a modified change of curvature tensor, which is a third-order polynomial in the displacements and their derivatives, while the boundary conditions are non-rational functions of the displacements. Associated variational principles can be found in [18,19], stability equations in [20] and stability equations and post-buckling equations in [21].
A Lagrangean shell theory for large rotations, which avoids any non-rational dependency of the governing equations and which is derivable from variational principles, had been presented in [22]. In [23] various beam and shell problems are calculated numerically by using moderate, large or unrestricted rotation shell theories. The results, which are compared with those published in the literature, yield statements about the range of applicability of various nonlinear shell theories.

In the first part of this paper the geometrically nonlinear shell theory for small strains and unrestricted rotations is considered. Using energetic considerations the stability and post-buckling equations are derived in a compact form. It is shown that the equations, describing the general eigenvalue problem of shell buckling, are the Gâteaux differential of the nonlinear governing shell equations.
For many shell problems of engineering interest the rotations of shell elements can be restricted such that the squares of the rotations are small of the same order as the strains, leading to the so-called moderate rotation shell theory. In this case all equations, derived for the unrestricted rotation shell theory in the first part, can be simplified essentially, what is shown in the second part of this paper.

Notations and Basic Shell Relations

In this paper the following notations and basic shell relations will be used, where geometric quantities in the deformed shell configuration are indicated by a bar and given quantities by an asterisk:

M	shell middle surface
θ^{α}, $\alpha = 1,2$	surface coordinates on M
$(\cdot)_{,\alpha}$	partial differentiation

$(\cdot)_{\|\alpha}$	covariant differentiation
$r(\theta^{\alpha})$	position vector
$a_{\alpha} = r_{,\alpha}$	covariant base vector on M
$n = \frac{1}{2}\,\epsilon^{\alpha\beta} a_{\alpha} \times a_{\beta}$	unit vector orthogonal to M
$a_{\alpha\beta} = a_{\alpha} \cdot a_{\beta}$	metric tensor
$a = \det(a_{\alpha\beta})$	determinant of the metric tensor
$\epsilon_{\alpha\beta}$	surface permutation tensor
$b_{\alpha\beta} = a_{\alpha,\beta} \cdot n$	curvature tensor
$p = p^{\alpha} a_{\alpha} + p n$	surface load
$u = u^{\alpha} a_{\alpha} + w n$	displacement vector
$\bar{a}_{\alpha} = l^{\lambda}_{\cdot\alpha} a_{\lambda} + \varphi_{\alpha} n$	base vector on $\bar{M}$
$\bar{n} = n^{\lambda} a_{\lambda} + n n$	unit vector orthogonal to $\bar{M}$
$\theta_{\alpha\beta} = \frac{1}{2}\,(u_{\alpha\|\beta} + u_{\beta\|\alpha}) - b_{\alpha\beta} w$	linearized strain tensor
$\omega_{\alpha\beta} = \frac{1}{2}\,(u_{\beta\|\alpha} - u_{\alpha\|\beta}) = \epsilon_{\alpha\beta}\varphi$	linearized rotation tensor
$\varphi_{\alpha} = w_{,\alpha} + b^{\lambda}_{\alpha} u_{\lambda}$	linearized rotation tangent to M
$\varphi = \frac{1}{2}\,\epsilon^{\alpha\beta}\omega_{\alpha\beta}$	linearized rotation normal to M
$l_{\alpha\beta} = a_{\alpha\beta} + \theta_{\alpha\beta} - \omega_{\alpha\beta}$	nonsymmetric tensor
$\gamma_{\alpha\beta} = \frac{1}{2}\,(\bar{a}_{\alpha\beta} - a_{\alpha\beta})$	membrane strain tensor, defined by (I.1)
$\chi_{\alpha\beta} = -(\sqrt{\frac{\bar{a}}{a}}\,\bar{b}_{\alpha\beta} - b_{\alpha\beta}) + b_{\alpha\beta}\gamma^{\lambda}_{\lambda}$	modified tensor of change of curvature, defined by (I.4)
$m^{\lambda} = \sqrt{\frac{\bar{a}}{a}}\, n^{\lambda} = -\epsilon^{\kappa\rho}\epsilon^{\lambda\mu}\varphi_{\kappa} l_{\mu\rho}$	modified normal vector components
$m = \sqrt{\frac{\bar{a}}{a}}\, n = \frac{1}{2}\,\epsilon^{\kappa\rho}\epsilon^{\lambda\mu} l_{\lambda\kappa} l_{\mu\rho}$	modified normal vector component
h	shell thickness
E	Young's modulus
ν	Poisson's ratio
$H^{\alpha\beta\lambda\mu}$	modified elasticity tensor
$N^{\alpha\beta}$	membrane stress tensor
$M^{\alpha\beta}$	stress couple tensor
f	strain energy density
J	total potential energy

$(.)^{(n)}$	n-th Gâteaux differential
C	boundary contour of M
C_f	boundary curve, static quantities given
C_u	boundary curve, geometric quantities given
s	length parameter on C
$\boldsymbol{t} = \boldsymbol{r}_{,s} = t^{\alpha}\boldsymbol{a}_{\alpha}$	unit vector tangent to C
$\underset{\sim}{\nu} = \boldsymbol{t} \times \boldsymbol{n} = \nu^{\alpha}\boldsymbol{a}_{\alpha}$	unit outward vector orthogonal to C
$\sigma_t = b_{\alpha\beta}t^{\alpha}t^{\beta}$	normal curvature of C
$\tau_t = -b_{\alpha\beta}\nu^{\alpha}t^{\beta}$	geodesic torsion of C
$\kappa_t = t^{\alpha}t^{\beta}\nu_{\alpha\|\beta}$	geodesic curvature of C
$T^*_{\nu} = T^*_{\nu\nu}\underset{\sim}{\nu} + T^*_{t\nu}\boldsymbol{t} + T^*_{n\nu}\boldsymbol{n}$	distributed boundary force
$H^*_{\nu} = H^*_{\nu\nu}\underset{\sim}{\nu} + H^*_{t\nu}\boldsymbol{t} + H^*_{n\nu}\boldsymbol{n}$	distributed static moment
$\boldsymbol{u} = u_{\nu}\underset{\sim}{\nu} + u_t\boldsymbol{t} + w\boldsymbol{n}\ (u_{\nu} = u^{\lambda}\nu_{\lambda},\ u_t = u^{\lambda}t_{\lambda})$	displacement vector at C
$\bar{\boldsymbol{n}} = n_{\nu}\underset{\sim}{\nu} + n_t\boldsymbol{t} + n\boldsymbol{n}\ (n_{\nu} = n^{\lambda}\nu_{\lambda},\ n_t = n^{\lambda}t_{\lambda})$	unit normal vector on $\bar{C}$

PART I: UNRESTRICTED ROTATION SHELL THEORY

Lagrangean Governing Shell Equations

In the theory of thin elastic shells the Lagrangean strain tensor $\gamma_{\alpha\beta}$ and the tensor of change of curvature $\kappa_{\alpha\beta}$ can be defined by:

$$\gamma_{\alpha\beta} = \frac{1}{2}(\bar{a}_{\alpha\beta} - a_{\alpha\beta}) = \frac{1}{2}(l^{\lambda}_{\cdot\alpha}l_{\lambda\beta} + \varphi_{\alpha}\varphi_{\beta} - a_{\alpha\beta}) \tag{I.1}$$

$$\kappa_{\alpha\beta} = -(\bar{b}_{\alpha\beta} - b_{\alpha\beta}) = -[n(\varphi_{\alpha|\beta} + b_{\lambda\beta}l^{\lambda}_{\cdot\alpha}) + n_{\lambda}(l^{\lambda}_{\cdot\alpha|\beta} - b^{\lambda}_{\beta}\varphi_{\alpha}) - b_{\alpha\beta}], \tag{I.2}$$

where $\gamma_{\alpha\beta}$ is a quadratic polynomial in u_{α}, w and their derivatives and $\kappa_{\alpha\beta}$ a non-rational function containing the square-root $\sqrt{\frac{a}{\bar{a}}}$ of the invariant:

$$\frac{\bar{a}}{a} = 1 + 2\gamma^{\alpha}_{\alpha} + 2(\gamma^{\alpha}_{\alpha}\gamma^{\beta}_{\beta} - \gamma^{\alpha}_{\beta}\gamma^{\beta}_{\alpha}) \ . \tag{I.3}$$

In the Lagrangean shell theory of [17] a modified tensor of change of curvature $\chi_{\alpha\beta}$ is introduced by:

$$\begin{aligned}\chi_{\alpha\beta} &= -(\sqrt{\frac{\bar{a}}{a}}\,\bar{b}_{\alpha\beta} - b_{\alpha\beta}) + b_{\alpha\beta}\gamma^{\lambda}_{\lambda} \\ &= (m^{\lambda}|_{\beta} - b^{\lambda}_{\beta}m)l_{\lambda\alpha} + (b_{\lambda\beta}m^{\lambda} + m|_{\beta})\varphi_{\alpha} + b_{\alpha\beta}(1 + \gamma^{\lambda}_{\lambda}) \ ,\end{aligned} \tag{I.4}$$

which is a third-degree polynomial of u_α, w and their derivatives. The tensor $\chi_{\alpha\beta}$ differs from the tensor $K_{\alpha\beta}$ defined in [1] only by the additional term $1/2(b^\lambda_\alpha\gamma_{\lambda\beta} + b^\lambda_\beta\gamma_{\lambda\alpha})$.

In the first approximation shell theory the strain energy density is given by:

$$f = \frac{h}{2} H^{\alpha\beta\lambda\mu}(\gamma_{\alpha\beta}\gamma_{\lambda\mu} + \frac{h^2}{12}\chi_{\alpha\beta}\chi_{\lambda\mu}), \quad H^{\alpha\beta\lambda\mu} = \frac{E}{2(1+\nu)}(a^{\alpha\lambda}a^{\beta\mu} + a^{\alpha\mu}a^{\beta\lambda} + \frac{2\nu}{1-\nu} a^{\alpha\beta}a^{\lambda\mu}) \tag{I.5}$$

with the associated stress-strain relations:

$$N^{\alpha\beta} = hH^{\alpha\beta\lambda\mu}\gamma_{\lambda\mu}; \quad M^{\alpha\beta} = \frac{h^3}{12} H^{\alpha\beta\lambda\mu}\chi_{\lambda\mu} . \tag{I.6}$$

If the boundary force T^*_ν and the static moment H^*_ν, per unit length of C_f, are prescribed resultants of the external stress vector distribution of dead-load type, the functional of total potential energy is defined by [17 - 19]:

$$J(u) = \iint_M [f(u) - p \cdot u]dA - \int_{C_f} [T^*_\nu \cdot u + H^*_\nu \cdot (\bar{n} - n)]\, ds \tag{I.7}$$

The deformation of the shell boundary can be described by using the three displacement components u_ν, u_t, w and the three components of the vector $\bar{n} = n_\nu \underset{\sim}{\nu} + n_t t + n\, n$ referred to the base vectors of the undeformed shell boundary. If a vector $\bar{a}_t$ tangent to the deformed boundary $\bar{C}$ is introduced:

$$\bar{a}_t = \bar{r}_{,s} = t + u_{,s} = c_\nu \underset{\sim}{\nu} + c_t t + c_n n$$
$$c_\nu = u_{\nu,s} + \tau_t w - \kappa_t u_t , \quad c_t = 1 + u_{t,s} + \kappa_t u_\nu - \sigma_t w , \quad c_n = w_{,s} + \sigma_t u_t - \tau_t u_\nu , \tag{I.8}$$

the Kirchhoff-Love constraints

$$\bar{a}_t \cdot \bar{n} = 0 \quad , \quad \bar{n} \cdot \bar{n} = 1 \tag{I.9}$$

yield the components n_t, n as function of u and $n_\nu = n_\lambda \nu^\lambda$ [5,17]:

$$n_t = - \frac{c_\nu c_t n_\nu - c_n D}{c_t^2 + c_n^2} , \quad n = - \frac{c_\nu c_n n_\nu + c_t D}{c_t^2 + c_n^2} , \quad D = -\sqrt{(c_\nu^2 + c_t^2 + c_n^2)(1-n_\nu^2) - c_\nu^2} . \tag{I.10}$$

To describe the nonlinear boundary value problem of thin elastic shells with unrestricted rotations the set of shell equations has to be completed by the equilibrium equations and the static and geometric boundary conditions:

Equilibrium equations

$$\left.\begin{aligned} T^{\alpha\beta}|_{\beta} - b^{\alpha}_{\beta}T^{\beta} + p^{\alpha} &= 0 \\ T^{\beta}|_{\beta} + b_{\alpha\beta}T^{\alpha\beta} + p &= 0 \end{aligned}\right\} \quad \text{in } M, \tag{I.11}$$

Static boundary conditions

$$\left.\begin{aligned} P_{\nu\nu}(u) &= P^{*}_{\nu\nu}(u) \\ P_{t\nu}(u) &= P^{*}_{t\nu}(u) \\ P_{n\nu}(u) &= P^{*}_{n\nu}(u) \\ M(u) &= M^{*}(u) \end{aligned}\right\} \quad \text{on } C_f \tag{I.12}$$

with static corner conditions

$$\left.\begin{aligned} F_{\nu}(u) &= F^{*}_{\nu}(u) \\ F_{t}(u) &= F^{*}_{t}(u) \\ F_{n}(u) &= F^{*}_{n}(u) \end{aligned}\right\} \quad \text{at } s_i \in C_f,\ i = 1,\ldots; \tag{I.13}$$

Geometric boundary conditions

$$\left.\begin{aligned} u_{\nu} &= u^{*}_{\nu} \\ u_{t} &= u^{*}_{t} \\ w &= w^{*} \\ n_{\nu} &= n^{*}_{\nu} \end{aligned}\right\} \quad \text{on } C_u \tag{I.14}$$

with geometric corner conditions

$$\left.\begin{aligned} u_{\nu} &= u^{*}_{\nu} \\ u_{t} &= u^{*}_{t} \\ w &= w^{*} \end{aligned}\right\} \quad \text{at } s_j \in C_u,\ j = 1,\ldots, \tag{I.15}$$

where the stress tensors, stress resultants, internal and external boundary loads as well as the corner forces in (I.11 - 15) are defined in appendix I.

Taylor Expansion of Total Potential Energy

Let $\bar{u}$ denote the displacement field in a fundamental equilibrium state and let $u = \bar{u} + \hat{u}$ be the displacement field of a geometrically admissible adjacent configuration such, that $\hat{u}$ satisfies homogenous geometric boundary conditions on C_u. Then the increments of the strain measures can be obtained in the form:

$$\Delta\gamma_{\alpha\beta} = \gamma_{\alpha\beta}^{(1)}(\bar{u};\hat{u}) + \frac{1}{2!}\,\gamma_{\alpha\beta}^{(2)}(\bar{u};\hat{u}^2)$$
$$\Delta\chi_{\alpha\beta} = \chi_{\alpha\beta}^{(1)}(\bar{u};\hat{u}) + \frac{1}{2!}\,\chi_{\alpha\beta}^{(2)}(\bar{u};\hat{u}^2) + \frac{1}{3!}\,\chi_{\alpha\beta}^{(3)}(\bar{u};\hat{u}^3), \tag{I.16}$$

where $\gamma_{\alpha\beta}^{(n)}(\bar{u};\hat{u}^n)$, $\chi_{\alpha\beta}^{(n)}(\bar{u};\hat{u}^n)$ denote the n-th Gâteaux differentials of $\gamma_{\alpha\beta}$, $\chi_{\alpha\beta}$ at $\bar{u}$ with respect to $\hat{u}$, given in detail in [21].
A Taylor expansion of the increment of the strain energy density leads to a representation by a finite series:

$$\Delta f = f^{(1)}(\bar{u};\hat{u}) + \frac{1}{2!}\,f^{(2)}(\bar{u};\hat{u}^2) + \frac{1}{3!}\,f^{(3)}(\bar{u};\hat{u}^3) + \frac{1}{4!}\,f^{(4)}(\bar{u};\hat{u}^4)$$
$$+ \frac{1}{5!}\,f^{(5)}(\bar{u};\hat{u}^5) + \frac{1}{6!}\,f^{(6)}(\bar{u};\hat{u}^6) \tag{I.17}$$

with the differentials:

$$f^{(1)}(\bar{u};\hat{u}) = N^{\alpha\beta}(\bar{u})\,\gamma_{\alpha\beta}^{(1)}(\bar{u};\hat{u}) + M^{\alpha\beta}(\bar{u})\chi_{\alpha\beta}^{(1)}(\bar{u};\hat{u}) \ ,$$
$$f^{(2)}(\bar{u};\hat{u}^2) = hH^{\alpha\beta\lambda\mu}[\gamma_{\alpha\beta}^{(1)}\gamma_{\lambda\mu}^{(1)} + \frac{h^2}{12}\,\chi_{\alpha\beta}^{(1)}\chi_{\lambda\mu}^{(1)}] + N^{\alpha\beta}(\bar{u})\gamma_{\alpha\beta}^{(2)} + M^{\alpha\beta}(\bar{u})\chi_{\alpha\beta}^{(2)} \ ,$$
$$f^{(3)}(\bar{u};\hat{u}^3) = 3hH^{\alpha\beta\lambda\mu}[\gamma_{\alpha\beta}^{(1)}\gamma_{\lambda\mu}^{(2)} + \frac{h^2}{12}\,\chi_{\alpha\beta}^{(1)}\chi_{\lambda\mu}^{(2)}] + M^{\alpha\beta}(\bar{u})\chi_{\alpha\beta}^{(3)} \ ,$$
$$f^{(4)}(\bar{u};\hat{u}^4) = hH^{\alpha\beta\lambda\mu}\{3\gamma_{\alpha\beta}^{(2)}\gamma_{\lambda\mu}^{(2)} + \frac{h^2}{12}[3\chi_{\alpha\beta}^{(2)}\chi_{\lambda\mu}^{(2)} + 4\chi_{\alpha\beta}^{(1)}\chi_{\lambda\mu}^{(3)}]\}, \tag{I.18}$$
$$f^{(5)}(\bar{u};\hat{u}^5) = \frac{5}{6}\,h^3H^{\alpha\beta\lambda\mu}\chi_{\alpha\beta}^{(2)}\chi_{\lambda\mu}^{(3)} \ ,$$
$$f^{(6)}(\bar{u};\hat{u}^6) = \frac{5}{6}\,h^3H^{\alpha\beta\lambda\mu}\chi_{\alpha\beta}^{(3)}\chi_{\lambda\mu}^{(3)} \ .$$

Because of the non-rational contribution of the boundary integral in (I.7) a Taylor expansion of the total potential energy increment ΔJ yields an infinite series:

$$\Delta J = J^{(1)}(\bar{u};\hat{u}) + \frac{1}{2!}\,J^{(2)}(\bar{u};\hat{u}^2) + \frac{1}{3!}\,J^{(3)}(\bar{u};\hat{u}^3) + \frac{1}{4!}\,J^{(4)}(\bar{u};\hat{u}^4) +$$
$$\frac{1}{5!}\,J^{(5)}(\bar{u};\hat{u}^5) + \frac{1}{6!}\,J^{(6)}(\bar{u};\hat{u}^6) + \ldots \tag{I.19}$$

with the differentials:

$$J^{(1)}(\bar{u};\hat{u}) = \iint_M [f^{(1)}(\bar{u};\hat{u}) - p\cdot\hat{u}]dA - \int_{C_f}[T_\nu^{*}\cdot\hat{u} + H_\nu^{*}\cdot n^{(1)}(\bar{u};\hat{u})]ds,$$
$$J^{(n)}(\bar{u};\hat{u}^n) = \iint_M f^{(n)}(\bar{u};\hat{u}^n)dA - \int_{C_f} H_\nu^{*}\cdot n^{(n)}(\bar{u};\hat{u}^n)ds, \quad n = 2,\ldots,6 \ ,$$
$$J^{(n)}(\bar{u};\hat{u}^n) = -\int_{C_f} H_\nu^{*}\cdot n^{(n)}(\bar{u};\hat{u}^n)ds, \quad n \geq 7 \ . \tag{I.20}$$

In (I.20) the Gâteaux differentials $n_t^{(n)}(\bar{u};\hat{u}^n)$, $n^{(n)}(\bar{u};\hat{u}^n)$ have to be expressed in terms of $\hat{u}_\nu, \hat{u}_t, \hat{w}$ and $n_\nu^{(n)}(\bar{u};\hat{u}^n)$ by solving the system of equations following from the Kirchhoff-Love conditions:

$$a_t(\bar{u}+\hat{u}) \cdot n(\bar{u}+\hat{u}) = 0 \quad ; \quad n(\bar{u}+\hat{u}) \cdot n(\bar{u}+\hat{u}) = 1 \tag{I.21}$$

Governing Shell Equations in Operator Description

To represent the governing shell equations in a compact form we introduce an operator description for the unrestricted rotation shell theory.
Let $(.,.)$ denote the integral over the undeformed shell middle surface M with:

$$(p,u) = \iint_M (p^\alpha u_\alpha + pw)dA \tag{I.22}$$

and let $[.,.]_C$ be the integral along the boundary contour C together with the contribution of the corner forces:

$$[b(u)\,,\,\hat{u}_b(u)]_C = \int_C [P_{\nu\nu}\hat{u}_\nu + P_{t\nu}\hat{u}_t + P_{n\nu}\hat{w} + Mn_\nu^{(1)}(u\,;\hat{u})]ds +$$

$$+\sum_k [(F_\nu\hat{u}_\nu + F_t\hat{u}_t + F_n\hat{w})_{(s_k+0)} - (F_\nu\hat{u}_\nu + F_t\hat{u}_t + F_n\hat{w})_{(s_k-0)}], \tag{I.23}$$

where at the boundary C the static quantity $b(u)$ and the geometric quantity $\hat{u}_b(u)$ are introduced by:

$$b(u) = \begin{pmatrix} P_{\nu\nu}(u) \\ P_{t\nu}(u) \\ P_{n\nu}(u) \\ M(u) \\ F_\nu(u) \\ F_t(u) \\ F_n(u) \end{pmatrix} ; \; b^*(u) = \begin{pmatrix} P^*_{\nu\nu}(u) \\ P^*_{t\nu}(u) \\ P^*_{n\nu}(u) \\ M^*(u) \\ F^*_\nu(u) \\ F^*_t(u) \\ F^*_n(u) \end{pmatrix} ; \; \hat{u}_b(u) = \begin{pmatrix} \hat{u}_\nu \\ \hat{u}_t \\ \hat{w} \\ n_\nu^{(1)}(u;\hat{u}) \\ \hat{u}_\nu(s_k) \\ \hat{u}_t(s_k) \\ \hat{w}(s_k) \end{pmatrix} . \tag{I.24}$$

In $(I.24)_3$ $n_\nu^{(1)}(u;\hat{u})$ is the first Gâteaux differential of $n_\nu(u) = n_\lambda(u)\nu^\lambda$.

Furthermore we define a nonlinear operator $B(u)$ on M:

$$B(u) = -\begin{pmatrix} T^{\alpha\beta}(u)|_\beta - b^\alpha_\beta T^\beta(u) \\ \\ T^\beta(u)|_\beta + b_{\alpha\beta}T^{\alpha\beta}(u) \end{pmatrix} \quad . \tag{I.25}$$

By partial integration and use of Gauss' divergence theorem the first differential $J^{(1)}(u;\hat{u})$ according to $(I.20)_1$ can be transformed into:

$$J^{(1)}(u;\hat{u}) = (B(u) - p,\hat{u}) + [b(u),\hat{u}_b(u)]_{C_u} + [b(u) - b^*(u),\hat{u}_b(u)]_{C_f} . \tag{I.26}$$

If we assume that u satisfies the geometric boundary conditions $u_b = u_b^*$ on C_u then the principle of stationary total potential energy yields the statement:

$$J^{(1)}(u;\hat{u}) = 0 \tag{I.27}$$

for all geometrically admissible $\hat{u}$ satisfying the homogeneous geometric boundary conditions $\hat{u}_b = 0$ on C_u. Introducing (I.26) into the stationarity condition (I.27) leads to the Euler-Lagrange equations

$$B(u) = p \quad \text{in } M \quad ; \quad b(u) = b^*(u) \quad \text{on } C_f \quad , \tag{I.28}$$

which are, in operator description, the equilibrium equations (I.11) and the static boundary conditions (I.12 - 13) of the unrestricted rotation shell theory.

Equations of Critical Equilibrium

To derive the equations defining critical equilibrium configurations, where snap-through or bifurcation buckling can occur, we have to consider the second differential of the total potential energy. If $\hat{u}$ and $\hat{\hat{u}}$ are two small deformations superimposed upon a fundamental equilibrium deformation u satisfying $u_b = u_b^*$ on C_u, the second differential of the total potential energy can be obtained in the following form:

$$J^{(2)}(u;\hat{u}\,\hat{\hat{u}}) = (B^{(1)}(u)\hat{\hat{u}},\hat{u}) + [b^{(1)}(u)\hat{\hat{u}},\hat{u}_b(u)]_{C_u} + [b(u),\hat{u}_b^{(1)}(u)\hat{\hat{u}}]_{C_u}$$
$$+ [b^{(1)}(u)\hat{\hat{u}} - b^{*(1)}(u)\hat{\hat{u}},\hat{u}_b(u)]_{C_f} + [b(u) - b^*(u),\hat{u}_b^{(1)}(u)\hat{\hat{u}}]_{C_f} \tag{I.29}$$

where $B^{(1)}(u)\hat{\hat{u}}$, $b^{(1)}(u)\hat{\hat{u}}$, $\hat{u}_b^{(1)}(u)\hat{\hat{u}}$ are the Gâteaux differentials of (I.25) and (I.24).

According to the energy criterion of stability an equilibrium configuration is a critical one, if the second differential of the total potential energy vanishes:

$$J^{(2)}(u;\hat{u}\,\hat{\hat{u}}) = 0 \tag{I.30}$$

for arbitrary $\hat{u}$ satisfying $\hat{u}_b = \hat{u}_b^{(1)}\hat{\hat{u}} = 0$ on C_u. Introducing (I.29) into (I.30) yields the stability equations of the unrestricted rotation shell theory:

$$\begin{aligned} B^{(1)}(u)\hat{\hat{u}} &= 0 \qquad \text{in } M \\ b^{(1)}(u)\hat{\hat{u}} - b*^{(1)}(u)\hat{\hat{u}} &= 0 \qquad \text{on } C_f \end{aligned} \tag{I.31}$$

Together with a well-chosen normalization of the buckling modes $\hat{\hat{u}}$ (see [15, 16]) equations (I.31) represent the general eigenvalue problem of shell buckling, depending nonlinearly on the prebuckling state $u = \bar{u}$. Comparing (I.31) with (I.28) it follows that the stability equations (I.31) are the Gâteaux differentials of the equilibrium equation $(I.28)_1$ and the static boundary conditions $(I.28)_2$.
If we are dealing with a one-parameter dependency of the shell deformation, a load-factor λ can be introduced, such that we have to determine the fundamental equilibrium path $u = \bar{u}(\lambda)$ with discrete critical loads λ_c. As solution of the eigenvalue problem (I.31) a series of buckling loads $\lambda_c(\lambda_1 \leq \leq \lambda_2 \leq \ldots)$ with associated buckling modes $\hat{\hat{u}}_c(\hat{\hat{u}}_1, \hat{\hat{u}}_2, \ldots)$ are obtained, where λ_c can be a singular or multiple eigenvalue with associated singular or multiple buckling modes.

Post-Buckling Behaviour

Let us assume that at a singular critical point u_c, λ_c a new equilibrium path $u^b(\lambda)$ bifurcates from the fundamental equilibrium path $\bar{u}(\lambda)$:

$$u^b(\lambda) = \bar{u}(\lambda) + v(\lambda) \ , \tag{I.32}$$

where $v(\lambda)$ is the differential displacement field. It is presumed that $\bar{u}(\lambda)$ is known and given by the following Taylor expansion:

$$\bar{u}(\lambda) = u_c + (\lambda - \lambda_c)u_c^{\diamond} + \frac{1}{2}(\lambda - \lambda_c)^2 u_c^{\diamond\diamond} + \frac{1}{6}(\lambda - \lambda_c)^3 u_c^{\diamond\diamond\diamond} + O((\lambda - \lambda_c)^4) \tag{I.33}$$

with $(.)^{\diamond} = \frac{d(.)}{d\lambda}$. For the differential displacement field $v(\lambda)$ we use a parametric representation

$$\upsilon(\lambda(t)) = t\upsilon_c^{\cdot} + \frac{t^2}{2}\upsilon_c^{\cdot\cdot} + \frac{t^3}{6}\upsilon_c^{\cdot\cdot\cdot} + O(t^4), \quad (.)^{\cdot} = \frac{d(.)}{dt} \tag{I.34}$$

together with an appropriate normalization [15,16]. A similar parametric expansion is chosen to represent the load factor λ:

$$\lambda(t) = \lambda_c + t\lambda_c^{\cdot} + \frac{t^2}{2}\lambda_c^{\cdot\cdot} + \frac{t^3}{6}\lambda_c^{\cdot\cdot\cdot} + O(t^4) . \tag{I.35}$$

With the procedure outlined in [3,15,16] the coefficients of the series (I.34) and (I.35) can be calculated. The two load factors $\lambda_c^{\cdot}$ and $\lambda_c^{\cdot\cdot}$ of (I.35), which define symmetric or unsymmetric bifurcation and also the stability behaviour at critical points u_c, λ_c, are obtained as:

$$\lambda_c^{\cdot} = -\frac{1}{2}\,\frac{J'''(u_c;\upsilon_c^{\cdot 3})}{J'''(u_c;\overset{\diamond}{u}_c\upsilon_c^{\cdot 2})} \tag{I.36}$$

and for $\lambda_c^{\cdot} = 0$:

$$\lambda_c^{\cdot\cdot} = -\frac{J'''(u_c;\upsilon_c^{\cdot 2}\upsilon_c^{\cdot\cdot}) + \frac{1}{3}J''''(u_c;\upsilon_c^{\cdot 4})}{J'''(u_c;\overset{\diamond}{u}_c\upsilon_c^{\cdot 2})} . \tag{I.37}$$

To determine the bifurcated equilibrium path (I.32) the coefficients of the parametric representation (I.34) can be calculated by solving a sequence of linear boundary value problems.

PART II: MODERATE ROTATION SHELL THEORY

For a wide area of shell problems of engineering interest the rotations of shell elements can be restricted such that the squares of the rotations are small of the same order as the strains leading to the so-called moderate rotation shell theory. In this case all results, presented in PART I of this paper, can be simplified essentially.

The strain measures $\gamma_{\alpha\beta}$ and $\chi_{\alpha\beta}$ are defined now by:

$$\begin{aligned} \gamma_{\alpha\beta} &= \theta_{\alpha\beta} + \frac{1}{2}\varphi_\alpha\varphi_\beta + \frac{1}{2}a_{\alpha\beta}\varphi^2 - \frac{1}{2}(\theta_\alpha^\lambda\omega_{\lambda\beta} + \theta_\beta^\lambda\omega_{\lambda\alpha}) \\ \chi_{\alpha\beta} &= -\frac{1}{2}[\varphi_{\alpha|\beta} + \varphi_{\beta|\alpha} + b_\alpha^\lambda(\theta_{\lambda\beta} - \omega_{\lambda\beta}) + b_\beta^\lambda(\theta_{\lambda\alpha} - \omega_{\lambda\alpha})] , \end{aligned} \tag{II.1}$$

where $\gamma_{\alpha\beta}$ is a quadratic and $\chi_{\alpha\beta}$ a linear function of the displacements and their derivatives.

Then we obtain the equilibrium equations and the static boundary conditions (I.28) in the form:

$$\mathcal{B}(u) = p \quad \text{in } M \quad ; \quad b(u) = b^* \quad \text{on } C_f \, , \tag{II.2}$$

where the nonlinear operators $\mathcal{B}(u)$, $b(u)$ and the prescribed statical boundary quantitiy b^* are given by (I.24 - 25) and (AII.1 - 4). It should be pointed out that in the moderate rotation shell theory b^* does not depend on the deformation u as it is the case in the unrestricted rotation shell theory.

A further essential simplification is achieved in the geometric boundary quantity $\hat{u}_b(u)$ according to $(I.24)_3$. If the shell deformations are restricted such that the assumption of the moderate rotation shell theory are satisfied, the first differential $n_\nu^{(1)}(u;\hat{u})$ can be expanded in a Taylor series with respect to u. Neglecting higher order terms the following result is obtained:

$$n_\nu^{(1)}(u;\hat{u}) = n_\lambda^{(1)}(u;\hat{u})\nu^\lambda = -\hat{\varphi}_\lambda \nu^\lambda = \hat{\beta}_\nu . \tag{II.3}$$

With (II.3) the geometric boundary quantitiy $\hat{u}_b$ is no more depending on the deformation u.

With (II.2) the general stability equations (I.31) lead to the stability equations of the moderate rotation shell theory:

$$\mathcal{B}^{(1)}(u)\hat{\hat{u}} = 0 \quad \text{in } M \quad ; \quad b^{(1)}(u)\hat{\hat{u}} = 0 \quad \text{on } C_f \quad . \tag{II.4}$$

Furthermore in the post-buckling equations (I.36 - 37) the differentials of the total potential energy can be simplified considerably. With (II.1) and (II.3) the functional of total potential energy is of order four and all higher order differentials vanish, which leads to further simplifications in the post-buckling analysis.

Finally it should be mentioned that for the moderate rotation shell theory the nonlinear operators $\mathcal{B}(.)$, $b(.)$ of equation (II.2) can be splitted into linear operators A, a and nonlinear operators $C(.)$, $c(.)$, where the linear operators A, a describe the linear shell theory [15,16].

References

1. Budiansky, B.: Notes on nonlinear shell theory, Trans. ASME, Ser. E, J. Appl. Mech. 35, 2 (1968) 393 - 401.

2. Koiter, W. T.and Simmonds J. G.: Foundations of shell theory, in: Theoretical and Applied Mechanics, 150 - 175, Proc. 13-th IUTAM Congr., Moscow 1972; Springer Berlin-New York (1974).

3. Budiansky, B.: Theory of buckling and post-buckling behavior of elastic structures, Adv. Appl. Mech. 14 (1974) 1 - 65.

4. Pietraszkiewicz, W.: Introduction to the non-linear theory of shells, Mitt. Inst. f. Mech. 10, Ruhr-Univ. Bochum (1977) 1 - 154.

5. Pietraszkiewicz, W.: On consistent approximations in the geometrically non-linear theory of shells, Mitt. Inst. f. Mech. 26, Ruhr-Univ. Bochum (1981) 1 - 39.

6. Stumpf, H.: Stationary and extremal variational principles in nonlinear moderate rotation shell theory, Proc. Symp. Mechanics of Inelastic Media and Structures, Warsaw 1978, Polish Scientific Publishers, Warsaw (1982).

7. Stein, E.: Variational functionals in the geometrical nonlinear theory of thin shells and finite-element-discretizations with applications to stability problems, in "Theory of Shells", Proc. 3-rd IUTAM Symp. on Shell Theory, Tbilisi 1978, 509 - 535, North-Holland Publ. Co., Amsterdam-New York-Oxford (1980).

8. Stumpf, H.: The derivation of dual extremum and complementary stationary principles in geometrical non-linear shell theory, Ing. Arch. 48 (1979) 221 - 237.

9. Schmidt, R. and Pietraszkiewicz, W.: Variational principles in the geometrically non-linear theory of shells undergoing moderate rotations, Ing.-Arch. 50 (1981) 187 - 201.

10. Stumpf H.: On the linear and nonlinear stability analysis in the theory of thin shells, Ing.-Arch. 51 (1981) 195 - 213.

11. Stumpf, H.: The stability equations of the consistent nonlinear elastic shell theory with moderate rotations, Proc. Second IUTAM Symp. on Stability in the Mechanics of Continua, Nümbrecht, Springer Publ. (1982) 89 -100.

12. Krätzig, W. B.; Bazar, Y.; Wittek, U.: Nonlinear behavior and elastic stability of shells - Theoretical concepts - numerical computations - results, in: "Buckling of Shells", Springer Berlin-New York (1982) 19 - 56.

13. Stein, E.; Berg, A.; Wagner, W.: Different levels of nonlinear shell theory in finite element stability analysis, in: "Buckling of Shells", Springer Berlin New-York (1982) 91 - 136.

14. Stumpf, H.: On the post-buckling analysis of thin elastic shells, ZAMM 63 (1981) T101 - T103.

15. Stumpf, H.: Unified operator description, nonlinear buckling and post-buckling analysis of thin elastic shells, Mitt. Inst. f. Mech. 34, Ruhr-Univ. Bochum (1982) 1 - 46.

16. Stumpf, H.: On the nonlinear buckling and post-buckling analysis of thin elastic shells, to appear in: Int. J. Non-linear Mech. 1984.

17. Pietraszkiewicz, W. and Szwabowicz, M. L.: Entirely Lagrangian nonlinear theory of thin shells, Arch. of Mech. 33 (1981) 273 - 288.

18. Schmidt, R.: On variational formulations in the general geometrically non-linear first approximation theory of thin elastic shells, ZAMM 62 (1982) T165 - T167.

19. Schmidt, R.: Variational principles for general and restricted geometrically non-linear Kirchhoff-Love type shell theories, Proc. Int. Conf. on Finite Element Methods, Shanghai (China), 621 - 626, New York: Gordon and Breach 1982.

20. Nolte, L.-P.: Stability equations of the general geometrically non-linear first approximation theory of thin elastic shells, ZAMM 63 (1983) T79 - T82.

21. Schmidt, R. and Stumpf, H.: On the stability and post-buckling of thin elastic shells with unrestricted rotations, to appear in: MRC 1984.

22. Nolte, L.-P. and Stumpf, H.: Energy-consistent large rotation shell theories in Langrangean description, Mech. Res. Comm., Vol. 10(4), 1983, 213 - 221.

23. Nolte, L.-P.: Beitrag zur Herleitung und vergleichende Untersuchung geometrisch-nichtlinearer Schalentheorien unter Berücksichtigung großer Rotationen, Diss. Ruhr-Univ. Bochum 1983.

APPENDIX I

In the equations (I.11 - 15) and in the operator definitions (I.24 - 25) of the unrestricted rotation shell theory the following formulas are used:

$$\begin{aligned} T^{\lambda\beta} &= l^{\lambda}_{\cdot\alpha}(N^{\alpha\beta} + a^{\alpha\beta} b_{\kappa\rho} M^{\kappa\rho}) + (m^{\lambda}|_{\alpha} - b^{\lambda}_{\alpha} m) M^{\alpha\beta} \\ &+ \epsilon^{\alpha\beta}\epsilon^{\lambda\mu}\{ l_{\mu\rho}[(\varphi_{\kappa} M^{\kappa\rho})|_{\rho} + l_{\gamma\kappa} b^{\gamma}_{\rho} M^{\kappa\rho}] - \varphi_{\alpha}[(l_{\mu\kappa} M^{\kappa\rho})|_{\rho} - \varphi_{\kappa} b_{\mu\rho} M^{\kappa\rho}]\} \\ T^{\beta} &= \varphi_{\alpha}(N^{\alpha\beta} + a^{\alpha\beta} b_{\kappa\rho} M^{\kappa\rho}) + (m_{|\alpha} + b^{\lambda}_{\alpha} m_{\lambda}) M^{\alpha\beta} + \\ &+ \epsilon^{\alpha\beta}\epsilon^{\lambda\mu} l_{\lambda\alpha}[(l_{\mu\kappa} M^{\kappa\rho})|_{\rho} - \varphi_{\kappa} b_{\mu\rho} M^{\kappa\rho}], \end{aligned} \tag{AI.1}$$

$$\begin{aligned} &P_{\nu\nu} = T^{\alpha\beta}\nu_{\alpha}\nu_{\beta} - Ck_{\nu} + F_{\nu,s} \; ; \; P_{t\nu} = T^{\alpha\beta} t_{\alpha}\nu_{\beta} - Ck_{t} + F_{t,s} \; ; \; P_{n\nu} = T^{\beta}\nu_{\beta} - Ck_{n} + F_{n,s} \\ &M = H_{\nu\nu} + FH_{t\nu} + KH_{n\nu} , \end{aligned} \tag{AI.2}$$

with

$$\begin{aligned} &H_{\nu\nu} = \sqrt{\frac{\bar{a}}{a}}\; M^{\alpha\beta} l^{\lambda}_{\cdot\alpha}\nu_{\lambda}\nu_{\beta} \; ; \; H_{t\nu} = \sqrt{\frac{\bar{a}}{a}}\; M^{\alpha\beta} l^{\lambda}_{\cdot\alpha} t_{\lambda}\nu_{\beta} \; ; \; H_{n\nu} = \sqrt{\frac{\bar{a}}{a}}\; M^{\alpha\beta}\varphi_{\alpha}\nu_{\beta} \\ &C = -\frac{1}{D}(H_{t\nu} n - H_{n\nu} n_{t}) \; ; \; F_{\nu} = Cn_{\nu} \; ; \; F_{t} = Cn_{t} \; ; \; F_{n} = Cn \end{aligned} \tag{AI.3}$$

$$k_\nu = \kappa_t n_t - \tau_t n \quad ; \quad k_t = \sigma_t n - \kappa_t n_\nu \quad ; \quad k_n = \tau_t n_\nu - \sigma_t n_t \quad ;$$

$$F = \frac{1}{D}(c_\nu n - c_n n_\nu) \quad ; \quad K = \frac{1}{D}(c_t n_\nu - c_\nu n_t) \quad ; D = c_n n_t - c_t n \ ,$$

and:

$$P^*_{\nu\nu} = T^*_{\nu\nu} - C^* k_\nu + F^*_{\nu,s} \ ; \ P^*_{t\nu} = T^*_{t\nu} - C^* k_t + F^*_{t,s} \ ; \ P^*_{n\nu} = T^*_{n\nu} - C^* k_n + F^*_{n,s}$$

$$M^* = H^*_{\nu\nu} + FH^*_{t\nu} + KH^*_{n\nu} \ ; \ F^*_\nu = C^* n_\nu \ ; \ F^*_t = C^* n_t \ ; \ F^*_n = C^* n \qquad \text{(AI.4)}$$

$$C^* = -\frac{1}{D}(H^*_{t\nu} n - H^*_{n\nu} n_t) \ .$$

APPENDIX II

For the moderate rotation shell theory the following simplified formulas have to be introduced into (I.24 - 25):

$$T^{\alpha\beta} = N^{\alpha\beta} - \frac{1}{2}(b^\alpha_\lambda M^{\lambda\beta} + b^\beta_\lambda M^{\lambda\alpha}) - \frac{1}{2}(b^\alpha_\lambda M^{\lambda\beta} - b^\beta_\lambda M^{\lambda\alpha}) - \frac{1}{2}\omega^{\alpha\beta} N^\lambda_\lambda -$$
$$- \frac{1}{2}(\omega^{\alpha\lambda} N^\beta_\lambda + \omega^{\beta\lambda} N^\alpha_\lambda) + \frac{1}{2}(\theta^{\alpha\lambda} N^\beta_\lambda - \theta^{\beta\lambda} N^\alpha_\lambda) \qquad \text{(AII.1)}$$

$$T^\beta = (M^{\alpha\beta}|_\alpha + \varphi_\alpha N^{\alpha\beta}) \ ,$$

$$P_{\nu\nu} = \nu_\alpha \nu_\beta [N^{\alpha\beta} - b^\alpha_\lambda M^{\lambda\beta} - \omega^{\alpha\lambda} N^\beta_\lambda] + \tau_t M_{t\nu}$$

$$P_{t\nu} = t_\alpha \nu_\beta [N^{\alpha\beta} - \frac{1}{2}(b^\alpha_\lambda M^{\lambda\beta} + b^\beta_\lambda M^{\lambda\alpha}) - \frac{1}{2}(b^\alpha_\lambda M^{\lambda\beta} - b^\beta_\lambda M^{\lambda\alpha}) - \frac{1}{2}\omega^{\alpha\beta} N^\lambda_\lambda -$$
$$- \frac{1}{2}(\omega^{\alpha\lambda} N^\beta_\lambda + \omega^{\beta\lambda} N^\alpha_\lambda) + \frac{1}{2}(\theta^{\alpha\lambda} N^\beta_\lambda - \theta^{\beta\lambda} N^\alpha_\lambda)] - \sigma_t M_{t\nu}$$

$$P_{n\nu} = \nu_\beta (M^{\alpha\beta}|_\alpha + \varphi_\alpha N^{\alpha\beta}) + M_{t\nu,s} \qquad \text{(AII.2)}$$

$$M = \nu_\alpha \nu_\beta M^{\alpha\beta}$$

$$F_\nu = F_t = 0$$

$$F_n = M_{t\nu}(s_k + 0) - M_{t\nu}(s_k - 0) \ ,$$

$$P^*_{\nu\nu} = T^*_{\nu\nu} + \tau_t M^*_{t\nu} \ ; \ P^*_{t\nu} = T^*_{t\nu} - \sigma_t M^*_{t\nu} \ ; \ P^*_{n\nu} = T^*_{n\nu} + M^*_{t\nu,s} \ ;$$
$$M^* = M^*_{\nu\nu} \ ; \ F^*_\nu = F^*_t = 0 \ ; \ F^*_n = M^*_{t\nu}(s_k + 0) - M^*_{t\nu}(s_k - 0) \ , \qquad \text{(AII.3)}$$

$$n^{(1)}_\nu(u;\hat{u}) = \hat{\beta}_\nu = -\hat{\varphi}_\alpha \nu^\alpha \ ,$$
$$\hat{u}_\nu(s_k) = \hat{u}_t(s_k) = 0 \qquad \text{(AII.4)}$$

On Entirely Lagrangian Displacemental Form of Non-Linear Shell Equations

W. PIETRASZKIEWICZ

Institute of Fluid-Flow Machinery of the
Polish Academy of Sciences
ul.Fiszera 14, 80-952 Gdańsk, Poland

Summary

Equations of equilibrium and corresponding four geometric and static boundary conditions are derived for an entirely Lagrangian non-linear theory of thin shells. In case of a linearly elastic material and conservative external forces all shell relations are exactly derivable as stationarity conditions of the Hu - Washizu free functional. The set of equations is consistently reduced in the case of the geometrically non-linear theory of thin elastic shells undergoing large/small rotations.

1. Introduction

In the numerical analysis of flexible shell structures it is desirable to apply shell relations which are referred entirely to the undeformed shell geometry. Such entirely Lagrangian theory of shells should also be derivable from appropriately constructed variational principles.

Some forms of Lagrangian equilibrium equations for thin shells, but without associated boundary conditions, were given already in [1,2] . In [3,4] the equilibrium equations and three force boundary conditions were also referred to base vectors of the undeformed shell middle surface. However, in the fourth static boundary condition of [3,4] the resulting boundary couple was measured per unit length of the undeformed boundary contour but its axial vector was still tangent to the unknown boundary contour of the deformed shell. This caused difficulties in the construction of corresponding variational functionals even in the simplest case of dead loads applied to the shell boundary. Only recently [5] a complete set of entirely Lagrangian shell equations was derived which allowed for a proper formulation of corresponding Lagrangian variational principles [6,7]. In [5] a new Lagrangian displacemental parameter n_ν was introduced at the shell boundary and a modified tensor of change of curvature $\chi_{\alpha\beta}$ was used, which by definition was a third-degree polynomial in displacements.

In this report a different but equivalent to [5] version of the entirely Lagrangian theory of shells is presented in terms of the usual tensor of change of curvature $\kappa_{\alpha\beta}$ which is a non-rational function of displacements. The main reason for the development of the theory are simple transformation properties of $\kappa_{\alpha\beta}$ under the change of the reference shell configuration. This feature becomes of primary importance when superposed deformations and incremental formulations of shell equations are discussed in the total Lagrangian and in the updated Lagrangian descriptions [8,9] . In case of a linearly elastic material and when conservative loads are applied to the shell middle surface and the shell lateral boundary surface, the entirely Lagrangian shell equations are shown to be derivable exactly as stationarity conditions of the Hu - Washizu type free functional.

When strains are assumed to be small everywhere the shell relations derived here reduce exactly to those given in [5] for the geometrically non-linear theory. Additionally, rotations of the shell material elements may be restricted to be small, moderate or large, according to the classification scheme suggested in [10-12]. As a result, several consistently simplified versions of the entirely Lagrangian non-linear theory of shells may be constructed [13]. Here two consistent versions of equations of the non-linear theory of shells undergoing large/small rotations are developed. This shell theory describes accurately the behaviour of a majority of elastic flexible shell structures.

The sets of Lagrangian shell equations presented here are supposed to be solved in displacements as basic independent field variables. Since our shell equations are derivable from variational principles, mixed hybrid finite element methods may also be applied in which the strain and/or the stress fields appearing in corresponding variational functionals are discretized independently of the discretization of the displacement field.

2. Lagrangian Shell Equations

Within the Kirchhoff - Love type theory of shells the deformation of the three-dimensional thin shell-like body is described by the deformation of its middle surface. During the surface deformation components of the Lagrangian strain tensor and of the tensor of change of curvature are given by [5,12]

$$\gamma_{\alpha\beta} = \frac{1}{2}(\bar{a}_{\alpha\beta} - a_{\alpha\beta}) = \frac{1}{2}(l^{\lambda}_{\cdot\alpha} l_{\lambda\beta} + \phi_\alpha \phi_\beta - a_{\alpha\beta}) \, ,$$
$$\kappa_{\alpha\beta} = -(\bar{b}_{\alpha\beta} - b_{\alpha\beta}) = l_{\lambda\alpha}(n^{\lambda}|_{\beta} - b^{\lambda}_{\beta} n) + \phi_\alpha (n_{,\beta} + b^{\lambda}_{\beta} n_\lambda) + b_{\alpha\beta} . \tag{2.1}$$

Here $a_{\alpha\beta}$, $\bar{a}_{\alpha\beta}$ and $b_{\alpha\beta}$, $\bar{b}_{\alpha\beta}$ are components of the surface metric and curvature tensors in its undeformed M and deformed $\bar{M}$ configurations, respectively,

$$l_{\alpha\beta} = a_{\alpha\beta} + \theta_{\alpha\beta} - \omega_{\alpha\beta} \, , \qquad \theta_{\alpha\beta} = \frac{1}{2}(u_{\alpha|\beta} + u_{\beta|\alpha}) - b_{\alpha\beta} w \, ,$$
$$\omega_{\alpha\beta} = \frac{1}{2}(u_{\beta|\alpha} - u_{\alpha|\beta}) = \epsilon_{\alpha\beta}\phi \, , \quad \phi_\alpha = w_{,\alpha} + b^{\lambda}_{\alpha} u_\lambda \, ,$$
$$n_\mu = \frac{1}{j} m_\mu = \frac{1}{j}\epsilon^{\alpha\beta}\epsilon_{\lambda\mu}\phi_\alpha l^{\lambda}_{\cdot\beta} \, , \qquad n = \frac{1}{j} m = \frac{1}{2j}\epsilon^{\alpha\beta}\epsilon_{\lambda\mu} l^{\lambda}_{\cdot\alpha} l^{\mu}_{\cdot\beta} \, , \tag{2.2}$$
$$j = \sqrt{\frac{\bar{a}}{a}} \, , \quad \frac{\bar{a}}{a} = 1 + 2\gamma^{\alpha}_{\alpha} + 2(\gamma^{\alpha}_{\alpha}\gamma^{\beta}_{\beta} - \gamma^{\alpha}_{\beta}\gamma^{\beta}_{\alpha}) \, ,$$

$$\bar{\underset{\sim}{a}}_\alpha = l^{\lambda}_{\cdot\alpha}\underset{\sim}{a}_\lambda + \phi_\alpha \underset{\sim}{n} \, , \qquad \bar{\underset{\sim}{n}} = n^{\lambda}\underset{\sim}{a}_\lambda + n\underset{\sim}{n} \tag{2.3}$$

and $\underset{\sim}{u} = u_\alpha \underset{\sim}{a}^\alpha + w\underset{\sim}{n}$ is the displacement vector.

The deformation of the shell boundary surface may be described [5] by two vectors defined at the boundary contour C of M

$$\underset{\sim}{u} = \bar{\underset{\sim}{r}} - \underset{\sim}{r} = u_\nu \underset{\sim}{\nu} + u_t \underset{\sim}{t} + w\underset{\sim}{n} \, ,$$
$$\underset{\sim}{\beta} = \bar{\underset{\sim}{n}} - \underset{\sim}{n} = n_\nu \underset{\sim}{\nu} + n_t \underset{\sim}{t} + (n-1)\underset{\sim}{n} \, . \tag{2.4}$$

When rotations of the shell boundary elements do not exceed $\pm\frac{\pi}{2}$ for $\bar{\underset{\sim}{n}}$ we have the unique vector representation

$$\bar{\underset{\sim}{n}} = \frac{1}{c_t^2 + c^2}\left[n_\nu \bar{\underset{\sim}{a}}_t \times (\underset{\sim}{\nu} \times \bar{\underset{\sim}{a}}_t) + \sqrt{(1+2\gamma_{tt})(1-n_\nu^2) - c_\nu^2}\; \underset{\sim}{\nu} \times \bar{\underset{\sim}{a}}_t \right],$$
$$\bar{\underset{\sim}{a}}_t = 1 + \frac{d\underset{\sim}{u}}{ds} = c_\nu \underset{\sim}{\nu} + c_t \underset{\sim}{t} + c\underset{\sim}{n} \, , \quad c_\nu = \frac{du_\nu}{ds} + \tau_t w - \kappa_t u_t \, , \tag{2.5}$$
$$c_t = 1 + \frac{du_t}{ds} + \kappa_t u_\nu - \sigma_t w \, , \qquad c = \frac{dw}{ds} + \sigma_t u_t - \tau_t u_\nu \, .$$

It follows from (2.5) that along the boundary contour $\bar{\underset{\sim}{n}}$ is described completely by $\underset{\sim}{u}$ and n_ν as independent parameters.

Consider now $\bar{M}$ to be a middle surface of a thin shell in an equilibrium state, under the Lagrangian surface force $\underset{\sim}{p} = p^\alpha \underset{\sim}{a}_\alpha + p\underset{\sim}{n}$, per unit area of M, and under the Lagrangian boundary force $\underset{\sim}{T} = T_\nu \underset{\sim}{\nu} + T_t \underset{\sim}{t} + T \underset{\sim}{n}$ and the boundary static moment $\underset{\sim}{H} = H_\nu \underset{\sim}{\nu} + H_t \underset{\sim}{t} + H \underset{\sim}{n}$, both per unit length of C, such that

$$\int_C \underset{\sim}{T} ds = \iint_{\partial B} \underset{\sim}{f} dA \quad , \quad \int_C \underset{\sim}{H} ds = \iint_{\partial B} \underset{\sim}{f} \zeta dA \quad , \tag{2.6}$$

where $\underset{\sim}{f}$ is the Lagrangian surface load, per unit area of the undeformed shell boundary surface ∂B, while ζ is the distance from M. Then for any additional virtual displacement field $\delta \underset{\sim}{u} = \delta u_\alpha \underset{\sim}{a}^\alpha + \delta w \underset{\sim}{n}$, which is subject to geometric constraints, the principle of virtual displacements can be presented in the Lagrangian description to be

$$\iint_M (N^{\alpha\beta} \delta\gamma_{\alpha\beta} + M^{\alpha\beta} \delta\kappa_{\alpha\beta}) dA = \iint_M \underset{\sim}{p} \cdot \delta \underset{\sim}{u} dA + \int_{C_f} (\underset{\sim}{T} \cdot \delta \underset{\sim}{u} + \underset{\sim}{H} \cdot \delta \underset{\sim}{\beta}) ds \ , \tag{2.7}$$

where $N^{\alpha\beta}$, $M^{\alpha\beta}$ are components of the symmetric second Piola - Kirchhoff stress resultant and stress couple tensors.
Taking into account that

$$\delta\left(\frac{1}{j}\right) = -\frac{1}{j}\frac{a}{\bar{a}}[(1 + 2\gamma^\sigma_\sigma) a^{\alpha\beta} - 2\gamma^{\alpha\beta}]\delta\gamma_{\alpha\beta} \quad , \tag{2.8}$$

the left-hand side of (2.7) can be transformed into

$$IVW = -\iint_M \underset{\sim}{T}^\beta\big|_\beta \cdot \delta \underset{\sim}{u} + \int_{C_f} (\underset{\sim}{T}^\beta \cdot \delta \underset{\sim}{u} + M^{\alpha\beta} \bar{\underset{\sim}{a}}_\alpha \cdot \delta \bar{\underset{\sim}{n}}) \nu_\beta ds \quad , \tag{2.9}$$

$$\underset{\sim}{T}^\beta = T^{\lambda\beta} \underset{\sim}{a}_\lambda + T^\beta \underset{\sim}{n} \ ,$$

$$\begin{aligned} T^{\lambda\beta} = \Big\{ & N^{\alpha\beta} + \frac{a}{\bar{a}}[(1 + 2\gamma^\sigma_\sigma) a^{\alpha\beta} - 2\gamma^{\alpha\beta}]\Big([(M^{\kappa\rho} l_{\mu\kappa})|_\rho - M^{\kappa\rho}\phi_\kappa b_{\mu\rho}]n^\mu + \\ & + [(M^{\kappa\rho}\phi_\kappa)|_\rho + M^{\kappa\rho} l_{\gamma\kappa} b^\gamma_\rho] n\Big)\Big\} l^\lambda_{\cdot\alpha} + M^{\alpha\beta}(n^\lambda|_\alpha - b^\lambda_\alpha n) + \\ & + \frac{1}{j}\epsilon^{\alpha\beta}\epsilon^{\lambda\mu}\Big\{[(M^{\kappa\rho}\phi_\kappa)|_\rho + M^{\kappa\rho} l_{\gamma\kappa} b^\gamma_\rho] l_{\mu\alpha} - [(M^{\kappa\rho} l_{\mu\kappa})|_\rho - M^{\kappa\rho}\phi_\kappa b_{\mu\rho}]\phi_\alpha\Big\} \ , \end{aligned}$$

$$\begin{aligned} T^\beta = \Big\{ & N^{\alpha\beta} + \frac{a}{\bar{a}}[(1 + 2\gamma^\sigma_\sigma) a^{\alpha\beta} - 2\gamma^{\alpha\beta}]\Big([(M^{\kappa\rho} l_{\mu\kappa})|_\rho - M^{\kappa\rho}\phi_\kappa b_{\mu\rho}]n^\mu + \\ & + [(M^{\kappa\rho}\phi_\kappa)|_\rho + M^{\kappa\rho} l_{\gamma\kappa} b^\gamma_\rho] n\Big)\Big\}\phi_\alpha + M^{\alpha\beta}(n_{,\alpha} + b^\lambda_\alpha n_\lambda) + \\ & + \frac{1}{j}\epsilon^{\alpha\beta}\epsilon^{\lambda\mu}[(M^{\kappa\rho} l_{\mu\kappa})|_\rho - M^{\kappa\rho}\phi_\kappa b_{\mu\rho}] l_{\lambda\alpha} \ . \end{aligned} \tag{2.10}$$

Since

$$\delta \bar{\underset{\sim}{n}} = \frac{1}{a_\nu}[(\underset{\sim}{\nu} \times \bar{\underset{\sim}{n}})(\bar{\underset{\sim}{n}} \cdot \frac{d}{ds}\delta \underset{\sim}{u}) + \bar{\underset{\sim}{a}}_\nu \delta n_\nu] \quad , \quad a_\nu = (\bar{\underset{\sim}{a}}_t \times \bar{\underset{\sim}{n}}) \cdot \underset{\sim}{\nu} \ , \tag{2.11}$$

we can reduce to four the number of independent variations of displacemental parameters at the shell boundary, transforming the last term in (2.9) as follows

$$\int\limits_{C_f} M^{\alpha\beta}\underset{\sim}{\bar{a}}_\alpha\cdot\delta\underset{\sim}{\bar{n}}\nu_\beta ds = \int\limits_{C_f}\left(-\underset{\sim}{F}\cdot\frac{d}{ds}\delta\underset{\sim}{u} + M\delta n_\nu\right)ds =$$
$$= \int\limits_{C_f}\left(\frac{d\underset{\sim}{F}}{ds}\cdot\delta\underset{\sim}{u} + M\delta n_\nu\right)ds + \sum_j \underset{\sim}{F}_j\cdot\delta\underset{\sim}{u}_j \quad , \tag{2.12}$$

where

$$\underset{\sim}{F} = -\frac{1}{a_\nu}[(\underset{\sim}{\bar{n}}\times\underset{\sim}{\bar{a}}_\alpha)\cdot\underset{\sim}{\nu}]M^{\alpha\beta}\nu_\beta\underset{\sim}{\bar{n}} =$$
$$= (g_\nu R_{t\nu} + r_\nu R_\nu)\underset{\sim}{\nu} + (g_t R_{t\nu} + r_t R_\nu)\underset{\sim}{t} + (gR_{t\nu} + rR_\nu)\underset{\sim}{n} \quad , \tag{2.13}$$
$$M = \frac{1}{a_\nu}(\underset{\sim}{\bar{n}}\times\underset{\sim}{\bar{a}}_\alpha)\cdot\underset{\sim}{\bar{a}}_t M^{\alpha\beta}\nu_\beta = R_{\nu\nu} + fR_{t\nu} + kR_\nu \quad ,$$
$$\underset{\sim}{F}_j = \underset{\sim}{F}(s_j + 0) - \underset{\sim}{F}(s_j - 0) \quad ,$$

$$R_{\nu\nu} = \nu^\lambda l_{\lambda\alpha}M^{\alpha\beta}\nu_\beta = (1+\theta_{\nu\nu})M_{\nu\nu} + (\theta_{\nu t} - \phi)M_{t\nu} \quad ,$$
$$R_{t\nu} = t^\lambda l_{\lambda\alpha}M^{\alpha\beta}\nu_\beta = (\theta_{\nu t} + \phi)M_{\nu\nu} + (1+\theta_{tt})M_{t\nu} \quad , \tag{2.14}$$
$$R_\nu = \phi_\alpha M^{\alpha\beta}\nu_\beta = \phi_\nu M_{\nu\nu} + \phi_t M_{t\nu} \quad ,$$

$$g_\nu = \frac{n_\nu n}{a_\nu} \ , \quad g_t = \frac{n_t n}{a_\nu} \ , \quad g = \frac{n^2}{a_\nu} \ ,$$
$$r_\nu = \frac{n_\nu n_t}{a_\nu} \ , \quad r_t = \frac{n_t^2}{a_\nu} \ , \quad r = \frac{n_t n}{a_\nu} \ , \tag{2.15}$$
$$f = \frac{1}{a_\nu}(cn_\nu - c_\nu n) \quad , \quad k = \frac{1}{a_\nu}(c_\nu n_t - c_t n_\nu) \ .$$

Now the Lagrangian principle of virtual displacements (2.7) takes the final form

$$-\iint\limits_M (\underset{\sim}{T}^\beta|_\beta + \underset{\sim}{p})\cdot\delta\underset{\sim}{u}dA + \int\limits_{C_f}[(\underset{\sim}{P} - \underset{\sim}{P}^*)\cdot\delta\underset{\sim}{u} + (M - M^*)\delta n_\nu]ds + \sum_j(\underset{\sim}{F}_j - \underset{\sim}{F}_j^*)\cdot\delta\underset{\sim}{u}_j = 0 \ , \tag{2.16}$$

where

$$\underset{\sim}{P} = \underset{\sim}{T}^\beta\nu_\beta + \frac{d\underset{\sim}{F}}{ds} \quad , \quad \underset{\sim}{P}^* = \underset{\sim}{T} + \frac{d\underset{\sim}{F}^*}{ds} \quad ,$$

$$\underset{\sim}{F}^* = -\frac{1}{a_\nu}[(\underset{\sim}{\bar{n}}\times\underset{\sim}{H})\cdot\underset{\sim}{\nu}]\underset{\sim}{\bar{n}} =$$
$$= (g_\nu H_t + r_\nu H)\underset{\sim}{\nu} + (g_t H_t + r_t H)\underset{\sim}{t} + (gH_t + rH)\underset{\sim}{n} \quad , \tag{2.17}$$
$$M^* = \frac{1}{a_\nu}(\underset{\sim}{\bar{n}}\times\underset{\sim}{H})\cdot\underset{\sim}{\bar{a}}_t = H_\nu + fH_t + kH \quad , \quad \underset{\sim}{F}_j^* = \underset{\sim}{F}^*(s_j + 0) - \underset{\sim}{F}^*(s_j - 0) \ .$$

From (2.16) follow the equilibrium equations and the corresponding static boundary conditions, together with already known geometric boundary conditions, of the entirely Lagrangian non-linear theory of thin shells

$$\underset{\sim}{T}^{\beta}\big|_{\beta} + \underset{\sim}{p} = \underset{\sim}{0} \quad \text{in} \quad M \ ,$$
$$P = P^* \ , \ M = M^* \ \text{on} \ C_f \ \text{and} \ \underset{\sim}{F}_j = \underset{\sim}{F}_j^* \ \text{at each corner} \ M_j \in C_f \ , \qquad (2.18)$$
$$\underset{\sim}{u} = \underset{\sim}{u}^* \ , \ n_\nu = n_\nu^* \ \text{on} \ C_u \ \text{and} \ \underset{\sim}{u}_i = \underset{\sim}{u}_i^* \ \text{at each corner} \ M_i \in C_u \ .$$

Note that, in general, the Lagrangian equilibrium equations and the Lagrangian static boundary conditions are linear in the stress measures $N^{\alpha\beta}$, $M^{\alpha\beta}$ but are non-rational in the displacemental parameters, since in the expressions (2.10), (2.13) and (2.17) there are square roots of polynomials of those parameters.

Within the first-approximation theory of thin isotropic and elastic shells the strain energy function may be approximated by the quadratic form [14]

$$\Sigma = \frac{h}{2} H^{\alpha\beta\lambda\mu}(\gamma_{\alpha\beta}\gamma_{\lambda\mu} + \frac{h^2}{12}\kappa_{\alpha\beta}\kappa_{\lambda\mu}) + O(Eh\eta^2\theta^2) \ , \qquad (2.19)$$

where $H^{\alpha\beta\lambda\mu}$ are components of the modified elasticity tensor. The error of Σ at any point of the shell is expressed through the small parameter θ defined by [15,16]

$$\theta = \max(h/d,\ h/L,\ h/L^*,\ \sqrt{h/R},\ \sqrt{\eta}) \ , \qquad (2.20)$$

where d is the distance of the point from the lateral shell boundary, L is the wave length of deformation patterns of M , L^* is the wave length of the curvature patterns of M , R is the smallest principal radius of curvature of M and η is the largest principal strain in the shell space. From (2.19) follow the constitutive equations

$$N^{\alpha\beta} = \frac{\partial\Sigma}{\partial\gamma_{\alpha\beta}} = \frac{Eh}{1-\nu^2}[(1-\nu)\gamma^{\alpha\beta} + \nu a^{\alpha\beta}\gamma^{\kappa}_{\kappa}] + O(Eh\eta\theta^2) \ ,$$
$$M^{\alpha\beta} = \frac{\partial\Sigma}{\partial\kappa_{\alpha\beta}} = \frac{Eh^3}{12(1-\nu^2)}[(1-\nu)\kappa^{\alpha\beta} + \nu a^{\alpha\beta}\kappa^{\kappa}_{\kappa}] + O(Eh^2\eta\theta^2) \ . \qquad (2.21)$$

In the case of external dead loads there exist potential functions $\Phi(\underset{\sim}{u}) = -\underset{\sim}{p}\cdot\underset{\sim}{u}$ and $\Psi(\underset{\sim}{u},\underset{\sim}{\beta}) = -(\underset{\sim}{T}\cdot\underset{\sim}{u} + \underset{\sim}{H}\cdot\underset{\sim}{\beta})$ which allow to transform (2.16) into the variational principle $\delta I = 0$ for the functional

$$I = \iint\limits_M [\Sigma(\gamma_{\alpha\beta},\kappa_{\alpha\beta}) - \underset{\sim}{p}\cdot\underset{\sim}{u}]dA - \int\limits_{C_f}[\underset{\sim}{T}\cdot\underset{\sim}{u} + \underset{\sim}{H}\cdot\underset{\sim}{\beta}]ds \qquad (2.22)$$

with (2.1), (2.5) and $(2.18)_3$ as subsidiary conditions. Eliminating at C_f the parameters n_t and n with the use of $(2.5)_1$

and introducing other subsidiary conditions into the functional I with the help of Lagrange multipliers $N^{\alpha\beta}$, $M^{\alpha\beta}$, $\underset{\sim}{P}$ and M we obtain the free Hu - Washizu type functional

$$I_1 = \iint_M \Big\{ \Sigma(\gamma_{\alpha\beta},\kappa_{\alpha\beta}) - \underset{\sim}{p}\cdot\underset{\sim}{u} - N^{\alpha\beta}[\gamma_{\alpha\beta} - \gamma_{\alpha\beta}(\underset{\sim}{u})] - M^{\alpha\beta}[\kappa_{\alpha\beta} - \kappa_{\alpha\beta}(\underset{\sim}{u})]\Big\} dA -$$
$$- \int_{C_f} \Big\{ \underset{\sim}{T}\cdot\underset{\sim}{u} + H\cdot[\underset{\sim}{\bar{n}}(\underset{\sim}{u},n_\nu) - \underset{\sim}{n}]\Big\} ds - \tag{2.23}$$
$$- \int_{C_u} [\underset{\sim}{P}\cdot(\underset{\sim}{u} - \underset{\sim}{u}^*) + M(n_\nu - n_\nu^*)] ds - \sum_i \underset{\sim}{F}_i\cdot(\underset{\sim}{u}_i - \underset{\sim}{u}_i^*) .$$

The associated Hu - Washizu variational principle $\delta I_1 = 0$ states that among all possible values of independent fields indicated in (2.23), which are not restricted by any subsidiary conditions, the solution values render the functional I_1 stationary. It can be shown by direct calculations that the stationarity conditions of the functional I_1 are exactly the Lagrangian shell equations (2.18), the strain-displacement relations (2.1) and the constitutive equations (2.21) together with relations which identify the Lagrange multipliers with the functions described already by the symbols used in (2.23). In analogy to [6,7,17] many other free or constrained functionals and associated with them variational principles may be constructed for the entirely Lagrangian non-linear theory of thin isotropic elastic shells. The functionals form a solid basis for a computerized analysis of flexible shell structures.

The shell relations derived above are two-dimensionally exact for the shell middle surface. However, the relations are meaningful for shells only within small strains, since by using the Kirchhoff - Love constraints the effect of change of the shell thickness was ignored in the description of shell deformation. Such simplified approach is consistent within the first-approximation theory of elastic shells used here, but it would not be permissible if large strains in the shell space were allowed [18] .

Within small strains some shell relations may be simplified by omitting strains with respect to unity. In particular,

$$j \simeq 1 + \gamma^\alpha_\alpha \simeq 1 \quad , \quad n \simeq m(1 - \gamma^\alpha_\alpha) \simeq m \quad , \quad n_\mu \simeq m_\mu \quad ,$$

$$\kappa_{\alpha\beta} \simeq l_{\lambda\alpha}(m^{\lambda}|_{\beta} - b^{\lambda}_{\beta}m) + \phi_{\alpha}(m_{,\beta} + b^{\lambda}_{\beta}m_{\lambda}) + b_{\alpha\beta}(1+\gamma^{\kappa}_{\kappa}) \quad , \tag{2.24}$$

$$\bar{\underset{\sim}{n}} \simeq \frac{1}{1-c^2_{\nu}}[n_{\nu}\underset{\sim}{\bar{a}}_t \times (\underset{\sim}{\nu} \times \underset{\sim}{\bar{a}}_t) + \sqrt{1-n^2_{\nu}-c^2_{\nu}}\;\underset{\sim}{\nu} \times \underset{\sim}{\bar{a}}_t] \ .$$

If $(2.24)_2$ is used in the left-hand side of (2.7) then [5] it generates the following reduced definitions of (2.10)

$$T^{\lambda\beta} = l^{\lambda}_{\cdot\alpha}(N^{\alpha\beta} + a^{\alpha\beta}b_{\kappa\rho}M^{\kappa\rho}) + (m^{\lambda}|_{\alpha} - b^{\lambda}_{\alpha}m)M^{\alpha\beta} +$$
$$+ \epsilon^{\alpha\beta}\epsilon^{\lambda\mu}\Big\{l_{\mu\alpha}[(\phi_{\kappa}M^{\kappa\rho})|_{\rho} + l_{\gamma\kappa}b^{\gamma}_{\rho}M^{\kappa\rho}] - \phi_{\alpha}[(l_{\mu\kappa}M^{\kappa\rho})|_{\rho} - \phi_{\kappa}b_{\mu\rho}M^{\kappa\rho}]\Big\} \ , \tag{2.25}$$

$$T^{\beta} = \phi_{\alpha}(N^{\alpha\beta} + a^{\alpha\beta}b_{\kappa\rho}M^{\kappa\rho}) + (m_{,\alpha} + b^{\lambda}_{\alpha}m_{\lambda})M^{\alpha\beta} + \epsilon^{\alpha\beta}\epsilon^{\lambda\mu}l_{\lambda\alpha}[(l_{\mu\kappa}M^{\kappa\rho})|_{\rho} - \phi_{\kappa}b_{\mu\rho}N^{\kappa\rho}].$$

As a result of the simplified expressions (2.24) and (2.25), for the geometrically non-linear theory of shells the Lagrangian equilibrium equations become linear in $N^{\alpha\beta}$, $M^{\alpha\beta}$ and quadratic in u_{α} , w while the Lagrangian static boundary conditions are linear in $N^{\alpha\beta}$, $M^{\alpha\beta}$ but still remain non-rational functions of u_{α} , w and n_{ν} since in the approximate formula $(2.24)_3$ for $\bar{\underset{\sim}{n}}$ there is still the square-root function of the displacemental parameters. When (2.24) and (2.25) are introduced into the Hu - Washizu functional (2.23) , then analytically derived stationarity conditions of I_1 will not exactly coincide with the shell boundary parameters (2.13) and (2.17). However, since the error of (2.24) and (2.25) lies within an error margin of the first-approximation theory, the Lagrangian shell equations and the Hu - Washizu free functional may be regarded as corresponding equivalent descriptions of the same version of the geometrically non-linear theory of shells.

It should be pointed out that the general definitions of the Lagrangian measures of change of curvature used in this paper and in [1-5,19] are equivalent to each other from the point of view of the error introduced into the strain energy within the first-approximation theory. However, the important qualitative differences appear in the displacemental forms of the measures. In the derivations of changes of curvature presented in [1-4,19] the representation $\bar{b}_{\alpha\beta} = \underset{\sim}{\bar{a}}_{\alpha}|_{\beta} \cdot \bar{\underset{\sim}{n}}$ was applied for the curvature tensor of $\bar{M}$, while in [5] and in deriving $(2.1)_2$ here a different (although mathematically equivalent) expression $\bar{b}_{\alpha\beta} = -\underset{\sim}{\bar{a}}_{\alpha} \cdot \bar{\underset{\sim}{n}}|_{\beta}$ was used. As a result, our line integral in (2.9) consisted of six terms containing $\delta\underset{\sim}{u}$ and $\delta\bar{\underset{\sim}{n}}$. Since $\bar{\underset{\sim}{n}} = \bar{\underset{\sim}{n}}(\underset{\sim}{u}, n_{\nu})$, those terms were reduced further to four terms containing only $\delta \underset{\sim}{u}$ and $\delta \bar{n}_\nu$ as independent variations along the shell boundary contour. This allowed to construct the two-dimensionally exact and variationally derivable four natural static boundary and corner conditions associated with the equilibrium equations (2.18). On the other hand, when the displacemental forms of changes of curvatures given in [1-4,19] are introduced into IVW and Stokes' theorem is applied, the resulting line integral consists of six terms containing $\delta \underset{\sim}{u}$ and $\frac{d}{ds_\nu}\delta \underset{\sim}{u}$. Those six terms cannot be reduced further to only four terms containing $\delta \underset{\sim}{u}$ and one of components of $\frac{d}{ds_\nu}\delta \underset{\sim}{u}$ as independent variations of displacemental variables. Let us remind that $\frac{d}{ds_\nu}$ above means differentiation at C performed in the direction of $\underset{\sim}{\nu}$ orthogonal to C , [10-12]. As a result, no four variationally derivable static boundary and corner conditions can be associated with the Lagrangian equilibrium equations given in [1-4]. This was the reason why no boundary conditions were given in [1,2] for the general bending theory of shells while the static boundary conditions suggested in [3,4] were derived by transforming corresponding Eulerian parameters into the undeformed reference configuration but not by the direct variational procedure. However, such transformed static boundary conditions are not entirely Lagrangian and do not allow to construct a free variational functional of the Hu - Washizu type even in the simplest case of dead loads applied to the shell lateral boundary surface.

3. Restricted Rotations

By the polar decomposition theorem of the deformation gradient tensor [10-12] strains and rotations of the shell material elements were exactly separated from each other. Therefore, further consistent simplifications of the geometrically non-linear entirely Lagrangian shell equations may be achieved by imposing some restrictions upon the rotations.

The basic parameter describing the magnitude of a rotation is the angle of rotation ω about the rotation axis defined by the unit vector $\underset{\sim}{e}$. The angle may be used to classify rotations in terms of the small parameter (2.20) as follows [10-12]: $\omega \leq O(\theta^2)$ - small rotations, $\omega = O(\theta)$ - moderate rotations, $\omega = O(\sqrt{\theta})$ - large rotations, $\omega \geq O(1)$ - finite rotations. Introducing the finite rotation vector $\underset{\sim}{\Omega} = \sin\omega\,\underset{\sim}{e}$ we may approximate it within small strains

$$\underset{\sim}{\Omega} \simeq \epsilon^{\beta\alpha}\,[\,\phi_\alpha(1+\tfrac{1}{2}\theta^\kappa_\kappa) - \tfrac{1}{2}\phi^\lambda(\theta_{\lambda\alpha}-\omega_{\lambda\alpha})]\underset{\sim}{a}_\beta + \phi\underset{\sim}{n}\;. \tag{3.1}$$

For any restriction imposed on ω from (3.1) follow estimates for ϕ_α, ϕ and from $(2.1)_1$ we obtain an estimate for $\theta_{\alpha\beta}$. Estimates of those linearized parameters allow to simplify consistently the shell strain measures (2.1) within the error of the strain energy function (2.19). When introduced into (2.7) the simplified strain measures generate corresponding entirely Lagrangian non-linear shell equations for each simplified version of the theory of shells.

Within small rotations $\gamma_{\alpha\beta} = \theta_{\alpha\beta} + O(\eta\theta^2)$, $\kappa_{\alpha\beta} = -\frac{1}{2}(\phi_{\alpha|\beta} + \phi_{\beta|\alpha}) + O(\eta\theta/\lambda)$, where $\lambda = h/\theta$, and the theory reduces to the bending linear theory of shells which is discussed in many monographs.

Within moderate rotations the shell strain measures (2.7) may be simplified [10,12] to the form

$$\begin{aligned}\gamma_{\alpha\beta} &= \theta_{\alpha\beta} + \tfrac{1}{2}\phi_\alpha\phi_\beta + \tfrac{1}{2}a_{\alpha\beta}\phi^2 - \tfrac{1}{2}(\theta^\lambda_\alpha\omega_{\lambda\beta} + \theta^\lambda_\beta\omega_{\lambda\alpha}) + O(\eta\theta^2)\,,\\ \kappa_{\alpha\beta} &= -\tfrac{1}{2}[\phi_{\alpha|\beta} + \phi_{\beta|\alpha} + b^\lambda_\alpha(\theta_{\lambda\beta}-\omega_{\lambda\beta}) + b^\lambda_\beta(\theta_{\lambda\alpha}-\omega_{\lambda\alpha})] + O(\eta\theta/\lambda).\end{aligned} \tag{3.2}$$

The complete set of Lagrangian equations of the theory of shells undergoing moderate rotations was given in [12] . The theory contains as special cases the equations of various simpler versions of the Lagrangian theory of shells which were proposed in the literature. A detailed review of those versions was given in [17] where also many free and constrained functionals and associated with them variational principles were constructed. Stability equations for flexible shells based on (3.2) were derived in [8,20].

Within large rotations $\gamma_{\alpha\beta}$ can not be simplified while for the tensor of change of curvature we obtain [13]

$$\begin{aligned}\kappa_{\alpha\beta} = {}& \tfrac{1}{2}(1^{\lambda}_{\cdot\alpha}m_{\lambda|\beta} + 1^{\lambda}_{\cdot\beta}m_{\lambda|\alpha}) + \tfrac{1}{2}(\phi_\alpha m_{,\beta} + \phi_\beta m_{,\alpha}) -\\ &-\tfrac{1}{2}[b^\lambda_\alpha(\theta_{\lambda\beta}-\omega_{\lambda\beta}) + b^\lambda_\beta(\theta_{\lambda\alpha}-\omega_{\lambda\alpha})] + \tfrac{1}{2}(b^\lambda_\alpha\omega_{\lambda\beta} + b^\lambda_\beta\omega_{\lambda\alpha})(\theta^\kappa_\kappa + \tfrac{1}{2}\omega^{\kappa\rho}\omega_{\kappa\rho}) -\\ &-\tfrac{1}{2}(b^\lambda_\alpha\phi_\beta + b^\lambda_\beta\phi_\alpha)(\phi_\lambda + \phi^\mu\omega_{\mu\lambda}) + \tfrac{1}{2}b_{\alpha\beta}\phi^\lambda\phi_\lambda + O(\eta\theta/\lambda)\,.\end{aligned} \tag{3.3}$$

The complete set of entirely Lagrangian shell equations based on $(2.1)_1$ and (3.3) was derived in [13] . Two special cases of (3.3) were also discussed in [13] in which rotations associated with in-surface deformation were allowed to be small or moderate. Even for such large/small or large/moderate rotation shell theories the resulting Lagrangian shell equations were still very complex and hardly readable. However, in all three cases it was possible to get rid of the non-rational expressions at the shell boundary approximating consistently the square-root functions by polynomials of the displacemental parameters.

4. Simplified Theories of Shells Undergoing Large/Small Rotations

In the two simplified theories discussed here within the large/small rotation range of shell deformation a greater error $O(Eh\eta^2\theta\sqrt{\theta})$ or $O(Eh\eta^2\theta)$ is allowed in the strain energy function (2.19). The scheme of derivation and equilibrium equations for such simplified versions of shell theory were given already in [10,12], but at that time we failed to construct variationally derivable Lagrangian static boundary and corner conditions. Only when entirely Lagrangian theory of shells was developed [5] it became possible to reduce it consistently also within the large/small rotation range of deformation [13,21,22] and to formulate properly the corresponding static boundary and corner conditions. In what follows the relations given in [13,21,22] are modified further and presented in what is believed to be their canonical form.

When rotations about tangents to M are allowed to be large while rotations about normals to M are supposed to be always small then from (3.1) it follows that $\phi = O(\theta^2)$, $\phi_\alpha = O(\sqrt{\theta})$ and from $(2.1)_1$ we obtain $\theta_{\alpha\beta} = O(\theta)$.

Within the error $O(Eh\eta^2\theta\sqrt{\theta})$ of the strain energy function the shell strain measures (2.1) of the large/small rotation theory take the consistently reduced form [13]

$$\begin{aligned}
\gamma_{\alpha\beta} &= \theta_{\alpha\beta} + \frac{1}{2}\phi_\alpha\phi_\beta + \frac{1}{2}\theta^\lambda_{\alpha\cdot}\theta_{\lambda\beta} - \frac{1}{2}(\theta^\lambda_\alpha\omega_{\lambda\beta} + \theta^\lambda_\beta\omega_{\lambda\alpha}) + O(\eta\theta\sqrt{\theta}) \,, \\
\kappa_{\alpha\beta} &= \frac{1}{2}[(\mathfrak{m}_{\alpha|\beta} + \mathfrak{m}_{\beta|\alpha}) - (\theta^\lambda_\alpha\phi_{\lambda|\beta} + \theta^\lambda_\beta\phi_{\lambda|\alpha}) + (\phi_\alpha\mathfrak{m}_{,\beta} + \phi_\beta\mathfrak{m}_{,\alpha}) - \\
&\quad - (b^\lambda_\alpha\theta_{\lambda\beta} + b^\lambda_\beta\theta_{\lambda\alpha}) - (b^\lambda_\alpha\phi_\beta + b^\lambda_\beta\phi_\alpha)\phi_\lambda + b_{\alpha\beta}\phi^\lambda\phi_\lambda] + O(\eta\sqrt{\theta}/\lambda) \,, \\
\mathfrak{m}_\lambda &= -(1+\theta^\kappa_\kappa)\phi_\lambda + \phi^\mu\theta_{\mu\lambda} \,, \qquad \mathfrak{m} = 1 + \theta^\kappa_\kappa \,. \qquad (4.1)
\end{aligned}$$

It follows from $(4.1)_1$ that within the same approximation $\theta_{\alpha\beta} = -\frac{1}{2}\phi_\alpha\phi_\beta + O(\theta^2)$, see [10,22], which introduced into $(4.1)_3$ allow to reduce the parameters into the simpler forms

$$\mathfrak{m}_{\lambda} = -\phi_{\lambda} + O(\theta^2\sqrt{\theta}) \quad , \quad \mathfrak{m} = 1 - \frac{1}{2}\phi^{\kappa}\phi_{\kappa} + O(\theta^2) \quad ,$$
$$\bar{\underset{\sim}{n}} \simeq -\phi_{\nu}\underset{\sim}{\nu} - \phi_t\underset{\sim}{t} + (1 - \frac{1}{2}\phi_{\nu}^2 - \frac{1}{2}\phi_t^2)\underset{\sim}{n} \quad , \quad \phi_t = \frac{dw}{ds} - \tau_t u_{\nu} + \sigma_t u_t \quad . \tag{4.2}$$

This allows to present $\kappa_{\alpha\beta}$ by a simpler equivalent expression

$$\kappa_{\alpha\beta} = -\frac{1}{2}\Big\{[(\delta_{\alpha}^{\lambda} + \theta_{\alpha}^{\lambda})\phi_{\lambda|\beta} + (\delta_{\beta}^{\lambda} + \theta_{\beta}^{\lambda})\phi_{\lambda|\alpha}] + \frac{1}{2}[\phi_{\alpha}(\phi^{\kappa}\phi_{\kappa})_{,\beta} + \phi_{\beta}(\phi^{\kappa}\phi_{\kappa})_{,\alpha}] +$$
$$+ (b_{\alpha}^{\lambda}\theta_{\lambda\beta} + b_{\beta}^{\lambda}\theta_{\lambda\alpha}) + (b_{\alpha}^{\lambda}\phi_{\beta} + b_{\beta}^{\lambda}\phi_{\alpha})\phi_{\lambda} - b_{\alpha\beta}\phi^{\lambda}\phi_{\lambda}\Big\} + O(\eta\sqrt{\theta}/\lambda) . \tag{4.3}$$

Using some identities and the estimate for $\theta_{\alpha\beta}$ given above (4.2) the expression (4.3) can be shown to be equivalent, within the assumed error, to the one proposed in [12] f.(5.3.7). However, for the reasons explained at the end of §3, the displacemental form used in [12] did not allow to construct four variationally derivable static boundary conditions. In [13,22] equivalent to (4.3) measures of the change of curvature were proposed as quadratic polynomials in displacements. However, four static boundary conditions were constructed in [13] approximating consistently the square-root functions of the exact theory [5] by polynomials of the displacemental parameters, while in the transformation of IVW performed in [22] an approximate expression $\theta_{\kappa}^{\kappa} = -\frac{1}{2}\phi^{\kappa}\phi_{\kappa} + O(\theta^2)$ had to be additionally applied in the corresponding line integral. On the other hand, our expression (4.3) is a third-degree polynomial in displacements but it allows to perform exactly all further transformations presented in the following part of this paper.

When $(4.1)_1$ and (4.3) are introduced into the principle of virtual displacements, after appropriate transformations it can be reduced exactly to the form (2.16), only now

$$T^{\lambda\beta} = (\delta_{\alpha}^{\lambda} + \theta_{\alpha}^{\lambda})N^{\alpha\beta} - \frac{1}{2}(\omega^{\lambda\alpha}N_{\alpha}^{\beta} + \omega^{\beta\alpha}N_{\alpha}^{\lambda}) - \frac{1}{2}[(b_{\alpha}^{\lambda} + \phi^{\lambda}|_{\alpha})M^{\alpha\beta} + (b_{\alpha}^{\beta} + \phi^{\beta}|_{\alpha})M^{\alpha\lambda}] ,$$
$$T^{\beta} = \phi_{\alpha}N^{\alpha\beta} + [(\delta_{\lambda}^{\beta} + \theta_{\lambda}^{\beta})M^{\lambda\alpha}]_{|\alpha} + (\phi_{\lambda}M^{\lambda\alpha})_{|\alpha}\phi^{\beta} - \phi^{\lambda}|_{\alpha}\phi_{\lambda}M^{\alpha\beta} -$$
$$- (b_{\alpha}^{\lambda}M^{\alpha\beta} + b_{\alpha}^{\beta}M^{\alpha\lambda})\phi_{\lambda} + b_{\alpha\lambda}\phi^{\beta}M^{\alpha\lambda} ,$$
$$R_{\nu\nu} = (1 + \theta_{\nu\nu})M_{\nu\nu} + \theta_{\nu t}M_{t\nu} , \quad R_{t\nu} = \theta_{t\nu}M_{\nu\nu} + (1 + \theta_{tt})M_{t\nu} ,$$
$$R_{\nu} = \phi_{\nu}M_{\nu\nu} + \phi_t M_{t\nu} , \quad M = R_{\nu\nu} + \phi_{\nu}R_{\nu} = (1 + \theta_{\nu\nu} + \phi_{\nu}^2)M_{\nu\nu} + (\theta_{\nu t} + \phi_{\nu}\phi_t)M_{t\nu} , \tag{4.4}$$
$$\underset{\sim}{F} = F\underset{\sim}{n} , \quad F = R_{t\nu} + \phi_t R_{\nu} = (\theta_{\nu t} + \phi_{\nu}\phi_t)M_{\nu\nu} + (1 + \theta_{tt} + \phi_t^2)M_{t\nu} ,$$
$$\underset{\sim}{F}^* = F^*\underset{\sim}{n} , \quad F^* = H_t + \phi_t H , \quad M^* = H_{\nu} + \phi_{\nu}H .$$

From (2.16) with (4.4) follow corresponding Lagrangian shell equations (2.18), in which $(4.2)_2$ and (4.4) should be used.

In some engineering applications we may be interested in the use of even simpler but consistently reduced shell relations which follow when a larger error $O(Eh\eta^2\theta)$ in the strain energy function is assumed to be permissible. Within this larger error the strain measures of such simplest large/small rotation theory of shells take the extremely simple form

$$\gamma_{\alpha\beta} = \theta_{\alpha\beta} + \frac{1}{2}\phi_\alpha\phi_\beta + \frac{1}{2}\theta^\lambda_\alpha\theta_{\lambda\beta} + O(\eta\theta) \quad ,$$
$$\kappa_{\alpha\beta} = -\frac{1}{2}[(\delta^\lambda_\alpha + \theta^\lambda_\alpha + \phi^\lambda\phi_\alpha)\phi_{\lambda|\beta} + (\delta^\lambda_\beta + \theta^\lambda_\beta + \phi^\lambda\phi_\beta)\phi_{\lambda|\alpha}] + O(\eta/\lambda). \tag{4.5}$$

Again, within the assumed accuracy the strain measures (4.5) are equivalent to those proposed in [12,13,21,22] where various different displacemental expressions for $\kappa_{\alpha\beta}$ were suggested. However, only the expression $(4.5)_2$ given here allows to perform exactly all further transformations. When introduced into (2.7) the measures (4.5) lead exactly to (2.16) with the following corresponding definitions of the static field parameters

$$T^{\lambda\beta} = N^{\lambda\beta} + \frac{1}{2}(\theta^\lambda_\alpha N^{\alpha\beta} + \theta^\beta_\alpha N^{\alpha\lambda}) - \frac{1}{2}(\phi^\lambda|_\alpha M^{\alpha\beta} + \phi^\beta|_\alpha N^{\alpha\lambda}) \quad ,$$
$$T^\beta = \phi_\alpha N^{\alpha\beta} + [(\delta^\beta_\lambda + \theta^\beta_\lambda)M^{\lambda\alpha}]|_\alpha + (\phi_\lambda M^{\lambda\alpha})|_\alpha\phi^\beta - \phi^\lambda\phi_{\lambda|\alpha}M^{\alpha\beta} \quad , \tag{4.6}$$

while corresponding static boundary parameters remain identical with those given in $(4.4)_{3-6}$ and $(4.2)_2$.

From (2.16) we obtain the following component form of the entirely Lagrangian equations for both simplified versions of the theory of shells undergoing large/small rotations:
the equilibrium equations in M

$$T^{\lambda\beta}|_\beta - b^\lambda_\beta T^\beta + p^\lambda = 0 \ , \quad T^\beta|_\beta + b_{\lambda\beta}T^{\lambda\beta} + p = 0 \quad ; \tag{4.7}$$

the static boundary conditions on C_f

$$T^{\lambda\beta}\nu_\lambda\nu_\beta + \tau_t(R_{t\nu} + \phi_t R_\nu) = T_\nu + \tau_t(H_t + \phi_t H) \quad ,$$
$$T^{\lambda\beta}t_\lambda\nu_\beta - \sigma_t(R_{t\nu} + \phi_t R_\nu) = T_t - \sigma_t(H_t + \phi_t H) \quad ,$$
$$T^\beta\nu_\beta + \frac{d}{ds}(R_{t\nu} + \phi_t R_\nu) = T + \frac{d}{ds}(H_t + \phi_t H) \quad ,$$
$$R_{\nu\nu} + \phi_\nu R_\nu = H_\nu + \phi_\nu H \quad ; \tag{4.8}$$

the static corner conditions at each corner $M_j \in C_f$

$$F(s_j + 0) - F(s_j - 0) = F^*(s_j + 0) - F^*(s_j - 0) \quad ; \tag{4.9}$$

the geometric boundary conditions on C_u

$$u_\nu = u^*_\nu \; , \quad u_t = u^*_t \; , \quad w = w^* \; , \quad \phi_\nu = \phi^*_\nu \quad ; \tag{4.10}$$

the geometric corner condition at each corner $M_i \in C_u$

$$w_i = w^*_i \; . \tag{4.11}$$

Introducing appropriate definitions $(4.4)_{1,2}$ or (4.6) into (4.7) and (4.8) extended representations of the equilibrium equations and the static boundary conditions in terms of $N^{\alpha\beta}, M^{\alpha\beta}$ and u_α, w, ϕ_ν may easily be derived.

The structure of the Lagrangian shell relations given above is relatively simple. The strain tensors $(4.1)_1$ or $(4.5)_1$ are quadratic in displacements while the tensors of change of curvature (4.3) or $(4.5)_2$ are cubic in displacements , where cubic terms are expressed only through ϕ_α . The equilibrium equations (4.7) are linear in $N^{\alpha\beta}, M^{\alpha\beta}$ and quadratic in u_α, w only through squares of ϕ_α . The static boundary (4.8) and corner (4.9) conditions are linear in $N^{\alpha\beta}, M^{\alpha\beta}$ and quadratic in displacemental parameters again through ϕ_ν and ϕ_t . All four geometric boundary conditions (4.10) and the geometric corner conditions (4.11) are linear in displacements, what is very important when a numerical solution of a non-linear shell problem is constructed using finite elements in order to discretize the displacement field.

The Hu - Washizu free functional corresponding to the shell relations presented above takes the form

$$\begin{aligned} I_1 = & \iint\limits_{M} \Big\{ \Sigma(\gamma_{\alpha\beta}, \kappa_{\alpha\beta}) - \underset{\sim}{p} \cdot \underset{\sim}{u} - N^{\alpha\beta}[\gamma_{\alpha\beta} - \gamma_{\alpha\beta}(\underset{\sim}{u})] - M^{\alpha\beta}[\kappa_{\alpha\beta} - \kappa_{\alpha\beta}(\underset{\sim}{u})] \Big\} dA - \\ & - \int\limits_{C_f} \Big\{ \underset{\sim}{T} \cdot \underset{\sim}{u} + \underset{\sim}{H} \cdot [\underset{\sim}{\bar{n}}(u, \phi_\nu) - \underset{\sim}{n}] \Big\} ds - \\ & - \int\limits_{C_u} [\underset{\sim}{P} \cdot (\underset{\sim}{u} - \underset{\sim}{u}^*) - M(\phi_\nu - \phi^*_\nu)] ds - \sum_i F_i (w_i - w^*_i) \quad , \end{aligned} \tag{4.12}$$

where (2.19), $(4.1)_1$ and (4.3) or (4.5), $(4.2)_2$, $(4.4)_{3-6}$ and left-hand sides of (4.8) should be used. As stationarity conditions of I_1 we obtain exactly all relations of the respective simplified versions of the entirely Lagrangian theory of shells undergoing large/small rotations. From I_1 , following [17,6,7], a number of other free or constrained functionals and associated variational principles may be constructed. Appropriate stability equations for the large/small rotation theory of shells may be derived by specialization of those given in [9,23,24].

Let us remind some simplified versions of the non-linear theory of thin shells for which non-linear expressions for changes of curvatures were suggested. Koiter [19] f.(12.2) proposed a quadratic expression for $\rho_{\alpha\beta} \equiv -\kappa_{\alpha\beta} - \frac{1}{2}(b^{\lambda}_{\alpha}\gamma_{\lambda\beta} + b^{\lambda}_{\beta}\gamma_{\lambda\alpha})$ in the case of "moderate deflections", Başar [25] derived a quadratic expression for $\kappa_{\alpha\beta}$ in the case of "moderately large rotations", from Galimov [4] f.(3.38) follows a quadratic expression for $\kappa_{\alpha\beta}$ in the case of "strong bending". When compared with corresponding expression (5.3.7) of [12] for $\kappa_{\alpha\beta}$ with greater error, which is equivalent to our (4.3) in the sense of error, the lack of terms $\frac{1}{4}\phi^{\lambda}\phi_{\lambda}(\phi_{\alpha|\beta} + \phi_{\beta|\alpha})$ was noted in the measure [19] and in transformed version of the measure [25] obtained using an identity (3.34) of [19], while terms $\phi^{\lambda}(\theta_{\lambda\alpha|\beta} + \theta_{\lambda\beta|\alpha} - \theta_{\lambda\alpha|\beta})$ were missed in the resulting measure of [4]. According to our estimates, those terms are $O(\theta\sqrt{\theta}/\lambda)$ and should be taken into account even within the simplest large/small rotation shell theory, see $(5.3.9)_2$ of [12]. Apart from that, for the reasons explained at the end of §3, the expressions suggested in [4,19, 25] for the changes of curvatures do not allow to construct variationally derivable four Lagrangian static boundary conditions. Shapovalov [26] f.(1.9) proposed an extremely simple quadratic theory of shells in which $\gamma_{\alpha\beta}$ contain two first terms of (4.1), κ_{12} is linear while $\kappa_{\alpha\alpha} = -\phi_{\alpha|\alpha} - \frac{1}{2}b_{\alpha\alpha}\phi^2_{\beta}$, $\alpha \neq \beta$. The quadratic terms in $\kappa_{\alpha\alpha}$ result from the second line of (4.3) and are $O(\theta^2/\lambda)$. Since other terms $O(\theta^2)$ were omitted in $\gamma_{\alpha\beta}$ and even more important terms $O(\theta\sqrt{\theta}/\lambda)$ were omitted in $\kappa_{\alpha\beta}$, the version of [26] can not be regarded as consistent within the large-rotation theory of shells. In a refined version [27] a theory of shells undergoing finite/small rotations was given. At the shell boundary contour a "vector of elastic rotation" was introduced which had no geometric meaning of a finite rotation vector [10-12]. It was assumed that only one of the three components of the vector was independent but explicit transformation formulae for other two components were not given. The work performed by the Eulerian boundary couple on the "vector of elastic rotation" was assumed as a scalar product of the vectors what may not be correct in the general case. As a result, the boundary conditions constructed in [27] can not be regarded as entirely Lagrangian and their physical and geometrical meaning as well as the range of applicability is open to discussion. Finally, let us remind that already in [28,29] it was suggested to take into account all quadratic terms in $\kappa_{\alpha\beta}$. However, the non-linear theory of shells generated by such formal quadratic strain measures can not be regarded as consistent from the point of view of an error introduced into the shell strain energy function. Besides, such measures would not allow to construct variationally derivable four Lagrangian static boundary conditions.

Acknowledgements. Partial support for this research was provided by the Ruhr-Universität Bochum /FRG/.

References

1. Sanders, J.L.: Nonlinear theories for thin shells. Quart. Appl. Math. 21 (1963) 21-36.

2. Budiansky, B.: Notes on nonlinear shell theory. J. Appl. Mech. Trans. ASME E35 (1968) 393-401.

3. Pietraszkiewicz, W.: Lagrangian nonlinear theory of shells. Arch. Mech. Stos. 26 (1974) 221-228.

4. Galimov, K.Z.: Foundations of the nonlinear theory of shells /in Russian/. Kazan': Kazan' Univ. Press 1975.

5. Pietraszkiewicz, W.; Szwabowicz, M.L.: Entirely Lagrangian nonlinear theory of thin shells. Arch. Mech. 33 (1981) 273 - 288.

6. Szwabowicz, M.: Równania podstawowe i zasady wariacyjne geometrycznie nieliniowej teorii ciemkich powłok sprężystych. Zesz. Nauk. WSI w Opolu Nr 89, Ser. Budownictwo z.18 (1982) 267-272.

7. Schmidt, R.: Variational principles for general and restricted geometrically non-linear Kirchhoff-Love type shell theories. Proc. Int. Conf. on FEM, Shanghai. New York: Gordon and Breach Sci. Publ. 1982.

8. Makowski, J.: Liniowa i nieliniowa analiza stateczności sprężystej cienkich powłok. Praca doktorska. Politechnika Gdańska, Instytut Konstrukcji Budowlanych: Gdańsk 1981.

9. Makowski, J.; Pietraszkiewicz, W.: Incremental formulation of the non-linear theory of thin shells in the total Lagrangian description. ZAMM 64 (1984) T65-T67.

10. Pietraszkiewicz, W.: Introduction to the non-linear theory of shells. Ruhr-Universität, Mitt. Inst. f. Mech. Nr 10: Bochum 1977.

11. Pietraszkiewicz, W.: Finite Rotations and Lagrangean description in the non-linear theory of shells. Warszawa, Poznań: Polish Scientific Publishers 1979.

12. Pietraszkiewicz, W.: Finite rotations in the nonlinear theory of thin shells. in: Olszak, W. (ed.) Thin shell theory, new trends and applications. CISM courses and lectures No 240. Wien, New York: Springer-Verlag 1980.

13. Pietraszkiewicz, W.: On consistent approximations in the geometrically non-linear theory of shells. Ruhr-Universität, Mitt. Inst. f. Mech. Nr 26: Bochum 1981.

14. Koiter, W.T.: A consistent first approximation in the general theory of thin elastic shells. in: Koiter, W.T. (ed.) Theory of thin shells. Proc. IUTAM Symp., Delft 1959. Amsterdam: North-Holland P.Co. 1960.

15. Koiter, W.T.; Simmonds, J.G.: Foundations of shell theory. in: Theoretical and applied mechanics. Proc. 13th IUTAM Congress, Moscow 1972. Berlin: Springer-Verlag 1974.

16. Koiter, W.T.: The intrinsic equations of shell theory with some applications. in: Nemat-Nasser, S. (ed.) Mechanics Today vol. 5. London: Pergamon Press 1980.

17. Schmidt, R.; Pietraszkiewicz, W.: Variational principles in the geometrically non-linear theory of shells undergoing moderate rotations. Ingenieur-Archiv 50 (1981) 187-201.

18. Chernykh, K.F.: Non-linear theory of thin isotropic elastic shells. Mechanics of Solids (1980) 148-159.

19. Koiter, W.T.: On the nonlinear theory of thin elastic shells. Proc. Konink. Ned. Akad. Wetensch. B69 (1966) 1-54.

20. Stumpf, H.: The stability equations of the consistent nonlinear elastic shell theory with moderate rotations. in: Schröder, F.H. (ed.) Stability in the mechanics of continua. Berlin, Heidelberg, New York: Springer-Verlag 1982.

21. Pietraszkiewicz, W.: A simplest consistent version of the geometrically non-linear theory of elastic shells undergoing large/small rotations. ZAMM 63 (1983) T200-T202.

22. Nolte, L.-P.; Stumpf, H.: Energy-consistent large rotation shell theories in Lagrangian description. Mech. Res. Comm. 10 (1983) 213-221.

23. Nolte, L.-P.: Stability equations of the general geometrically non-linear first approximation theory of thin elastic shells. ZAMM 63 (1983) T79-T82.

24. Schmidt, R.; Stumpf, H.: On the stability and post-buckling of thin elastic shells with unrestricted rotations. Mech. Res. Comm. 11 (1984)(in print).

25. Başar, Y.: Eine geometrisch nichtlineare Schalentheorie. Konstruktiver Ingenieurbau-Berichte Nr. 38/39, Ruhr-Universität Bochum 1981.

26. Shapovalov, L.A.: On a simplest variant of equations of geometrically non-linear theory of thin shells /in Russian/. Inzh. Zhurnal, Mekh. Tv. Telȧ (1968) 56-62.

27. Shapovalov, L.A.: Equations of thin elastic shells for nonsymmetric deformations. Mechanics of Solids (1976) 62-72.

28. Novozhilov, V.V.: Foundations of the non-linear theory of elasticity /in Russian/. Moscow, Gostekhizdat 1948. English transl.: Baltimore: Graylock 1953.

29. Tsao, C.H.: Strain-displacement relations in large displacement theory of shells. AIAA J. 2 (1964)

Shallow Caps with a Localized Pressure Distribution Centered at the Apex[1]

FREDERIC Y. M. WAN[2]

Department of Mathematics and
Institute of Applied Mathematics and Statistics
The University of British Columbia
Vancouver, B.C. V6T 1W5
Canada

Introduction

Under favourable loading conditions, dome-shaped thin elastic shells of revolution are known to exhibit a predominantly inextensional bending deformation in the form of a finite axisymmetric dimple centered at the pole. For example, it has been shown in [1,2] that polar dimpling is possible when a spherical shell is subject to an axisymmetric normal pressure distribution which is directed inward near a pole and outward in an adjacent region[3]. To a good approximation, the dimple base radius, which characterizes the location of the dimple base and therefore the dimple size, was shown to depend on the external loading in a simple way. To bring out the essential idea behind the asymptotic method for constructing the simple solution, results for a spherical cap with a clamped edge were first presented in [1] for a quadratically varying pressure distribution along the shell meridian. Analogous and more general results were reported for a complete spherical shell in [2] for a meridionally sinusoidal pressure distribution.

The two approximate analyses in [1] and [2] both consist of showing the existence of two types of inextensional bending deformation which, when

---

(1) The research is partly supported by NSERC Operating Grant No. A9259. The author gratefully acknowledges Messrs. Derek Lee and Joseph Pang for their assistance in machine computation and computer graphics.

(2) The author is currently a Professor of Applied Mathematics and Mathematics in the Applied Mathematics Program, FS-20, University of Washington, Seattle, WA 98195, USA.

(3) This type of load distributions is of interest to the designers of fuel containers for space crafts and to biomechanicians and physiologists interested in biological structures involving fluid-filled thin shells (see [3] and [4] for examples).

pieced together by a bending layer solution, give an axisymmetric dimple centered at the pole. In the case of a shallow cap, the conditions of a support at the cap edge are satisfied by an edge bending layer solution. In the language of singular perturbations, (e.g., [19]), the inextensional bending solutions and the layer solutions correspond to the outer and inner asymptotic expansions, respectively, of the exact solution of the boundary value problem (BVP) governing the elastostatics of the shell. The main result in [1] and [2] is the simple determination of an accurate, first approximation dimple radius by way of the outer solutions of the BVP alone without any reference to the inner solution(s). Consistent with the spirit of this main result, the analyses in [1] and [2] also avoid the final matching of the inner and outer solutions. They rely on accurate numerical solutions of the original BVP to show the adequacy of the approximate location of the dimple base and of the outer solutions away from the dimple base and shell edge, for sufficiently thin shells and for a maximum applied inward pressure at least comparable to the classical buckling pressure for a spherical shell with the same radius of curvature. In a sequel to [1] soon to appear [5], the corresponding polar dimpling solution for general cap-like shallow shells of revolution under more general axisymmetric loadding conditions is obtained for a much wider range of maximum applied load magnitude. More precise conditions for the matching of inner and outer solutions are also given there.

To the extent that a detailed study of the more general polar dimpling solution obtained in [5] is presented there again only for a spherical cap with a quadratically varying pressure distribution (concentrating on the sub-buckling pressure magnitude range where dimples of two different sizes are possible), one of the purposes of this report is to complement the work of [5] and analyze in some detail the dimple solution for internally pressurized shallow caps with two types of localized axisymmetric inward applied loads important in engineering applications. These two types of localized loads have as their respective limiting case a point force at the pole and a ring load along a given lattitude. Beyond the dimple type deformations appropriate for the aforementioned applied loads, we will also discuss some typical load distributions for which nonlinear membrane action, instead of the inextensional bending action, dominates throughout the shallow cap except in some narrow interior and/or edge layers. The report concludes with some comments on a class of axisymmetric load distributions of interest to engineers which may give rise to one of the two kinds of nonlinear membrane behaviours, or to a dimple type inextensional bending behaviour, depending on the relative magnitude of the several load and geometric parameters.

Formulation

The elastostatics of shells of revolution, which have undergone an axisymmetric finite deformation (with infinitesimal strain), may be formulated as a boundary value problem for a pair of coupled nonlinear second order ordinary differential equations for a stress function and a meridional slope variable [6]. Let x ($0 \le x \le 1$) denote the dimensionless radial distance from the axis of revolution to a point on the middle surface of a cap-like shallow shell of revolution having uniform thickness h. We consider here shells subjected to a general axisymmetric (positive inward) normal load distribution $p_n = p_v p(x)$, with p_v chosen so that the dimensionless resultant axial force P (defined in (2.3)) is $O(1)$(4), and a genereral axisymmetric (positive outward) radial surface load distribution $p_r = p_H q(x)$ (with $|q|_{max} = 1$). With an undeformed meridional slope $\xi_0 \Phi_0(x)$ ($\Phi_0(1) = 1$), we may take these equations in the following dimensionless form [7]:

$$\varepsilon^2 x[\Psi'' + \frac{1}{x}\Psi' - \frac{1}{x^2}\Psi] + \frac{1}{2}(\Phi^2 - \Phi_0^2) = -4\mu x[xq' + (2+\nu)q] \tag{2.1}$$

$$\varepsilon^2 x[\Phi'' + \frac{1}{x}\Phi' - \frac{1}{x^2}\Phi] - \Phi\Psi = 4\kappa x P(x) + \varepsilon^2 x Q(x) \tag{2.2}$$

with

$$P(x) = \frac{1}{x}\int_0^x tp(t)dt, \tag{2.3}$$

$$Q(x) = \Phi_0'' + \frac{1}{x}\Phi_0' - \frac{1}{x^2}\Phi_0. \tag{2.4}$$

(4) p may be unbounded in the limiting case of a point load or ring load.

In (2.1)-(2.4), a prime indicates differentiation with respect to x, $\xi_0\Phi(x)$ denotes the meridional slope of the deformed middle surface of the shell, and the dimensionless independent variable x is related to the radial distance from the axis of revolution $r(0\le r\le r_0)$ by $x = r/r_0$ (see Fig. 1). In terms of r_0, the shell thickness h, Young's modulus E, Poisson's ratio ν, and the radius of curvature measure of the shell $a = r_0/\xi_0$, we have

$$\varepsilon^2 = \frac{ha}{2r_0^2\sqrt{3(1-\nu^2)}}, \qquad \mu = \frac{h}{2r_0\sqrt{3(1-\nu^2)}}\frac{p_H}{p_c}$$

$$\kappa = \frac{p_v}{p_c}, \qquad p_c = \frac{2Eh^2}{a^2\sqrt{3(1-\nu^2)}} \tag{2.5}$$

We note also that the dimensionless stress function Ψ is the conventional stress function normalized by $p_c a r_0/4$ with $p_c a\Psi/4x$ and $p_c a\Psi'/4$ being the radial and hoop stress resultant, respectively.

Equations (2.1) and (2.2) are supplemented by the regularity conditions at the apex:

$$x = 0: \qquad \Phi = 0, \qquad \Psi = 0, \tag{2.6}$$

and appropriate edge conditions depending on the type of support for the shell at x = 1. For example, we consider in sections (3) - (5) shells with a clamped edge so that

$$x = 1: \qquad \Phi = 1, \qquad \Psi' - \nu\Psi = 0, \tag{2.7}$$

though our analysis applied to shells with other types of edge support, possibly after suitable (but straightforward) modifications. The second condition in (2.7) corresponds to a requirement of no radial midsurface displacement. A discussion of other stress and deformation measures of the shallow shell can be found in [1,2,6,7].

For the most part of this report, we shall be concerned mainly with the simpler problem of a spherical cap, $\Phi_0(x) = x$, subjected only to a dimensionless axisymmetric normal pressure load distribution of the form

$$p(x) = \begin{cases} -p_2 & (0 \le x < x_1) \\ -p_2 + \dfrac{P_1}{(x_2^2-x_1^2)} & (x_1 < x < x_2) \\ -p_2 & (x_2 < x \le 1) \end{cases} \tag{2.8}$$

with the corresponding axial resultant given by

$$P(x) = \begin{cases} -\frac{1}{2} p_2 x & (0 \le x \le x_1) \\ -\frac{1}{2} p_2 x + \dfrac{P_1}{2x} \dfrac{x^2-x_1^2}{x_2^2-x_1^2} & (x_1 \le x \le x_2) \\ -\frac{1}{2} p_2 x + \dfrac{P_1}{2x} & (x_2 \le x \le 1) \end{cases} \tag{2.9}$$

For this special case, the ODEs (2.1) and (2.2) become (with $q(x)\equiv 0$)

$$\varepsilon^2 x[\Psi'' + \frac{1}{x}\Psi' - \frac{1}{x^2}\Psi] + \frac{1}{2}(\Phi^2 - x^2) = 0, \tag{2.10}$$

$$(0<x<1)$$

$$\varepsilon^2 x[\Phi'' + \frac{1}{x}\Phi' - \frac{1}{x^2}\Phi] - \Phi\psi = 4\kappa x P(x), \tag{2.11}$$

where $P(x)$ is as given by (2.9). Note that $\pi r_o p_v P_1$ is the dimensionless resultant force associated with the annular normal pressure distribution when $p_2 = 0$. Upon specializing the four parameters at our disposal, namely, p_2, P_1, x_1 and x_2, we get from (2.8) and (2.9) a variety of load conditions useful in engineering applications. We shall analyze the behaviour of shells under these external loads in the next few sections.

Internally Pressurized Caps with a Patch Load at the Apex

For $x_1 = 0$ and $0<x_2<<1$, the applied load (2.8) corresponds to a uniform internal pressure (of dimensionless magnitude $-p_2$) throughout the shell and a localized (axi-symmetric) uniform external pressure centered at the pole with a magnitude inversely proportional to its area of application.

For the net pressure to be inward near the pole, we take $P_1 > p_2 x_2^2$. Note that the magnitude of the resultant force associated with the external pressure is independent of the area of application. In the limit as $x_2 \to 0$, we have an internally (uniformly) pressurized cap with a point force at its pole. In the absence of internal pressure ($p_2=0$) and radial load ($q(x)\equiv 0$), the problem has been investigated in [8-16] and elsewhere. In this section, we are concerned with the case $q(x) = 0$ but $p_2 \neq 0$ for shells with a clamped edge and sufficiently thin so that $\varepsilon^2 \ll 1$.

When κ is O(1) or larger with $\kappa\varepsilon^2 \ll 1$, we know from [1,2,5] that polar dimpling is a possible mode of deformation. The dimple solution consists of the inextensional bending solutions

$$[\Phi,\Psi] \sim \begin{cases} [-\Phi_0,\ 4\kappa x P/\Phi_0] & (0 \leq x < x_T) \\ [\Phi_0,\ -4\kappa x P/\Phi_0] & (x_T < x < 1) \end{cases} \tag{3.1}$$

in two disjoint regions of the shell, an interior layer solution in the neighborhood of x_T to bridge them and a boundary layer solution adjacent to the shell edge to satisfy the clamped edge conditions (2.7). The leading term outer solutions (3.1) are obtained by setting to zero the small parameter ε in the two ODEs (2.1) and (2.2), assuming $xP(x)/\Phi_0(x)$ is well defined in [0,1] except at $x = x_T$. Cases where the shell has a horizontal tangent and/or a flat point have already been treated separately in [17]. We also do not concern ourselves here with the layer solutions, though their presence will be seen from the graphs of the numerical solutions to be shown and discussed at the end of the section. Note that our choice of outer solutions in the two disjoint regions of the shallow cap anticipates the occurrence of the dimpling phenomenon.

The dimensionless dimple base radius x_T is determined to a first approximation by the requirement of vanishing axial resultant to be a positive root $\bar{x}_t$ of $P(x) = 0$. This requirement is a necessary condition for the existence of the interior layer (or the inner) solution for bridging the two types of inextensional bending (or outer) solutions [5]. Physically, it renders the radial stress resultant continuous across the dimple base. For the $P(x)$ under consideration, there is always a single positive root (since we consider only the range $P_1 \geq p_2 x_2^2$) with

$$x_T \sim \bar{x}_t \equiv \sqrt{\frac{P_2}{p_2}} \geq x_2 \qquad (3.2)$$

($P(x)$ never vanishes except at $x = 0$ if $P_1 \leq p_2 x_2^2$ so that polar dimpling is not possible in that range of parameter values.) It is rather remarkable that to leading order, the dimple size depends only on the applied load and not on the shape, thickness or material properties of the shell, at least when the shell is sufficiently thin (and $\kappa >> \varepsilon$).

Accurate numerical solution for the BVP of (2.1), (2.2), (2.6) and (2.7) have been obtained by the BVP solver COLSYS [18], a general purpose computer code for solving ODEs with prescribed boundary values at the end (and/or intermediate) points of the solution domain. Based on spline-collocation at Gaussian points with error estimates, the codes solves a given problem on a sequence of meshes, automatically increasing the number of mesh points in region of abrupt changes, until the solution obtained meets a prescribed (combination of absolute and relative) error tolerance, or until the prescribed maximum mesh is reached without meeting the error tolerance. Some of these accurate numerical solutions are presented here for comparison with the asymptotic solution (3.1) and (3.2). For this purpose, we show only results for a shallow spherical cap with $\Phi_0(x) = x$ and with $\nu = 0.3$.

In Figure (2), we show distributions of the deformed slope $\Phi(x)$ for $\varepsilon^2 = 10^{-4}$, $\kappa = 1$, and $\bar{x}_t \equiv \sqrt{P_1/p_2} = 3/4$, each curve corresponding to a different value of x_2, i.e., a different size of application area (and a different magnitude) of the localized external pressure. These numerical results, generated by COLSYS for the BVP defined by (2.6), (2.7), (2.9)-(2.11), clearly show in all cases a dimple type deformation pattern with an identical dimple radius independent of x_2, a phenomenon predicted by our asymptotic solution (see (3.2)). The (approximate) dimple radius $\bar{x}_T$ as given by COLSYS, i.e. the location where the COLSYS solution for Φ vanishes, agrees with the asymptotic solution $\bar{x}_t$ to within 5%. It should be noted that all COLSYS results presented in this paper have met a prescribed error tolerance of 10^{-4} (and often smaller).

In Figure (3), three COLSYS solutions for $\Phi(x)$ are given for $\kappa = 1$, $x_2 = 0.1$ and $\bar{x}_t = \sqrt{P_1/p_2} = 3/4$; they correspond to three different thickness parameter values, $\varepsilon^2 = 10^{-3}$, 10^{-4} and 10^{-5}. We see from these plots that, in all cases, the dimple sharpens and the layer width decreases as ε decreases. They also show that as ε decreases $\bar{x}_t$ approaches $\bar{x}_T$, the COLSYS solution for x_T.

In Figure (4), three COLSYS solutions for $\Phi(x)$ are given for $x_2 = 0.05$, $\varepsilon^2 = 10^{-4}$ and $\kappa = 1$; they correspond to $\sqrt{P_1/p_2} = 0.4$, 0.6 and 0.75. The figure shows that the dimple size depends only on the ratio P_1/p_2 according to (3.2) to a first approximation.

Other numerical solutions not presented here show that the sharpness of the dimple solution deteriorates as κ increases. It is known from [5] that nonlinear membrane shell action becomes more significant as κ becomes much larger than unity. Our numerical solutions show that it completely dominates the inextensional bending shell action when $\kappa\varepsilon^2 \gtrsim 1$ for the present problem.

At the other end of the spectrum, we also know from [5] that, when κ is small of order ε, say $\kappa = k\varepsilon$ with $k = O(1)$, the necessary condition for the existence of an interior layer solution takes the form

$$4\kappa P(x_T) = \varepsilon I_0 + O(\varepsilon^2) \tag{3.3}$$

where I_0 is a pure number for a spherical cap ($I_0 \doteq 1.6674\ldots$). Therefore, we have to a first approximation

$$P_1 - p_2\bar{x}_t^2 = \frac{I_0}{2k}\,\bar{x}_t \tag{3.4}$$

in the case of $\kappa = k\varepsilon$. In contrast to the case of a quadratically varying pressure load distribution [5] which allows for two different dimple radii, the size of the dimple remains unique in this "sub-buckling" range of the localized load at the pole. Furthermore, for the localized load at the pole, the condition (3.3) always has a unique positive solution for all $k > 0$. We should note, however, that (3.3) ceases to be valid if $k = O(\varepsilon)$ (see the derivation of (3.3) in [5]) and is raplaced by $I_0 + O(\varepsilon) = 0$ instead. Therefore, polar dimpling is not possible for the type of applied loads considered in this section when $\kappa = O(\varepsilon^2)$ since the necessary condition for the existence of an interior layer solution (which bridges the two outer solutions given in (3.1)) cannot be satisfied. From Figure (5), we see that the dimple radius shrinks continuously from $O(1)$ to $O(\varepsilon)$ as k decreases from $O(1)$ to $O(\varepsilon)$. However, the form of $P(x)$ requires $\bar{x}_t \geq x_2$ giving a lower bound to the possible size of the dimple.

Finally, we note that all COLSYS solutions for $\Phi(x)$ display a layer phenomenon in the neighborhood of the load discontinuity at x_2 (see (2.8) with $x_1 = 0$). The bending layer is generated as neither membrane nor inextensional bending shell action could absorb the abrupt changes caused by the load discontinuity.

Internally Pressurized Caps with an Axi-symmetric Ring Load

For $0 < x_2 - x_1 << 1$ with $x_1 > 0$, the load distribution (2.8) corresponds to a uniform internal pressure throughout the shell and a localized uniform external pressure distributed axisymmetrically over an annular frustum of the shell (extending from x_1 to x_2) with a magnitude inversely proportional to the area of application. In the limit as $x_2 - x_1 \to 0$, we have an internally pressurized cap with a ring load at the lattitude $x = x_1$ $(=x_2)$. In the absence of the internal pressure ($p_2 = 0$) and radial load ($q(x) \equiv 0$), the problem has been investigated previously in [14] and elsewhere. In this section, we are concerned with the case $q(x) \equiv 0$ but $p_2 \neq 0$ for sufficiently thin shells so that $\varepsilon^2 << 1$.

For κ not small compared to unity but $\kappa\varepsilon^2 << 1$, we expect the possibility of polar dimpling. The condition $P(x) = 0$, determining x_T to a first approximation, now implies the following:

(1) Polar dimpling is not possible for $P_1 < p_2 x_2^2$, similar to the case of a patch load at the apex.
(2) There is a unique dimple size when $P_2/p_2 = x_2^2$, namely $\bar{x}_t = x_2$.
(3) For $P_1 > p_2 x_2^2$, $P(x) = 0$ has two distinct positive roots given by

$$\bar{x}_t = \begin{cases} x_1/[1 - p_2(x_2^2 - x_1^2)/P_1]^{1/2} \equiv \bar{x}_t^{(1)} & (x_1 < \bar{x}_t^{(1)} < x_2) \\ \\ \sqrt{P_1/p_2} \equiv \bar{x}_t^{(2)} & (>x_2) \end{cases} \tag{4.1}$$

For $P_1/p_2 > x_2$, the existence of two dimple base radii offers a number of possible configurations for the deformed shells. For example, one may have a large dimple corresponding to $\bar{x}_t^{(2)}$ or a small dimple corresponding to $\bar{x}_t^{(1)}$. From an analytical viewpoint, the admissable configurations are ultimately determined by matching the inextensional bending solutions and the various layer solutions. We will not carry out this matching process here since the following outer solution is readily suggested by our particular load distribution:

$$\Phi \sim \begin{cases} \Phi_o(x) & (0 \le x < x_T{}^{(1)}) \\ -\Phi_o(x) & (x_T{}^{(1)} < x < x_T{}^{(2)}) \\ \Phi_o(x) & (x_T{}^{(2)} < x < 1) \end{cases} \tag{4.2}$$

with the corresponding Ψ given by $-4\kappa x P/\Phi$. We also expect that, to a first approximation, we have

$$x_T{}^{(1)} \sim \bar{x}_t^{(1)} \quad , \quad x_T^{(2)} \sim \bar{x}_t^{(2)} \tag{4.3}$$

Note that $\bar{x}^{(1)}$ decreases from x_2 as P_1 increases (or p_2 decreases), tending to its lower bound x_1. On the other hand $\bar{x}_t^{(1)}$ approaches x_2 as P_1 decreases; $\bar{x}_t^{(1)}$ and $\bar{x}_t^{(2)}$ coalesce as P_1 tends to $p_2 x_2^2$ from above.

The expression (4.2) is confirmed to be the leading term outer solution for $\varepsilon << 1$ by the COLSYS solutions for $\Phi_o(x) = x$ and $\nu = 0.3$ given in Figures (6) and (7). In Figure (6), we have three COLSYS solutions for $\Phi(x)$ for $\kappa = 1$, $x_1 = 0.3$, $x_2 = 0.4$, $P_1 = 1$ and $p_2 = 16/9$ (so that $\bar{x}_t^{(2)} = 0.75$ and $\bar{x}_t^{(1)} = 0.3206$); they correspond to $\varepsilon^2 = 10^{-3}, 10^{-4}$, and 10^{-5}. While the COLSYS solution is numerically close to the solution (4.2) only for the smallest ε, the two locations of horizontal tangent in the deformed shell shape, $\bar{x}_T^{(1)}$ and $\bar{x}_T^{(2)}$, are nearly indistinguishable from $\bar{x}_t^{(1)}$ and $\bar{x}_t^{(2)}$, respectively, in all three cases. Figure (7) also displays three COLSYS solutions for $\Phi(x)$ for $\kappa = 1$, $x_1 = 0.4$, $\varepsilon^2 = 10^{-4}$, $P_1 = 1$ and $p_2 = 16/9$; they correspond to $x_1 = 0.1$, 0.25 and 0.39 (with $\bar{x}_t^{(1)} = 0.1168$, 0.2750, and 0.3928, respectively). They show that (4.2) continues to be a good approximation for (the COLSYS) $\Phi(x)$ whether the spread of the external pressure is wide or narrow and that $x_T^{(2)}$ is unaffected by the width of the spread. In general, COLSYS always converges to a solution with the qualitative features of (4.2) if $\varepsilon^2 \leq 10^{-2}$, whatever the initial guess of the solution may be.

Whenever polar dimpling is possible, the qualitative features of the dimple solution are similar to the point load case. The dimple base radius $x_T^{(2)} \sim \bar{x}_t^{(2)} \equiv \sqrt{P_1/p_2}$ is asymptotically independent of the size of the area over which the external pressure distributes. The nonlinear membrane action again dominates when $\kappa\varepsilon^2$ is <u>not</u> small compared to unity, with the dimple deteriorates gradually as κ tends to ε^{-2}. At the other end of the spectrum, the dependence of $\bar{x}_t^{(2)}$ on κ in the range $\kappa = O(\varepsilon)$, say $\kappa = k\varepsilon$, is again similar to the point force case discussed previously. In particular, the asymptotic value of $\bar{x}_t^{(2)}$ decreases to x_2 as k decreases from O(1) to o(1), while $x_t^{(1)}$ increases to x_2; the dimple disappears as $x_t^{(1)}$ and $x_t^{(2)}$ coalesce. Polar dimpling is again not possible if k is of order ε or smaller.

<u>Shallow Caps with a Ring of Internal Pressure</u>

Consider now the case of $p_2 = 0$, and $q(x) \equiv 0$ and $P_1 < 0$ so that the shell is subject only to an internal pressure distributed over an annular region of the shell. The magnitude of the applied pressure is sufficiently high so that $\kappa\varepsilon^2$ is <u>not</u> small compared to unity. In this range, the nonlinear membrane action dominates throughout the shell except for some narrow layers. Anticipating this non-linear membrane behaviour, we set

$$\Psi = \frac{\kappa^{2/3}}{\varepsilon}\psi \ , \qquad \Phi = (\kappa\varepsilon^2)^{1/3}\phi \tag{5.1}$$

and write the ODE (2.1), (2.2) and the boundary conditions (2.6), (2.7) as

$$x[\psi'' + \frac{1}{x}\psi' - \frac{1}{x^2}\psi] + \frac{1}{2}[\phi^2 - (\kappa\varepsilon^2)^{-2/3}\phi_0^2] = 0 \tag{5.2}$$

$$\varepsilon^4(\kappa\varepsilon^2)^{-2/3}x[\phi'' + \frac{1}{x}\phi' - \frac{1}{x^2}\phi] - \phi\psi = 4xP(x) + \varepsilon^4(\kappa\varepsilon^2)^{-1}xQ(x) \tag{5.3}$$

with

$$x = 0: \qquad \phi = 0, \qquad \psi = 0, \tag{5.4}$$

$$x = 1: \qquad \phi = (\kappa\varepsilon^2)^{-1/3} \qquad \psi' - \nu\psi = 0 \tag{5.5}$$

With $(\kappa\varepsilon^2)^{-2/3} = O(1)$ at most, the leading term outer solution of the BVP for $\varepsilon^2 << 1$, denoted by ϕ_0 and ψ_0, is determined by

$$\phi_0\psi_0 = -4xP(x) \tag{5.6}$$

along with (5.2), (5.4) and (5.5) (in which ϕ_0 and ψ_0 take the place of ϕ and ψ, respectively). For simplicity, we shall discuss the solution process only for a spherical cap with $x_1 = x_2$. The annular pressure load thus takes its limiting form of a ring load with

$$-4xP(x) = \begin{cases} 0 & (0 \leq x < x_1) \\ -2P_1 & (x_1 < x \leq 1) \end{cases} \tag{5.7}$$

where $P_1 < 0$. The method, with some straightforward modifications, also applies to the case $x_1 < x_2$ and more general shell shape $\Phi_0(x)$.

In the range $0 < x < x_1$, the algebraic equation (5.6) with the right hand side given by (5.7) may be satisfied by either $\phi_0(x) \equiv 0$ or $\psi_0(x) \equiv 0$. For the solution to be applicable for both moderate and large values of $\kappa\varepsilon^2$, we take

$$\phi_0(x) \equiv 0 \quad (\equiv \hat{\phi}_p(x)) \quad (0 \leq x < x_1) \tag{5.8}$$

which automatically satisfies the first boundary condition in (5.4). With (5.8), the shell is flattened out in the neighborhood of the pole (as we would expect for a sufficiently high pressure magnitude). Corresponding to this mode of deformation, we have from (5.2) (with $\Phi_0^2 = x^2$)

$$[\psi_0'' + \frac{1}{x}\psi_0' - \frac{1}{x^2}\psi_0] = \frac{1}{2}(\kappa\varepsilon^2)^{-2/3}x, \quad (0 \leq x < x_1). \tag{5.9}$$

The solution of (5.9) satisfying the second condition of (5.4) is

$$\psi_0(x) = A_1 x + \frac{1}{16}(\kappa\varepsilon^2)^{-2/3}x^3 \equiv \hat{\psi}_p(x) \quad (0 \leq x < x_1). \tag{5.10}$$

(With both regularity conditions at $x = 0$ satisfied, there will be no layer solution in the neighborhood of the apex.) The remaining constant of integration A_1 is to be determined by the matching of $\hat{\psi}_p(x)$ and $\hat{\phi}_p(x) \equiv 0$

with an appropriate layer (inner) solution in the neighborhood of x_1.

For the range $x_1 < x < 1$, we have from (5.6)

$$\phi_0 = -\frac{2P_1}{\psi_0}, \quad (x_1 < x < 1) \tag{5.11}$$

so that (5.2) (with $\Phi_0^2(x) = x^2$) may be written as

$$x[\psi_0'' + \frac{1}{x}\psi_0' - \frac{1}{x^2}\psi_0] + \frac{2P_1^2}{\psi_0^2} = \frac{1}{2}(\kappa\varepsilon^2)^{-2/3}x^2, \quad (x_1 < x < 1). \tag{5.12}$$

The solution of this second order ODE, denoted by $\hat{\psi}_e(x)$, cannot be made to satisfy the two boundary conditions at $x = 1$ given by (5.5) as well as matching with the layer solution in the neighborhood of x_1. A boundary layer (inner) solution adjacent to the shell edge is needed to satisfy the boundary conditions. The two constants of integration of $\hat{\psi}_e$ and the arbitrary constant A_1 in $\hat{\psi}_p$ are determined by matching the outer solutions $\hat{\psi}_p$ and $\hat{\psi}_e$ with the inner solutions, one in the neighborhood of x_1 and the other adjacent to the shell edge.

Analyses of the structure of the interior layer at $x = x_1$, and the boundary layer at $x = 1$ (details of which will not be included here) show that only Φ changes abruptly over a small distance along a shell meridian in these layers. The corresponding matching of the inner and outer solutions effectively requires $\hat{\psi}_e$ to satisfy

$$\hat{\psi}_e'(1) - \nu\hat{\psi}_e(1) = 0, \tag{5.13}$$

i.e., the "no radial displacement" condition in (5.6), and

$$x = x_1: \quad x\hat{\psi}_e' - \hat{\psi}_e = \frac{1}{8}(\kappa\varepsilon^2)^{-2/3}x^3, \tag{5.14}$$

with the only unknown constant A_1 in $\hat{\psi}_p(x)$ given by

$$A_1 = \hat{\psi}_e'(x_1) - \frac{3}{16}(\kappa\varepsilon^2)^{-2/3}x_1^2 \tag{5.15}$$

(The conditions (5.14) and (5.15) effectively ensure $\hat{\psi}_p(x_1) = \hat{\psi}_e(x_1)$ and $\hat{\psi}_p'(x_1) = \hat{\psi}_e'(x_1)$.) The solution procedure at this point is to solve the two point BVP defined by (5.12)-(5.14) to determine $\hat{\psi}_e(x)$, $x_1 \le x \le 1$ and then use (5.15) to calculate A_1, thereby completely determining $\hat{\psi}_p(x)$, $0 \le x \le x_1$.

There still remains the formulation of appropriate BVP for the two layer solutions. We shall not be concerned with this aspect of the problem as the stresses associated with the layer solutions are at most comparable to membrane stresses associated with $\hat{\psi}_p(x)$ and $\hat{\psi}_e(x)$. We merely note that the interior layer solution mainly serves to smooth out the discontinuity of the two different outer solutions for $\Phi(x)$ across x_1, while the boundary layer solution ensures the condition of no rotation, $\Phi(1) = \Phi_0(1)$, at the shell edge.

When $\Phi_0(x) \equiv 0$, the nonlinear ODE (5.12) for $\hat{\psi}_e$ is the governing equation in the Föppl-Hencky (small finite-deflection) nonlinear theory of an originally flat elastic membrane [20,21]. Existence, uniqueness and numerical methods of the solution for this ODE in the case of annular membranes with various stress and displacement boundary conditions, including (5.13), have been reported in [22] and other references cited therein. Numerical solutions of (5.12) with (5.14) taking the place of a conventional boundary condition at the inner edge pose no new difficulties. These solutions along with (5.11), (5.8), (5.10) and (5.15) (henceforth called nonlinear membrane solutions) give the limiting behaviour of the shell for very small values of ε.

Figure (8) shows three COLSYS solutions for $\Phi(x)$ for shells with a ring of uniform internal pressure extending from $x_1 = 0.5$ to $x_2 = 0.51$; they are for $\varepsilon^2 = 10^{-1}$, 10^{-2} and 10^{-4} with the corresponding pressure magnitude given by $P_1 = -1$ and $\kappa = \varepsilon^{-2}$ (or $\kappa\varepsilon^2 = 1$). These distributions of $\Phi(x)$ show clearly that the shells exhibit mainly nonlinear membrane action and that their deformed meridional slope $\Phi(x)$ tends to $(\kappa\varepsilon^2)^{1/3}\phi_0(x)$ as ε becomes smaller and smaller. For $\varepsilon^2 = 10^{-4}$, the COLSYS solution differs by less than 5% from the limiting nonlinear membrane solution throughout the shell except in the two narrow layers of abrupt changes located around $x = x_1$ and $x = 1$. It is remarkable that the rather complex and nearly discontinuous distribution of $\Phi(x)$ for $\varepsilon^2 << 1$ can be so accurately generated by COLSYS. It is even more remarkable that a simple outer asymptotic solution, which completely describes the limiting behaviour for shells with vanishing thickness, may be obtained from (2.1) and (2.2) by setting $\varepsilon = 0$ after re-scaling these ODEs (and then using technique of matched asymptotic expansions or physical reasoning).

The agreement between the COLSYS solution for the shell and the limiting nonlinear membrane solution is even better and extends to the edge of the shell ($x = 1$) if the shell is hinged (so that the condition $\Phi = \Phi_0$ in (2.7) is replaced by $\Phi' + \nu\Phi = \Phi_0' + \nu\Phi_0$) at the edge instead (as shown in Figure (9)). We show in Figure (9) two solutions for $\Phi(x)$ for shells with a clamped edge and with $\varepsilon^2 = 10^{-2}$, $\kappa\varepsilon^2 = 1$, $P_1 = -1$; they correspond to a ring pressure load at the two different locations, namely ($x_1 = 0.5$, $x_2 = 0.51$) and ($x_1 = 0.75$, $x_2 = 0.76$). Also shown there are the corresponding two solutions when the shell is hinged at $x = 1$ instead. The qualitative features of $\Phi(x)$ (outside the narrow layers of abrupt changes) as shown in Figure (8) remain unaffected by the change in the location of the ring load and the type of edge conditions.

Finally, Figure (10) shows the effect of the ring load spread $(x_2 - x_1)$ by the COLSYS solutions for $(x_2 - x_1) = 0.01$, 0.1 and 0.25 and for $x_1 = 0.5$. The change from a flat region in the deformed configuration (given by (5.8), (5.10) and (5.15)) to a nearly conical region (given by (5.11) - (5.14)) is much more gradual for a wider load spread. Over the region of the pressure ring ($x_1 \leq x \leq x_2$), the deformed shape of the shell continues to be spherical but with a smaller spherical radius.

Concluding Remarks

In this report, we have considered only shells subject to axial loads. For the more general situation of combined axial and radial loads, the possible modes of shell behaviour generally depends on the relative magnitude of the two types of loadings. This dependence on load magnitude is evidently more complicated than what we have encountered in this report for shells subject only to axial loads. A complete analysis of this dependence can be found in [23]. When subjected only to radial loads, however, we know from [24] that a very thin shell behaves effectively as a membrane except near an edge support and that a nonlinear membrane solution is often needed for a correct description of the qualitative behaviour of the shell even for very small radial load magnitudes.

References

1. Wan, F.Y.M.: The dimpling of spherical caps. Mechanics Today 5: The Eric Reissner Anniversary Volume (1980) 495-508.

2. _____________; Polar dimpling of complete spherical shells, Theory of Shells, Proc. of the Third IUTAM Shell Symp. (Tbilisi, USSR; August, 1978), Ed. W.T. Koiter & G.K. Mikhailov, North-Holland, 1980, 589-605.

3. Bucciarelli, L.L.: The analysis of propellant expulsion bladders and diaphrams. Dept. of Aero. & Astro. Tech. Rep., M.I.T., Cambridge, MA., 1967.

4. Taber, L.A.: Large deflection of a fluid-filled spherical shell under a point load. J. Appl. Mech. 49 (1982) 121-128.

5. Parker, D.F.; Wan, F.Y.M.: Finite polar dimpling of shallow caps under sub-buckling pressure loading. (I.A.M.S. Tech. Rep. No. 80-10, U.B.C., Vancouver, Canada and) SIAM J. Appl. Math (1984) to appear.

6. Reissner, E.: On Axisymmetric deformations of thin shells of revolution. Proc. Symp. in Appl. Math., III, Amer. Math. Soc., Providence (1950) 27-52.

7. _______________: Symmetric bending of shallow shells of revolution. J. Math. Mech. 7 (1958) 121-140.

8. Ashwell, D.G.: On the large deflexion of a spherical shell with an inward point load. Proc. First Symp. on Thin Elastic Shells (Delft, The Netherlands; August, 1959), Ed. W.T. koiter, North-Holland, Amsterdam (1960) 43-63.

9. Archer, R.R.: On the numerical solution of the nonlinear equations of shells of revolution. J. Math. & Phys. 41 (1962) 165-178.

10. Evan-Iwanowski, R.M.; Cheng, H.S.; Loo, T.C.: Experimental investigations of deformations and stability of spherical shells subjected to concentrated load at the apex. Proc. 4th U.S. Nat'l Congr. Appl. Mech. (1962) 563-575.

11. Penning, F.A.; Thurston, G.A.: The stability of shallow spherical shells under concentrated load. NASA CR-265, July, 1965.

12. Mescall, J.F.: Large deflections of spherical shells under concentrated loads. J. Appl. Mech. 32 (1965) 936-938.

13. Penning, F.A.: Experimental buckling modes of clamped shallow shells under concentrated load. J. Appl. Mech. 33 (1966) 297-304.

14. Bushnell, D.: Bifurcation phenomena in spherical shells under concentrated and ring loads. AIAA J. 5 (1967) 2034-2040.

15. Fitch, J.R.: The buckling and post-buckling behavior of spherical caps under concentrated load. Int'l J. Solids & Structures 4 (1968) 421-446.

16. Ranjan, C.V.; Steele, C.R.: Large deflection of deep spherical shells under concentrated load, Proc. AIAA/ASME 18th Structures, Structural Dynamics and Materials Conf. (San Diego, Calif.; March, 1977), 269-278.

17. Wan, F.Y.M.; Ascher, U.: Horizontal and flat points in shallow shell dimpling. Proc. of BAIL I Conf. (Dublin, 1980), Ed. J. Miller, Boole Press, Dublin, 1980, 659-679.

18. Ascher, U; Christiansen, J; Russell, R.D.: A collocation solver for mixed order systems of boundary value problems. Math. Comp. 33 (1979) 659-679.

19. Kevorkian, J.; Cole, J.D.: Perturbation methods in Applied Mathematics, Springer-Verlag, New York-Heidelberg-Berlin, 1981.

20. A. Föppl, Vorlesungen über technische Mechanik, Vol. III, Teubner, Leipzig, 1907.

21. Hencky, H.: Über den spannungzustaud in kreisrunden platten mit verschwindender biegungssteifigkeit. Z.f. Math. u. Phys. 63 (1915) 311-317.

22. Weinitschke, H.J.: On axisymmetric deformations of nonlinear elastic membranes. Mech. Today 5: E. Reissner Anniversary Volume (1980) 523-542.

23. Lin, Y.H.; Wan, F.Y.M.: Asymptotic solutions of steadily spinning shallow shells of revolution under uniform pressure. I.A.M.S. Tech. Rep. No. 83-25, U.B.C., Vancouver, Canada.

24. Reissner, E.; Wan, F.Y.M.: Rotating shallow elastic shells of revolution. J. of S.I.A.M. (now S.I.A.M. J. of Appl. Math.) 13 (1965) 333-352.

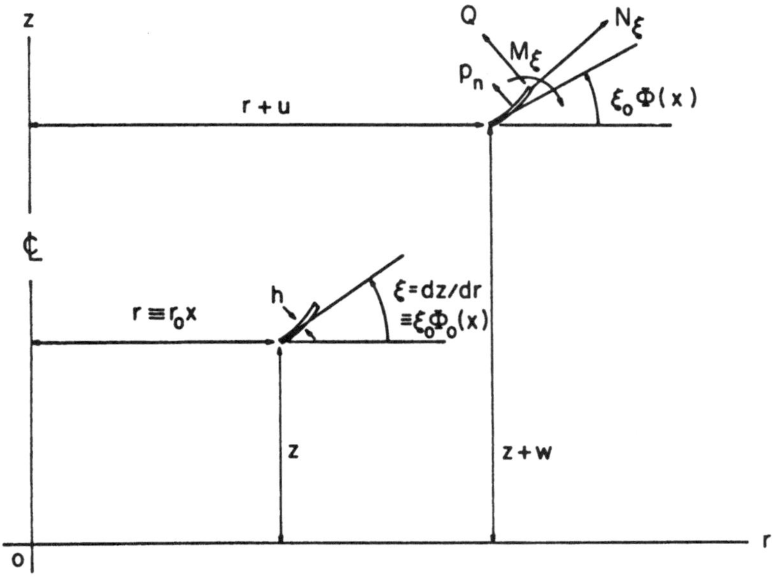

Figure 1. Dome-shaped Shallow Shells of Revolution

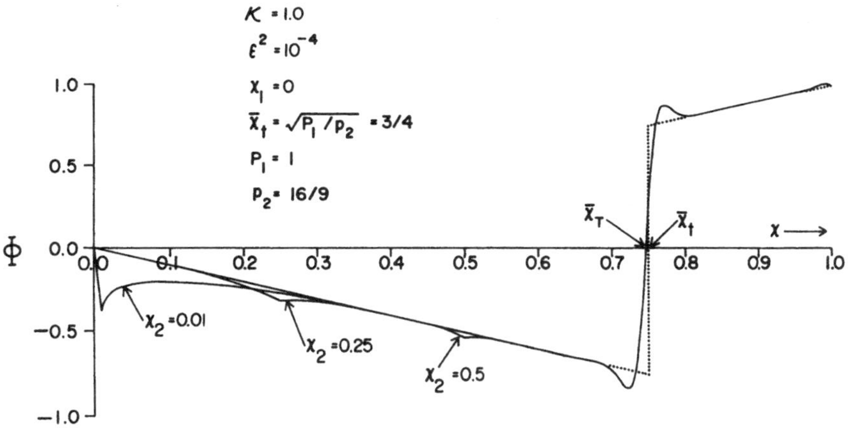

Figure 2. Distributions of Deformed Meridional Slope for Different Localized External Pressure Distributions with the Same Resultant

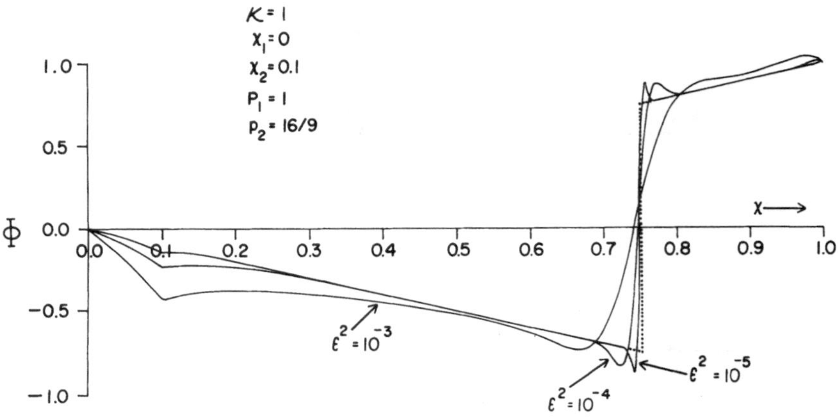

Figure 3. Distributions of Deformed Meridional Slope of Shells with Different Thickness Parameter Values Subjected to the Same Localized Load

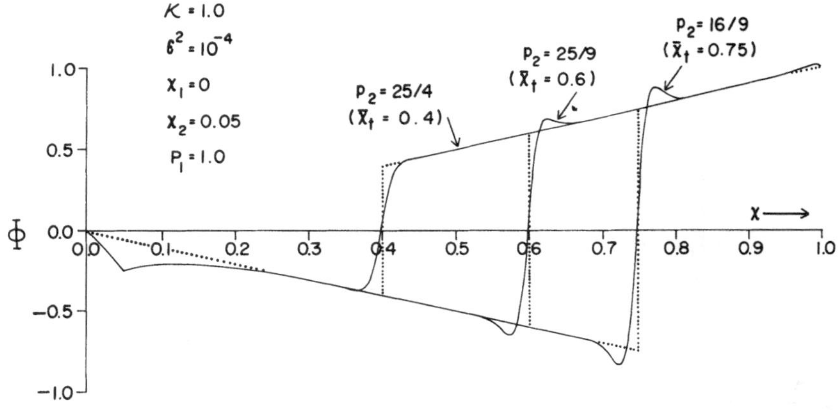

Figure 4. Distributions of Deformed Meridional Slope for Shells with the Same Localized Distribution External Pressure but with Different Internal Pressures

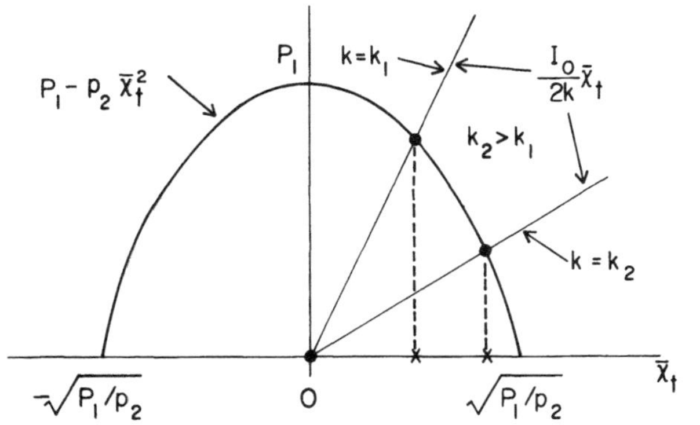

Figure 5. The Variation of Dimple Radius with Load Magnitude for $\kappa = \kappa\varepsilon$

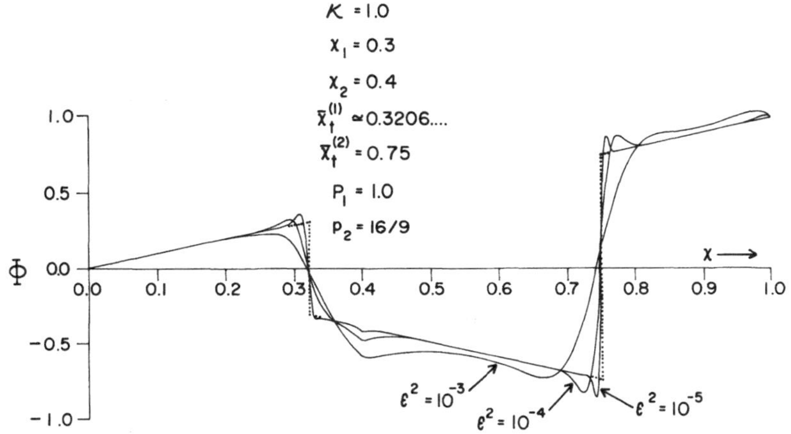

Figure 6. Distributions of Deformed Meridional Slope for Shells with the Same External Ring Pressure but Different Thickness Parameter Values

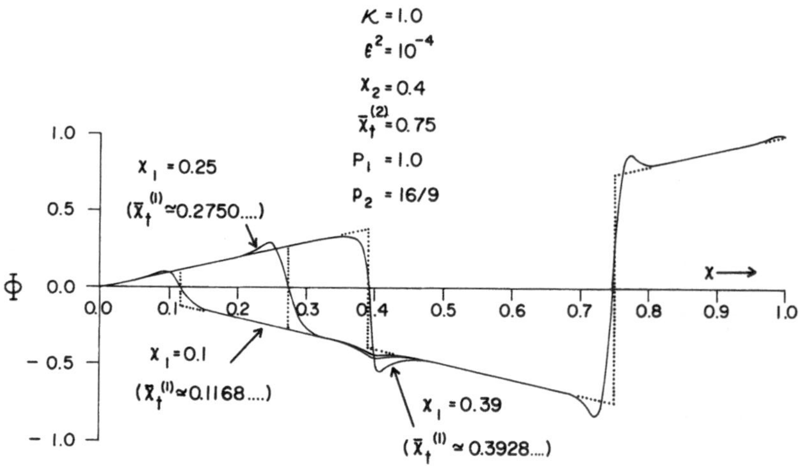

Figure 7. Distributions of Deformed Meridional Slope for Different Spreads of the External Ring Pressure Distribution

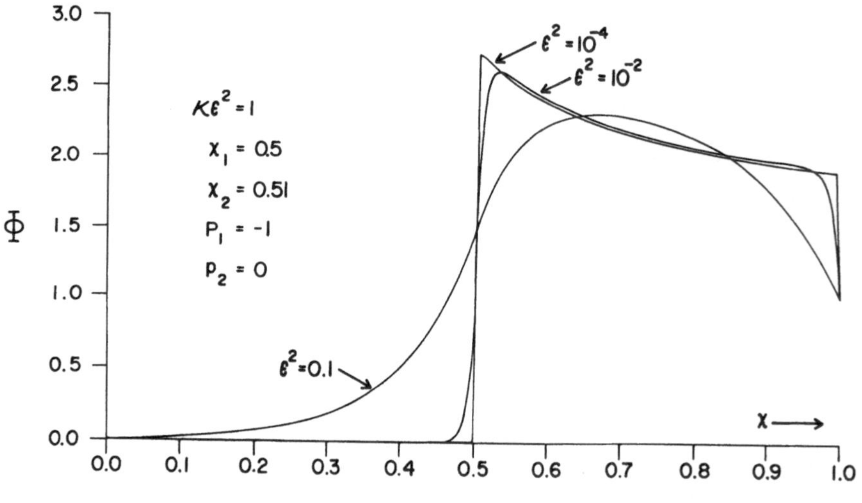

Figure 8. Distributions of Deformed Meridional Slope for Shells with Different Thickness Parameter Values Subjected to the Same Internal Ring Pressure Distribution

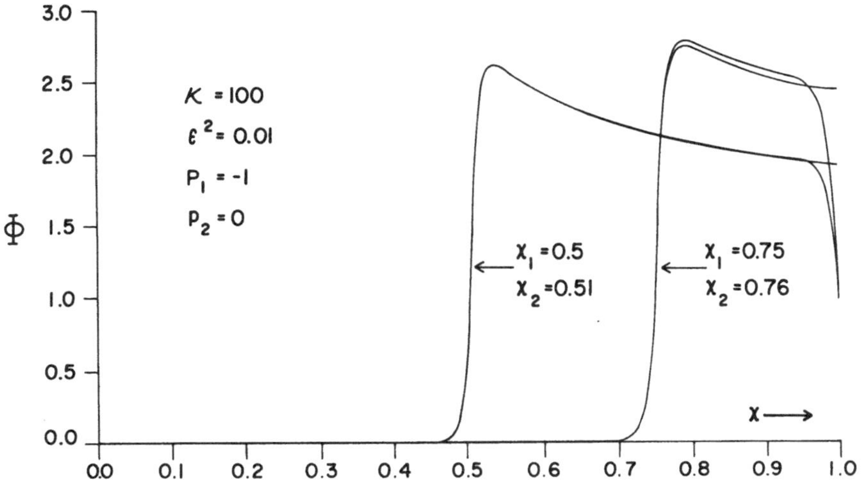

Figure 9. Distributions of Deformed Meridional Slope for Shells with a Clamped or Hinged Edge Subjected to an Internal Ring Pressure Distribution at Different Locations

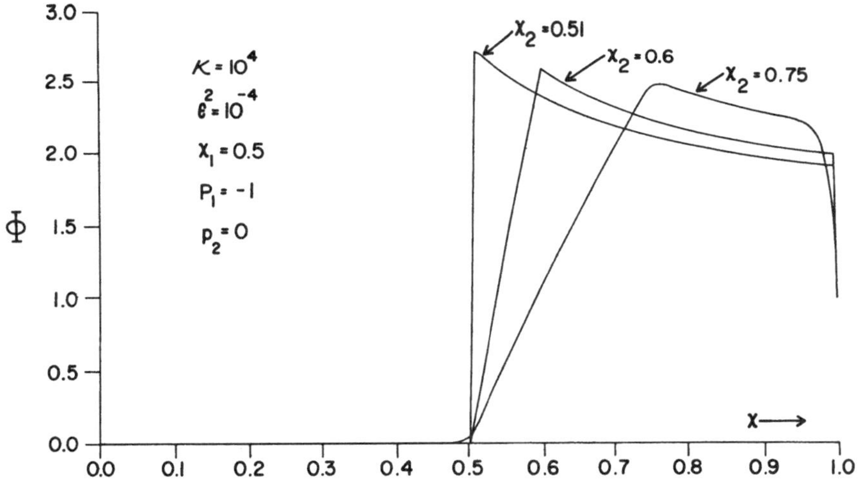

Figure 10. Distributions of the Deformed Meridional Slope for Shells Subjected to Internal Ring Pressure Distributions with the Same Resultant but Different Spreads

On the Buckling and Postbuckling of Spherical Shells

R.Scheidl and H.Troger
Institute of Mechanics, Department of Mechanical Engineering
Technical University Vienna, Austria.

Summary

For the axisymmetric buckling problem of a complete spherical shell we are able to give a complete imperfection sensitivity analysis for the most complicated case to occur, namely that of a double eigenvalue, if we restrict our analysis to nonlinear terms of quadratic order, which is correct mathematically as the bifurcation system turns out to be two determinate. As we make a classical bifurcation analysis our results are strictly local and therefore are of restricted practical importance in two respects. Firstly we do not take care of the fact that we have closely spaced eigenvalues. We comment on this in chapter 7. Secondly we are not able to show that the experimentally observed single dimple solution (Fig. 8) is one of the stable solutions found from our bifurcation equations up to terms of third order.

1. Introduction

The buckling and post-buckling problem of complete spherical shells constitutes a major challenge to engineers (|7|,|1|,|5|,|13|) and mathematicians (|6|,|8|). In a short review in the next chapter we shall try to show that the problem with arbitrary deformations poses a very difficult problem in mathematical respect which, as we believe it, has not yet been completely solved. This and two more reasons which we mention below have motivated us to restrict our analysis to the simpler case of axisymmetric deformations. One of these two reasons is that the industrial production of spheres is done by deep-drawing of hemispheres and by welding them together. This way of production favours the occurence of axisymmetric imperfections which create a preference for axisymmetric buckling patterns. The other is that we want to give an application of some mathematical concepts for the handling of bifurcation problems for engineers (|6|,|9|). Treating bifurcation problems one has to do basically two steps. One is to reduce the infinite dimensional system to a finite dimensional system of bifurcation equations by means of the Liapunov-Schmidt method. The second step then is to embed the bifurcation system in a versal parameter family in order to study the solutions of the bifurcation equations under arbitrary perturbations, which is nothing else than the study of the imperfection sensitivity of the problem. This last task can be done completely only if the bifurcation system is at most two dimensional and determined by low order nonlinear terms. This is in fact the case for the axisymmetric buckling problem and we shall treat it in detail in chapters 3,4, 5 and 6. However, it will turn out in the course of our analysis that though we are able to give the complete local solution of the bifurcation problem, the results are of limited practical importance because in our problem we have the case of closely spaced eigenvalues. We shall shortly comment on this point in chapter 7.

2. A short review of the general buckling problem of spherical shells

As it is well known (|7|,|6|), it is possible to reduce the system of equations describing the deformation of a spherical shell to a set of two nonlinear partial differential equations in two variables. One variable w is a generalized displacement and the other f a stress function (see also (7) below). An important difference in treating the problem in mathematical respect is whether one uses (a) shallow shell theory as it is done in |5| and |10| or whether one uses (b) the nonlinear shell equations on the complete sphere (|7|,|6|).

In case (a) one obtains for the buckling pattern doubly periodic functions (|5|)

$$w(x,y) = A \cos(k_x \frac{x}{a})\cos(k_y \frac{y}{a}). \tag{1}$$

a is the shell radius (Fig. 1) and x,y are the coordinates usually used in shallow shell theory (|5|). For the determination of k_x and k_y only a relation

$$k^2 = k_x^2 + k_y^2 \tag{2}$$

can be obtained, where k^2 is given by (h is shell thickness (Fig. 1))

$$k^4 = 12(1-\nu^2)(\frac{a}{h})^2, \tag{3}$$

and is related to the critical pressure p_c by

$$p_c = \frac{4Eh}{ak^2} . \tag{4}$$

E and ν are Youngs modulus and the Poison number respectively and are given constants. The problem now is that (2) can be fulfilled by an infinite number of possibilities. This means that one has an infinite number of different possible buckling modes.

However, if we consider case (b) the buckling pattern will be represented by spherical harmonics. Using the angles ξ and θ as variables (Fig. 1) we get for w

$$w(\xi,\theta) = \sum_{m=0}^{n} (\alpha_m \cos m\theta + \beta_m \sin m\theta)\, P_n^m(\cos\xi) , \tag{5}$$

where P_n^m are Legendre polynomials. From (5) we see that there exists only a finite number k = 2n+1 of possible different buckling modes. Furthermore this formulation possesses the advantage that n depends on a/h (|7|) and is given by

$$n(n+1) = \frac{a}{h} [12(1-\nu^2)]^{\frac{1}{2}}. \tag{6}$$

From (6) it can be seen that the shallow shell result is contained in (5) as an asymptotic limit. Furthermore it follows form (6) that for ratios h/a in the range usually for thin walled shells we obtain the number k = 2n+1 in an order of magnitude of about 10^2. This means that after reduction to the bifurcation equations (see chapter 4) we still have to solve a system of k nonlinear algebraic equations in k unknowns, which is a formidable task. A further reduction of the order k of the system is therefore essential, especially if one wants to study the imperfection sensitivity, which is an important part of shell buckling theory in engineering applications.

We only want to say that a further reduction of the order k of the system of bifurcation equations can be done by the use of concepts from group theory. The application, however, is not straight forward because the symmetry of the buckling pattern must be known beforehand, for example from experiments. Then for example a reduction to a number of about 10 bifurcation equations in the case of hexagonal symmetry can be achieved (|6|,|12|). But for a systematic study of parameter influences a system of 10 bifurcation equations is much too complicated. Therefore we shall restrict ourselves to the axisymmetric case where we have at most two variables and if we exclude the case that quadratic nonlinearities do not determine the problem we can give a complete description of the imperfection sensitivity using Catastrophe Theory.

We want to mention that in |5| and |10| it is shown that using shallow shell theory certain two and three mode solutions can be used to obtain the initial post-buckling path. The basic idea is that the nonlinear equations for the mode amplitudes decouple into seperate sets of equations corresponding to the interaction between either two or three of these critical modes. Furthermore it is shown in |7| p. 67, that the corresponding buckling pattern is hexagonal.

Finally we believe it to be interesting to remark that similar problems of pattern formation arise when considering a fluid layer heated from below (|3|,|2|). This is the so-called Benard Problem which has striking analogies with the shell buckling problem. The reason is that if one considères a plane fluid layer one has again an infinite dimensional degeneracy for the linearized problem whereas for a spherical fluid layer one obtains only a finite degeneracy. A group theoretic treatment of the latter problem is given in |4|. The convection problem is easier to handle mathematically than the shell buckling problem because for the fluid shell also thick walled shells are physically meaningful. Therefore the multiplicity of the critical eigenvalue is much lower than that for thin walled shells. A nice description of the solutions of the convection problem on the sphere is given in |2|.

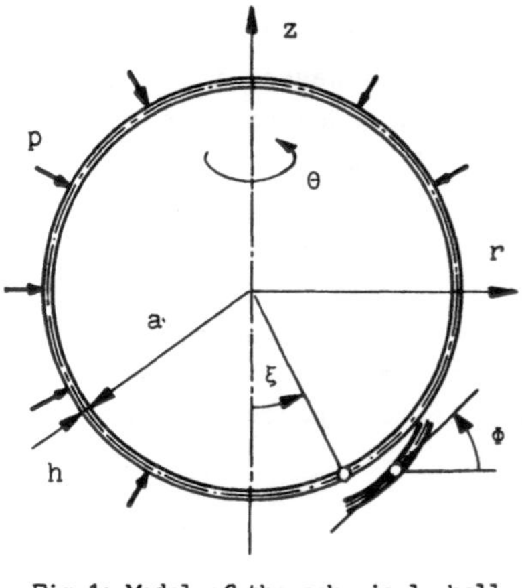

Fig.1: Model of the spherical shell.

3. Governing shell equations

We shall use shell equations derived in |11| for small but finite deflections of thin elastic shells of revolutions with axisymmetric states of deformation. As coordinate system we use spherical coordinates $r = a = \text{const}$, ξ and θ (Fig. 1). In |11| it is shown that it is possible to reduce the set of shell equations to two equations in two variables $\beta^*(\xi)$ and $\tilde{\psi}(\xi)$, which we write down up to third order nonlinear terms

$$
\begin{aligned}
&-\beta^{*''} - \beta^{*'}\operatorname{ctg}\xi + \beta^*\left(-\frac{a^3}{D}\frac{p}{2} + \nu + \operatorname{ctg}^2\xi\right) + \tilde{\psi}\frac{a}{D} = \\
&\beta^{*2}\left(\frac{3}{2} - \frac{\nu}{2}\right)\operatorname{ctg}\xi - \beta^*\tilde{\psi}\frac{a}{D}\operatorname{ctg}\xi + \\
&\beta^{*3}\left[\frac{2}{3}(\operatorname{ctg}^2\xi - 1) + \frac{\sin\xi}{6}(1+\nu) - \frac{a^3}{D}\frac{p}{12}\right] + \frac{\beta^{*2}}{2}\tilde{\psi}\frac{a}{D} \\
&\tilde{\psi}'' + \tilde{\xi}'\operatorname{ctg}\xi + \tilde{\psi}(\nu - \operatorname{ctg}^2\xi) + \frac{a}{A}\beta^* - \frac{a^2 p}{2}\beta^*(1-\nu) = \\
&-\tilde{\psi}[\beta^*(2+\nu)\operatorname{ctg}\xi + \beta^{*'}\nu] - \frac{\beta^{*2}}{2}\frac{a}{A}\operatorname{ctg}\xi + \frac{a^2 p}{2}(\nu\beta^*\beta^{*'} + \beta^{*2}\operatorname{ctg}\xi) \\
&-\tilde{\psi}\left[\beta^{*2}\left(\operatorname{ctg}\xi - 1 - \frac{\nu}{2}\right) + \beta^*\beta^{*'}\nu\operatorname{ctg}\xi\right] + \frac{\beta^{*3}}{6}\frac{a}{A} + \beta^{*3}\frac{a^2 p}{2}\left(\frac{\nu}{6} - \frac{2}{3}\right).
\end{aligned}
\tag{7}
$$

Here $\beta^*(\xi) = \Phi - \xi$ is the deformation variable, measuring the angular difference between the tangents to the buckled shell and the perfect spherical shell at the meridian ξ. $\tilde{\psi}(\xi) = \psi^* - \psi_0^*$ is a stress variable, giving the difference of ψ^* in the buckled and ψ_0^* in the trivial spherically deformed state. Physically ψ is proportional to the horizontal component of the section force in ξ.

The constants A and D are given by

$$A = \frac{1}{Eh} \quad \text{and} \quad D = \frac{Eh^3}{12(1-\nu^2)} \quad . \tag{8}$$

(7) is a set of two nonlinear ordinary differential equations. The corresponding boundary conditions are

$$\beta^*(0) = \beta^*(\pi) = 0 \qquad \text{and} \qquad \tilde{\psi}(0) = \tilde{\psi}(\pi) = 0 \;. \tag{9}$$

4. Reduction to bifurcation equations

Equations (7) represent a bifurcation problem which we can write in the form

$$G(w,\lambda') = 0, \tag{10}$$

where w is given by β^* and ψ and λ' is the load parameter. (10) is defined on an appropriate function space with suitable boundary conditions. We shall call (10) the perfect or one parameter buckling problem. We see immediately that for all physically reasonable (elastic behavior must be guaranteed) values of λ' the perfect spherical shell configuration which is given by $\beta^* = 0$ and $\psi^* =$

ψ_o^*, and which we shall call w_o, is a solution of (7). From (10) we can write $G(w_o,\lambda') = 0$ for all λ'. From the implicit function theorem (|12|) it now follows that if the Frechet derivative of G with respect to w at w_o and λ_o', which we denote by $G_w(w_o,\lambda_o')$ and which is a linear map, is invertible, then there exists locally a smooth unique curve $w = w(\lambda')$ through the point w_o, λ_o'. Consequently a bifurcation can only happen if $G_w(w_o,\lambda_o')$ is not invertible at a specific value $\lambda_o' = \lambda_c'$. In order to evaluate this critical parameter value λ_c', we have to solve the linear eigenvalue problem

$$G_w(w_o,\lambda_c')v = 0, \tag{11}$$

where v are the eigenfunctions of the linearized problem. Calculating the Frechet derivative of (7) we obtain the left hand side of (7) written in the variables $v^T = (\beta^*,\tilde{\psi})$. Before proceeding we introduce dimensionless quantities in (7) by means of the following relations

$$\begin{aligned} &\beta = -\beta^* , \quad \psi = \tilde{\psi}/a^2 p_k , \quad p_k = \frac{E(h/a)^2}{\sqrt{12(1-\nu^2)}} \\ &\lambda = \frac{p}{4p_k} , \quad \delta = \frac{h}{a}\frac{1}{\sqrt{12(1-\nu^2)}} . \end{aligned} \tag{12}$$

We want to point out that relations (12) and (4) are related by $p_c = 4p_k$ and that we are a little bit sloppy in our notation, because we designate the components of v with the same letters as those of w. Under consideration of (12) the linear eigenvalue problem (11) reads

$$\begin{aligned} &\beta'' + \beta'\mathrm{ctg}\xi + \beta\,(\, 2\lambda/\delta - \underline{\nu} - \mathrm{ctg}^2\xi \,) + \psi/\delta = 0 \\ &\psi'' + \psi'\mathrm{ctg}\xi + \psi[\underline{\nu} - \mathrm{ctg}^2\xi] - \beta/\delta + \underline{2\lambda\beta[1-\nu]} = 0 \end{aligned} \tag{13}$$

with the boundary conditions (9). Those terms in (13) which are underlined can be savely omitted, because for shells only the case δ small ($\delta = 0(10^{-2})$) is physically meaningful.

To solve (13) it is natural to look for a solution in spherical harmonics (|6|,|7|,|8|). We set

$$\begin{aligned} &\beta = A_n P_n^1(\cos\xi) \\ &\psi = B_n P_n^1(\cos\xi). \end{aligned} \tag{14}$$

where $P_n^m = (1-x^2)^{m/2}\frac{d^m}{dx^m} P_n(x)$. $P_n(x)$ is the Legendre polynomial of n-th order. The P_n^m form a complete set of eigenfunctions. Using the following relations $x = \cos\xi$, $\mathrm{ctg}\xi = x/\sqrt{1-x^2}$, $d^2/d\xi^2 = (1-x^2)d^2/dx^2 - x\,d/dx$, $d/d\xi = -\sqrt{1-x^2}\,d/dx$ and introducing (14) into (13) we obtain

$$\begin{aligned} &A_n(1-x^2)P_n^{1''} - 2A_n x P_n^{1'} + A_n P_n^1(2\lambda/\delta - x^2/(1-x^2)) + B_n 1/\delta P_n^1 = 0 \\ &B_n(1-x^2)P_n^{1''} - 2B_n x P_n^{1'} - B_n x^2/(1-x^2)P_n^1 - A_n/\delta P_n^1 = 0. \end{aligned} \tag{15}$$

A simple calculation shows that the P_n^m are solutions of the differential equations

$$(1-x^2)P_n^{m''} - 2xP_n^{m'} + (n(n+1) - \frac{m^2}{1-x^2})P_n^m = 0. \tag{16}$$

If we set in (16) m = 1, multiply (16) with A_n and B_n respectively and substract these equations from each equation (15) we get

$$\begin{aligned} &(2\lambda/\delta - n(n+1)+1)\ A_n + B_n/\delta = 0 \\ &A_n/\delta + (-n(n+1) + 1)\ B_n = 0. \end{aligned} \tag{17}$$

From (17) we obtain

$$\lambda_n = (\delta\mu_n + \frac{1}{\delta\mu_n})/2 \tag{18}$$

with

$$\mu_n = n(n+1) - 1. \tag{19}$$

Therefore λ_n is a function of δ and μ_n. Practically δ is fixed and we have to find the smallest value of λ_n as a function of μ_n. If, for a moment, we treat μ_n as a continuous variable we obtain from

$$\begin{aligned} &\frac{d\lambda_n}{d\mu_n} = \frac{1}{2}(\delta - \frac{1}{\delta\mu_n^2}) = 0 \\ &\mu_n = 1/\delta. \end{aligned} \tag{20}$$

From (20) and (18) we see that for small values of δ, which are the practically important ones, n will be rather big. In Fig. 2 the eigenvalue curves λ_n as a function of δ for different values of n are shown. One can see from Fig. 2, that in general one obtains simple eigenvalues. This is the generic case. However there are also special values of δ for which double eigenvalues occur. The corresponding values of δ are obtained from

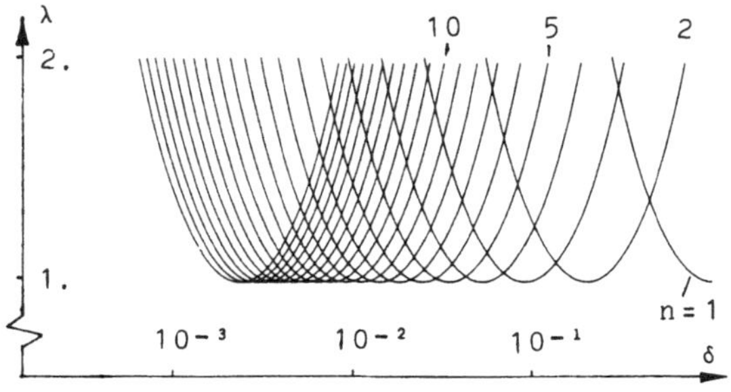

Fig.2: Eigenvalue curves in a load (λ) and thickness/radius (δ) diagram showing closely spaced eigenvalues.

$$\lambda_{n+1} = \lambda_n. \tag{21}$$

After introducing (18) into (21) we have

$$\delta^2 = 1/(\mu_n\ \mu_{n+1})\ . \tag{22}$$

Fig. 2 also shows that for small δ we obtain the already mentioned case of closely spaced eigenvalues.

The next step in our analyses is the replacement of (7) by a finite dimensional system of algebraic equations, which gives all solutions of (7) in a local neighborhood of the bifurcation point. The dimension of this set of bifurcation equations is equal to the multiplicity of the critical eigenvalue. As only the cases of a simple or a double eigenvalue can happen we shall obtain either a one dimensional or a two dimensional system of bifurcation equations.

For the reduction process we use the Liapunov Schmidt method (|6|,|4|,|12|) which is a mathematical formalization of the ingenious stability theory of W.T.Koiter (|7|). As we want to keep the functional analytic formalism at a minimum level we shall proceed in an engineering style. First we represent the two unknowns β and ψ by infinite series expansions

$$\beta = \sum_k a_k P_k^1(\cos\xi)$$
$$\psi = \sum_k b_k P_k^1(\cos\xi). \tag{23}$$

To obtain equations for the a_k and b_k we introduce (23) into (7) and form the inner product with $P_i^1(\cos\xi)$. Geometrically speaking we are projecting (7) on the eigenfunctions. We write (7) in the form

$$L_j(\beta,\psi,\lambda) = N_j(\beta,\psi,\lambda) \qquad j = 1,2 \tag{24}$$

with L_j and N_j designating the linear and nonlinear parts. For the projection of (24) on $P_i^1(\cos\xi)$ we get

$$\int_0^\pi L_j(\beta,\psi,\lambda)P_i^1(\cos\xi)\sin\xi d\xi = \int_0^\pi N_j(\beta,\psi,\lambda)P_i^1(\cos\xi)\sin\xi d\xi. \tag{25}$$

From (24) result two equations for a_i and b_i

$$\left(\frac{2\lambda}{\delta} - \mu_i\right)a_i + \frac{1}{\delta}b_i = \sum_{k,l} r_{ikl}\, a_k a_l - \sum_{k,l,m} r_{iklm}\left(\frac{1}{2}a_k a_l b_m + \frac{\lambda}{3}a_k a_l a_m\right)$$
$$-\frac{1}{\delta}a_i - \mu_i b_i = -\frac{1}{2}\sum_{k,l} r_{ikl}a_k a_l - \frac{1}{6}\sum_{k,l,m} r_{iklm}a_k a_l a_m \tag{26}$$

The coefficients r_{ikl} and r_{iklm} are given by

$$r_{ikl} = \frac{1}{\delta}\frac{2i+1}{2i(i+1)}\int_0^\pi P_k^1(\cos\xi)P_l^1(\cos\xi)P_i^1(\cos\xi)\cos\xi d\xi$$
$$r_{iklm} = \frac{1}{\delta}\frac{2i+1}{2i(i+1)}\int_0^\pi P_k^1(\cos\xi)P_l^1(\cos\xi)P_m^1(\cos\xi)P_i^1(\cos\xi)\sin\xi d\xi\ . \tag{27}$$

From the second equation of (26) we can calculate b_i and introduce it into the first equation of (26) which yields with

$$\omega_i = 2\lambda/\delta - \mu_i - 1/\delta^2\mu_i \tag{28}$$

one equation for a_i

$$\omega_i a_i = -\frac{1}{\delta}\sum_{k,l} r_{ikl}(1/2\mu_i + 1/\mu_l)a_k a_l + \sum_{k,l,m} a_k a_l a_m[-r_{iklm}/6\delta\mu_i +$$
$$\frac{1}{2}\sum_j r_{ikj}r_{jlm}/\mu_j - r_{iklm}/2\delta\mu_m + r_{iklm}\lambda/3] \quad , \; i = 1,2,3... \tag{29}$$

The elimination of one variable (b_K corresponds to ψ) is a basic way in treating problems of this kind and is also found in |6|, where it is already done in the system of differential equations corresponding to (7). From (28) we see that for those values of λ given by (18) the left hand side of (29) vanishes. This is the bifurcation case. Whether λ is a simple or double eigenvalue we obtain either one or two equations with vanishing linear parts (left hand side). These are the bifurcation equations. The corresponding variables are called critical variables and will be designated as a_α or a_α and $a_{\alpha+1}$ respectively. All other $a_i(i\neq\alpha,\alpha+1)$ are called noncritical variables. If we set $i=\alpha,\alpha+1$ in (29) which yields the bifurcation equations we see that on the right hand side of the bifurcation equations also the noncritical variables $a_i(i\neq\alpha,\alpha+1)$ are present. These still have to be eliminated. To eliminate them we express the noncritical variables $a_i(i\neq\alpha,\alpha+1)$ as functions of the critical variables from those equations of (29) for which the left hand sides do not vanish. By the implicit function theorem locally in the neighborhood of the bifurcation point these solutions

$$a_i = f_i(a_\alpha, a_{\alpha+1}) \qquad i \neq \alpha,\alpha+1 \tag{30}$$

are guaranteed. Introducing (30) into the bifurcation equations (equations (29) with index $i=\alpha,\alpha+1$) we obtain one (simple eigenvalue) or two (double eigenvalue) equations which contain only the critical variables and which we call the one-parameter bifurcation equations

$$F_i(a_\alpha, a_{\alpha+1}, \lambda) = 0 \qquad i = \alpha,\alpha+1. \tag{31}$$

For the practical calculation of (30) the question of determinacy is of great importance. One calls a system of bifurcation equations (31) n-determinate if the local behavior of the solutions of (31) is not influenced by the addition of terms of order n+1 or higher (|9|).

A second important observation in connection with (30) is that (|7|)

$$a_i = O(|a_\alpha + a_{\alpha+1}|^2) \qquad i \neq \alpha,\alpha+1. \tag{32}$$

Eq. (32) has the important consequence that if (31) is 2-determinate, then (30) only could have an influence on third order terms in (31) and in this case we can set all $a_i(i\neq\alpha,\alpha+1)$ to zero in (29). This will be the case which we shall treat for a double eigenvalue.

If, however, second order terms in (31) do not determine the problem we have to calculate (30). As long as we restrict ourself to third order terms in (31) we only have to retain quadratic terms from (29) and obtain

$$a_i = \frac{1}{\omega_i}\sum_{k,l=\alpha}^{\alpha+1} - \frac{1}{\delta} r_{ikl}\left(\frac{1}{2\mu_i} + \frac{1}{\mu_i}\right)a_k a_l + O(|a_k + a_l|^3) \ . \tag{33}$$

To prove whether (31) is n-determinate is simple in the case of one variable but a nontrivial problem in the case of several variables (|9| p. 59).

5. Unfolding of the bifurcation equations

One does not only want to solve (31) but one also wants to know how do the solutions of (31) change if one sligthly perturbs (31). From the mechanical standpoint this is the problem of the imperfection sensitivity of the shell buckling problem. The main question in this respect is how many parameters must be included in the model to obtain all qualitatively different solutions possible from (31). The answer gives the universal unfolding (|9|) of (31). By a universal unfolding of the bifurcation equations (31) we understand an embedding of (31) into a versal parameter family with the minimum number of parameters. A versal family is one, which contains every possible case of a given class. Depending on the dimension and on the determinacy of (31) the appropriate number of parameters and how they enter in (31) is supplied by singularity theory (|9| p. 320). Universal unfoldings which include at most four parameters are classified in Elementary Catastrophe Theory and we shall treat now those two cases which will become relevant in the following chapter.

In the case of a simple eigenvalue we obtain for (31) one equation in one variable x. Let us consider here for example the 3-determinate case

$$F(x) = x^3 + O(x^4). \tag{34}$$

From Catastrophe Theory (|9|) we know that (34) needs two parameters for a universal unfolding. This leads to

$$x^3 + qx + r = 0, \tag{35}$$

which is called the cusp-catastrophe (|9| p. 174). In Fig. 3 the complete qualitative behavior in the neighborhood of the bifurcation point is given. The semicubical parabola in the parameter plane is the bifurcation set which gives a partition of the parameter space according to the qualitative properties of the system. This is called a bifurcation diagramm. The task now is, to identify in (35) the mathematical unfolding parameters (q,r) physically, which will be given in the next chapter.

In the case of the double eigenvalue we shall meet a hyperbolic umbilic catastrophe which has two variables x,y

$$\begin{aligned} F_1 &= 3y^2 + x^2 = 0 \\ F_2 &= 2xy \qquad\;\; = 0 \end{aligned} \tag{36}$$

and the universal unfolding of which needs three parameters q,r,t (|9| p. 186)

$$\begin{aligned} &3y^2 + x^2 + 2ty + q = 0 \\ &2xy + 2tx + r \qquad = 0. \end{aligned} \tag{37}$$

To represent the full behavior of (37) would require a five dimensional space (two variables and three parameters). If we restrict ourselves to represent only the bifurcation diagram, we can do this in R^3 and obtain Fig. 4 which shows the bifurcation set in the parameter space q,r,t. We shall discuss Fig. 3 and 4 in the next chapter.

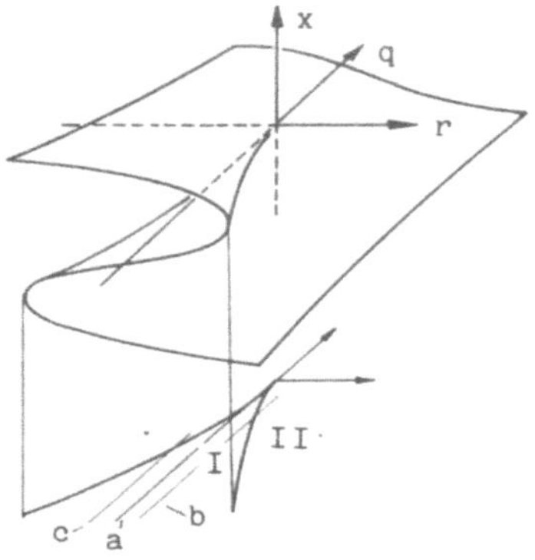

Fig.3: Complete behavior of the dual cusp catastrophe. For parameter values in domain I we have 3 solutions, one stable, two unstable whereas in domain II we have one unstable solution.

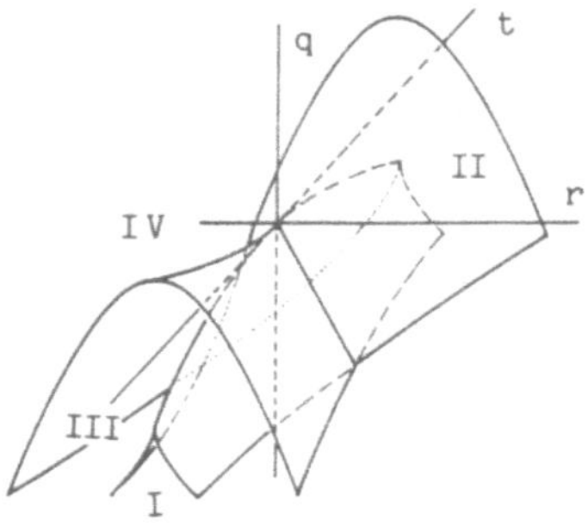

Fig.4: Bifurcation set of the hyperbolic umbilic catastrophe in the three dimensional parameter space (q,r,t), which is divided into four domains with qualitative different types of solutions: (I) 1 stable, 3 unstable (II) 2 unstable (III) 1 stable, 1 unstable (IV) no solutions.

6. Results

a) Simple eigenvalues.

As we shall need it, we write down the bifurcation equations up to nonlinear terms of third order. As we have a distinguished parameter in our problem, which is the external pressure p and which is included in (28) by means of (12) ω_α is a suitable choice for the parameter q of (35). We obtain from (29) with (33)

$$t + \omega_\alpha a_\alpha = k_2 a_\alpha^2 + k_3 a_\alpha^3$$

$$\text{with} \quad k_2 = -r_{\alpha\alpha\alpha} 3/2\delta\mu_\alpha \tag{38}$$

$$\text{and} \quad k_3 = \sum_{\substack{i \\ i\neq\alpha}} r_{\alpha\alpha i} r_{i\alpha\alpha}(1/2\mu_i + 1/\mu_\alpha)(2/\mu_\alpha + 1/\mu_i)/\delta^2\omega_i$$

$$+ \frac{1}{2}\sum_i r_{\alpha\alpha i} r_{i\alpha\alpha}/\mu_i - r_{\alpha\alpha\alpha\alpha}(2/3\delta\mu_\alpha - \lambda/3) .$$

The physical interpretation of the second parameter t which is simply a constant in (38) will be given at the end of this chapter after the discussion of the case of the double eigenvalue. Treating (38) we have to distinguish two cases:

(i) α even

In this case $r_{\alpha\alpha\alpha} \neq 0$ and (38) is 2-determinate. Therefore (38) reduces to

$$t + a_\alpha \omega_\alpha = k_2 a_\alpha^2 . \tag{39}$$

If t = 0 we get from (39) two solutions

$$a_\alpha = 0 \quad \text{and} \quad a_\alpha = - \frac{2\omega_\alpha \mu_\alpha \delta}{3r_{\alpha\alpha\alpha}} . \tag{40}$$

In terms of Catastrophe Theory (39) is the fold catastrophe. In Fig. 5 (40) and imperfect solutions are given. We see that depending on the sign of the imperfection the system has a physically quite different behavior.

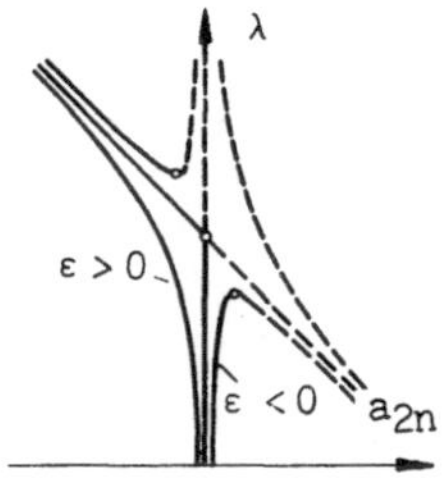

Fig.5: Bifurcation solutions for a buckling pattern of even order at a simple eigenvalue. The imperfect solutions correspond to Fig. 11. (Full lines stable, dotted lines unstable solutions, ε=-t in (39)).

(ii) α odd

In this case $r_{\alpha\alpha\alpha} = 0$ and we have to retain the third order term in (38). The bifurcation equation reads

$$t + \omega_\alpha a_\alpha = k_3 a_\alpha^3 . \tag{41}$$

The sums in (38) are finite, because the $r_{i\alpha\alpha}$, $r_{\alpha\alpha i}$ given by (27) are zero for $i > 2\alpha$. This follows from the properties of the Legendre polynomials, because we have

$$(P_\alpha(x))^2 = \sum_{j=1}^{2\alpha} a_j P_j(x).$$

Furthermore it is important to notice that k_3 is always negative only as consequence of the contribution of the noncritical variables, which result from the Liapunov Schmidt reduction process, in the bifurcation equation. We want to emphasize that the use of the Galerkin method where the a_i from (30) with $i \neq \alpha$, $\alpha + 1$ are always set to zero would yield a qualitatively incorrect result.

Considering the signs of the coefficients in (41) we get a dual cusp catastrophe (|9|). In Fig. 6 three sections from Fig. 3 (lines a,b,c) are shown. Both cases (i) and (ii) are critical cases concerning the post buckling behavior, as small imperfections can drastically reduce the carrying

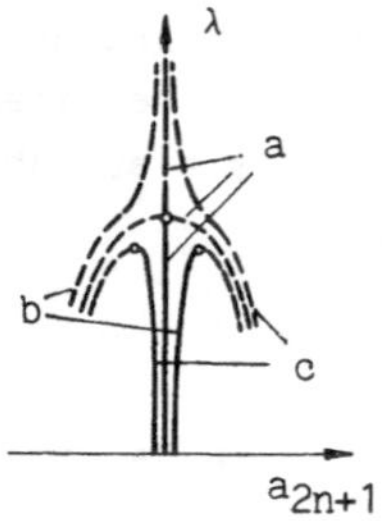

Fig.6: Bifurcation solutions for a buckling pattern of odd order at a simple eigenvalue. These solutions are obtained from the sections a,b,c of Fig. 3. (Full lines stable, dotted lines unstable solutions).

capacity of the shell.

b) Double eigenvalue

If the shell thickness has certain special values according to (22) we obtain a double eigenvalue. The consequence is that we now have two critical variables a_α, a_β where we have set $\beta = \alpha + 1$ and two bifurcation equations from (29), which are pretty simple as we only retain second order nonlinear terms

$$\begin{aligned} a_\alpha \omega_\alpha = &- \frac{1}{\delta} r_{\alpha\alpha\alpha}\left(\frac{1}{2\mu_\alpha} + \frac{1}{\mu_\alpha}\right)a_\alpha^2 - \frac{1}{\delta} r_{\alpha\alpha\beta}\left(\frac{1}{2\mu_\alpha} + \frac{1}{\mu_\beta}\right)a_\alpha a_\beta \\ &- \frac{1}{\delta} r_{\alpha\beta\alpha}\left(\frac{1}{2\mu_\alpha} + \frac{1}{\mu_\alpha}\right)a_\beta a_\alpha - \frac{1}{\delta} r_{\alpha\beta\beta}\left(\frac{1}{2\mu_\alpha} + \frac{1}{\mu_\beta}\right)a_\beta^2 \\ a_\beta \omega_\beta = &- \frac{1}{\delta} r_{\beta\alpha\alpha}\left(\frac{1}{2\mu_\beta} + \frac{1}{\mu_\alpha}\right)a_\alpha^2 - \frac{1}{\delta} r_{\beta\alpha\beta}\left(\frac{1}{2\mu_\beta} + \frac{1}{\mu_\beta}\right)a_\alpha a_\beta \\ &- \frac{1}{\delta} r_{\beta\beta\alpha}\left(\frac{1}{2\mu_\beta} + \frac{1}{\mu_\alpha}\right)a_\beta a_\alpha - \frac{1}{\delta} r_{\beta\beta\beta}\left(\frac{1}{2\mu_\beta} + \frac{1}{\mu_\beta}\right)a_\beta^2 \end{aligned} \tag{44}$$

We now assume $\alpha = 2n$, $\beta = 2n+1$. Then we get

$$r_{\alpha\alpha\alpha} \neq 0, \quad r_{\alpha\beta\beta} \neq 0, \quad r_{\beta\alpha\beta} = r_{\beta\beta\alpha} \neq 0.$$

All other r_{ijk} in (44) are zero. Thus (44) simplifies to

$$\begin{aligned} a_\alpha \omega_\alpha &= - \frac{1}{\delta\mu_\alpha}\frac{3}{2} r_{\alpha\alpha\alpha} a_\alpha^2 - \frac{1}{\delta\mu_\alpha}\frac{3}{2} r_{\alpha\beta\beta}\left(\frac{1}{3} + \frac{2}{3}\frac{\mu_\alpha}{\mu_\beta}\right)a_\beta^2 \\ a_\beta \omega_\beta &= - \frac{2}{\delta\mu_\alpha}\frac{3}{2} r_{\beta\alpha\beta}\left(\frac{2}{3}\frac{\mu_\alpha}{\mu_\beta} + \frac{1}{3}\right)a_\alpha a_\beta \ . \end{aligned} \tag{45}$$

(45) are the one parameter (external pressure) bifurcation equations which as follows from (36) are two-determinate. The corresponding bifurcation solutions are given in Fig. 7 which also can be found in |13| (Fig. 104). The physical interpretation of the solutions is that each of the coupled modes is a superposition of an even ($\alpha = 2n$) and an odd ($\alpha = 2n+1$) Legendre polynomial. This could lead to a deflection pattern given in Fig. 8, where a strong dimple occurs at the north pole and almost no deflection at the south pole. Though such solutions have been observed experimentally (|13|) we must say that they are unstable as long as we retain only second and third order terms in the bifurcation equations. The corresponding solutions for third order terms in the bifurcation equations are given in Fig. 9. They again are all unstable and form a double cusp catastrophe (|9|). In sketching Fig. 9 we have neglected the second order terms which are numerically of no influence. To give an unfolding of the double cusp catastrophe would require at least eight parameters (|9| p.320) and we do not try to do this here.

In order to study the physical significance of the imperfection parameters in (37) we make some reasonable simplifying assumptions for (45) which follow from (19), (27) and (20) ($\beta = \alpha + 1$) to

$$\mu_\alpha/\mu_\beta \simeq 1, \quad \frac{r_{\alpha\beta\beta}}{r_{\alpha\alpha\alpha}} \simeq \frac{r_{\beta\alpha\beta}}{r_{\alpha\alpha\alpha}} \simeq 1 \qquad \delta\mu_\alpha \simeq 1 \tag{46}$$

With (46) we obtain from (45)

$$\begin{aligned} a_\alpha\overline{\Delta\lambda} &= -a_\alpha^2 - a_\beta^2 \\ a_\beta\overline{\Delta\lambda} &= -2a_\alpha a_\beta \end{aligned} \tag{47}$$

where

$$\overline{\Delta\lambda} = \frac{4(\lambda-\lambda_c)}{3\delta\, r_{\alpha\alpha\alpha}} \tag{48}$$

and λ_c is given by (12) for p given by (4). To obtain from (47) equations (37) we have first to obtain the imperfection parameters and secondly we must make a transformation of coordinates in order to obtain the normal form of (37).

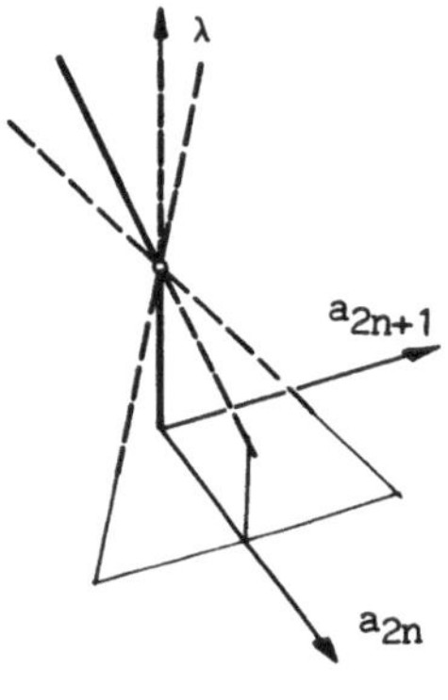

Fig.7: Bifurcation solutions at a double eigenvalue for the one parameter bifurcation equation with second order terms. (Full lines stable, dotted lines unstable solutions).

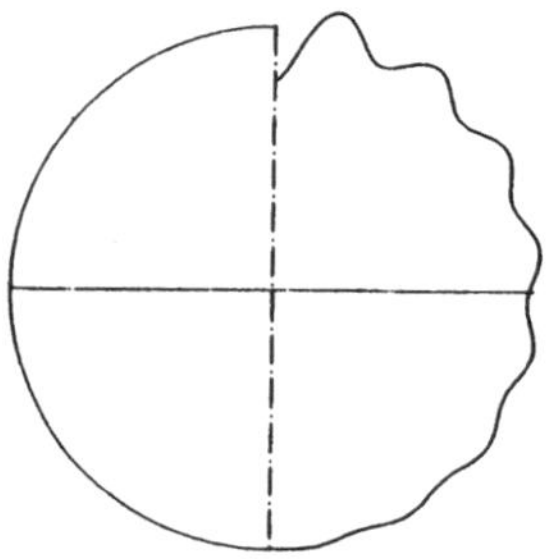

Fig.8: Buckling pattern of a single dimple solution, which is obtained by superposition of two solutions of order 2n and 2n+1.

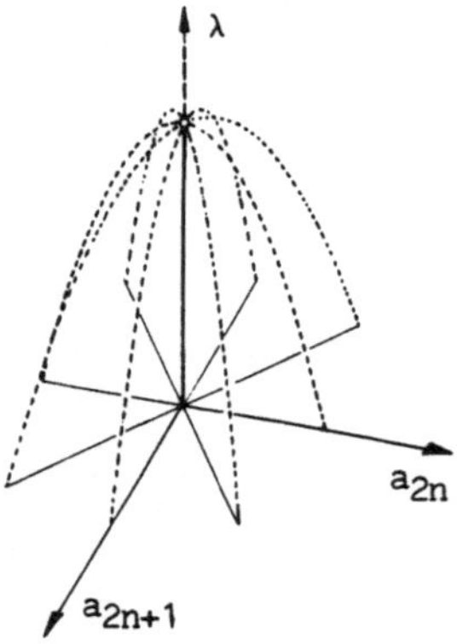

Fig.9: Bifurcation solutions at a double eigenvalue for third order nonlinear terms (Full lines stable, dotted lines unstable solutions).

Let us first identify the imperfection parameters. Of course it is possible to realize these in different ways. As we have already mentioned in the first chapter we want to be able to take into account manufacturing imperfections which manifest themselves by a non-uniform thickness $h(\xi)$ of the shell. Consequently A and D given by (9) are functions of ξ. From the derivation of the shell equations in |11| one easily sees what is changed in the equations. We obtain up to second order nonlinear terms the following shell equations

$$\beta'' + \beta'\mathrm{ctg}\xi + \beta(2\lambda/\delta - \mathrm{ctg}^2\xi) + \psi/\delta = \beta\psi\mathrm{ctg}\xi/\delta - 3(\delta'/\delta)(\beta' - \nu\beta\,\mathrm{ctg}\xi)$$
$$\psi'' + \psi'\mathrm{ctg}\xi - \psi\mathrm{ctg}^2\xi - \beta/\delta = -\beta^2\mathrm{ctg}\xi/2\delta + 2\lambda(\delta'/\delta)(1-\nu\cos\xi)\ . \tag{49}$$

Using (23) and projecting (49) on the eigenfunctions in the same way as we did with (24) we obtain

$$a_i(2\lambda/\delta - \mu_i) + b_i/\delta = \sum_{k,l} r_{ikl}\, a_k b_l$$
$$a_i/\delta \qquad b_i\mu_i = -\frac{1}{2}\sum_{k,l} r_{ikl}\, a_k a_l + \varepsilon_i \tag{50}$$

where

$$\varepsilon_i = \frac{1}{\delta}\,\frac{2i+1}{2i(i+1)}\int_0^\pi \delta'(\xi)2\lambda(1-\nu\cos\xi)P_i^1(\cos\xi)\sin\xi\, d\xi \tag{51}$$

and where we have used $1/\delta \approx 1/\delta_o$, which follows from $\delta(\xi) = \delta_o + d\delta(\xi)$ with $\| d\delta(\xi)\| \ll \delta_o$. Reducing now (50) to a system in the variables a_i as we did with (26) we get

$$a_i(2\lambda/\delta - \mu_i - 1/\delta^2\mu_i) = -\frac{1}{\delta}\sum_{k,l} r_{ikl}(2/\mu_i + 1/\mu_l)a_k a_l + \frac{1}{\delta\mu_i}\varepsilon_i\ . \tag{52}$$

We can say that the imperfection parameters ε_i express the contribution of the projection of the thickness variations $\delta'/\delta = h'/h$ on the corresponding eigenfunction.

Before proceeding we simplify (28) by introducing the deviations $\Delta\lambda = \lambda - \lambda_c$ and $\Delta\delta = \delta - \delta_c$ where λ_c is given by (12) and δ_c by (22). Introducing $\Delta\lambda$ and $\Delta\delta$ into (28) and making use of $\alpha \gg 1$ we get

$$\omega_\alpha = (2\Delta\lambda + \alpha\Delta\delta)/\delta$$
$$\omega_\beta = (2\Delta\lambda - \alpha\Delta\delta)/\delta\,. \tag{53}$$

We are able now to transform (52) exactly into the normal form (37). From (53) follows

$$\omega_\beta = \omega_\alpha - 2\alpha\Delta\delta/\delta \tag{54}$$

Making use of (48) we get from (52)

$$a_\alpha\overline{\Delta\lambda} = -a_\alpha^2 - a_\beta^2 + \frac{2\varepsilon_\alpha}{3r_{\alpha\alpha\alpha}}$$
$$a_\beta(\overline{\Delta\lambda} - \overline{\Delta\rho}) = -2a_\alpha a_\beta + \frac{2\varepsilon_\beta}{3r_{\alpha\alpha\alpha}} \tag{55}$$

where

$$\overline{\Delta\rho} = \frac{4\alpha\Delta\delta}{3\delta r_{\alpha\alpha\alpha}} \tag{56}$$

Setting

$$y = \sqrt{3}\,a_\alpha - \overline{\Delta\lambda}/2 - \overline{\Delta\rho}/4$$
$$x = a_\beta$$

we obtain from (55) the normal form (37) of the hyperbolic umbilic with the following coefficients

$$2t = -\overline{\Delta\rho}\,\sqrt{3}/2$$

$$r = \frac{1}{\sqrt{3}}\,\frac{2\varepsilon_\beta}{3r_{\alpha\alpha\alpha}} \tag{57}$$

$$q = \frac{2\varepsilon_\alpha}{3r_{\alpha\alpha\alpha}} - \frac{(\overline{\Delta\lambda})^2}{4} + \frac{(\overline{\Delta\rho})^2}{16}\,.$$

The main advantage of these cumbersome transformations leading to (57) is that we can immediately make use of the bifurcation set given in Fig. 4 where we have four distinct regions with qualitatively different solutions. Starting in domain I we have four different local solutions, which correspond to two saddles one maximum and one minimum of the potential

$$V = x^2y + y^3 + t(x^2+y^2) + rx + qy \tag{58}$$

from which (37) can have been derived. For a generic bifurcation the number of solutions changes by two if we cross the surface separating region I from regions II or III. We have one maximum and one saddle in region II and one minimum and one saddle in region III. Crossing from II or III the surface into domain IV we loose again two solutions, which means that there exist no solutions in IV. Hence we see that only in region I and III there exists one stable solution which is the solution a_{2n}. In Fig. 10 a plot of imperfect solutions is given in comparison to the perfect case of Fig. 7 representing parameters of domain I.

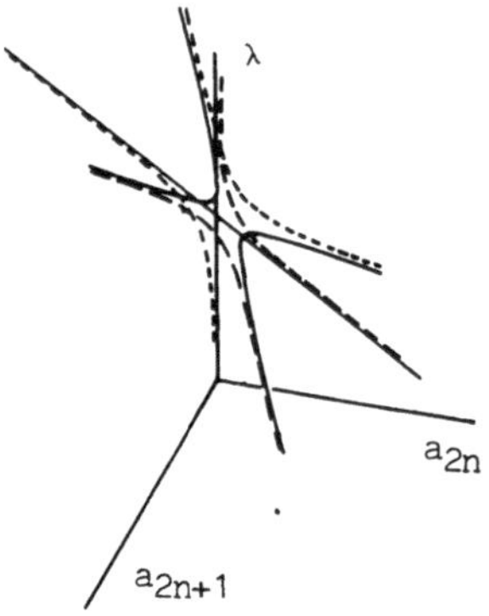

Fig.10: Imperfect solution paths for a parameter value from domain I of Fig. 4. (Dotted lines for imperfect shell, full lines for perfect shell).

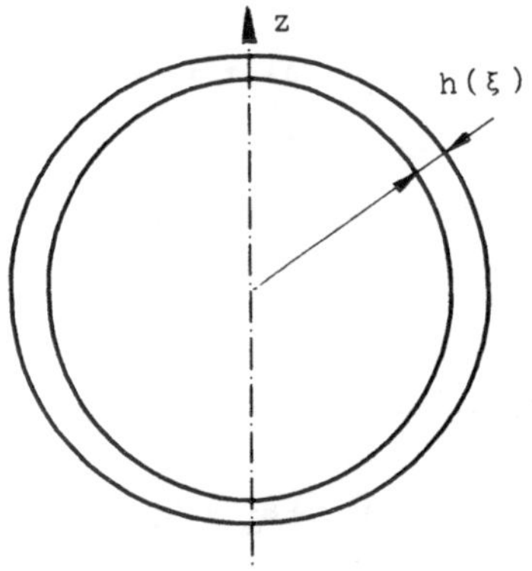

Fig.11: Imperfect shell, which gives a solution with $\varepsilon > 0$ in Fig. 5.

Finally we return to Fig. 5 in order to explain the physical difference of an imperfection $\varepsilon > 0$ and $\varepsilon < 0$. A possible imperfection $\varepsilon < 0$, which leads to a noncritical deformation is given in Fig. 11. The shell is thinner at the poles and fater at the equator. An imperfection $\varepsilon > 0$ could be a shell thinner at the equator and thicker at the poles and would lead to a critical deformation path.

7. Some comments on the case of closely spaced eigenvalues

As we have already mentioned the analysis we gave in strictly local. In this respect one has to make clear what is understood by local, what restrictions are given for the practical application and how one can overcome them. As it easily can be calculated from (18) and (22)

$$\lambda_{n+1} - \lambda_n = O(\delta) \qquad \text{for} \qquad \delta \to 0. \tag{59}$$

Therefore the range of validity ($|\lambda-\lambda_c| < \delta$) of our results is very small for small δ. In |8| by means of singular perturbation theory based on the smallness of δ a way is given to extend the solution branches from the local neighborhood of the bifurcation point to a range of the parameter variation $\lambda - \lambda_c = O(1)$. We do not want to go into detail here, but we only want to mention that the results of the classical bifurcation analysis as we have obtained are approached asymptotically by the results obtained in |8| in the close vicinity of the bifurcation point. Departing from the bifurcation point for more than it is allowed from (59) a much steeper post-bifurcation path is obtained (Fig. 1 in |8|). However also in |8| only unstable solutions are calculated.

8. Acknowledgement.

This research program was partly supported by a grant of the Fonds zur Förderung der Wissenschaftlichen Forschung in Austria, of the project "Datenbank Angewandte Mathematik".

L i t e r a t u r e

|1| Berke L. and R.L.Carlson, Experimental studies of the postbuckling behavior of complete spherical shells, Experimental Mech. 8 (1968) 548 - 553.

|2| Busse F.H., Patterns of convection in spherical shells, J.Fluid Mech. 72 (1975) 67 - 85.

|3| Chandrasekhar S., Hydrodynamic and Hydromagnetic Stability, Clarendon Press, London 1961.

|4| Golubitsky M. and D.Schaeffer, Bifurcations with O(3) symmetry including applications to the Benard problem, Comm.Pure Appl.Math. 35 (1982) 81 - 111.

|5| Hutchinson J.W., Imperfection sensitivity of externally pressurized spherical shells, J.Appl.Mech. 34 (1967) 49 - 55.

|6| Knightly G. and D.Sather, Buckled states of a spherical shell under external pressure, Arch.Rat.Mech.Anal. 72 (1980) 315 - 380.

|7| Koiter W.T., The nonlinear buckling problem of a complete spherical shell under uniform external pressure, Proc.Kon.Nederl.Akad.Wet Amsterdam B 72 (1969) 40 - 123.

|8| Lange C.G. and G.A.Kriegsmann, The axisymmetric branching behavior of complete spherical shells, Quart.Appl.Math. 39 (1981) 145 - 178.

|9| Poston T. and I.Stewart. Catastrophe Theory and its Applications, Pitman, London 1978.

|10| Reissner E., A note on postbuckling behavior of pressurized shallow spherical shells, J. Appl. Mech. 37 (1970) 533 - 534.

|11| Reissner E., On axisymmetrical deformations of thin shells of revolution, Proc. Symp. in Appl. Math., vol. 3 (1950) 27 - 52.

|12| Sattinger D.H., Bifurcation and symmetry breaking in applied mathematics, Bull. Amer. Math. Soc. 3 (1980) 779 - 819.

|13| Thompson J.M.T. and G.W.Hunt, A General Theory of Elastic Stability, J.Wiley and Sons, London 1973.

Implicit Relaxation Applied to Postbuckling Analysis of Cylindrical Shells

B. KRÖPLIN

Anwendung Numerischer Methoden im Bauwesen
Universität Dortmund

Summary

The paper deals with a viscous approach, where an artificial damping is used in order to arrive at points of equilibrium in the pre- or postbuckling range of shell structures.
The artificial damping is based on a creep type procedure with the time playing a purely ficticious role. The required numerical stability is obtained by using an implicit operator in time and a variable viscosity and time step. The viscosity and the time step are combined as a variable damping parameter which depends on the unbalanced forces of the structure.
The procedure allows to find states of static equilibrium with the load applied all at once. Incrementation is not necessary unless the path has to be traced in small steps. With a perturbation different branches of the nonlinear solution can be obtained.
Examples are given for elastic thin walled cylindrical shells in the postbuckling range. The used finite elements are based on a functional with displacements, axial forces and bending moments as nodal variables.

1. Introduction

The calculation of the postbuckling behaviour of thin shell structures is a very difficult task. The structure passes through a sequence of rapidly changing buckling patterns until a stable state of static equilibrium is reached. The actual sequence of buckling patterns in the static unstable region depends often on small structural or dynamic perturbations. The behaviour has been studied extensively by experimental investigators [1]. Cylindrical shells under axial load are known as extremely sensitive to structural imperfections.

The frequent limit and bifurcation points in the region of rapid changing buckling patterns makes the tracing of the history extremely difficult and costly.

Several strategies are proposed by different authors. For example: to find a limit or bifurcation point by accompanying measures exactly, to determine the lowest eigenvalue and eigenvector in the critical point and to superimpose the lowest eigenvector and the displacement field such, that the iteration converges to the next equilibrium state [2]. In this strategy is the information about the static unstable region incomplete. Different branches of the solution path may exist. Another possibility is to follow a solution path in the unstable region by imposing a contraint in the solution space. One way of doing is so well known as "Constant arc length method". Accompanying measures are necessary in order to find bifurcations of the solution path.

The here presented strategy consists of two steps:
First a numerically stable algorithm is presented in order to arrive at possible states of static equilibrium in the post-buckling range of the structure with the load applied all in one step. This is performed by the artificial damping concept, which is similar to dynamic relaxation [3].

Second the obtained state of equilibrium is tested against instability by a perturbation of a predetermined size and direction. On the reached state of equilibrium a perturbation is imposed. If the equilibrium state is stable, the iteration converges to the same state of static equilibrium if not, another equilibrium state is obtained. In the iteration the artificial damping is used as device for convergence.

The procedure is repeated several times until the obtained solution point is stable against a perturbation of the given size.

The strategy is applied together with a four node curved shell element with displacements, axial forces and bending moments as nodal variables [4]. The theory accounts for moderate large rotations.
In the following first the relaxation method is outlined.

Second some comments are made on the perturbation technique and third some examples are shown.

2. The Implicit Relaxation Procedure

The creep relaxation is based on the viscous equation (1),

$$\underline{D}\dot{\underline{x}} + \underline{C}\underline{x} = \bar{\underline{p}}, \tag{1}$$

where $\underline{x}$ means the vector of the nodal unknowns, $\underline{C}$ is a structural (secant) matrix of the static problem, $\underline{D}$ means an artificial damping and $\bar{\underline{p}}$ the applied load. $(\dot{\ })$ describes the derivatives with respect to the time. Using linear shape functions in time t, see fig. 1 and equation (2),

$$\underline{x}(t) = (\xi,\ 1-\varepsilon)\ (\underline{x}_{n+1}, \underline{x}_n)^T \tag{2}$$

and their derivatives with respect to the time

$$\dot{\underline{x}}(t) = \frac{1}{\Delta t}\ (1,\ -1)\ (\underline{x}_{n+1}, \underline{x}_n)^T \tag{3}$$

for $\xi = 1$ the backward difference formular

$$\left(\frac{1}{\Delta t}\ \underline{D} + \underline{C}\right)\ \underline{x}_{n+1} = \bar{\underline{p}} + \frac{1}{\Delta t}\ \underline{D}\ \underline{x}_n \tag{4}$$

is obtained. $\bar{\underline{p}}$ is taken as time independent load. With the incremental relationship

$$\Delta\underline{x} = \underline{x}_{n+1} - \underline{x}_n \tag{5}$$

the basis equation for the numerical application follows as

$$\left(\frac{1}{\Delta t}\underline{D} + \underline{C}\right)\Delta\underline{x} = \bar{\underline{p}} - \underline{C}\ \underline{x}_n. \tag{6}$$

It should be noted, that no assumptions have been made for the matrix $\underline{D}$.

For nonlinear structural analysis the right hand side of (6) is interpreted as unbalanced force $\underline{p}^u$, where

$$\underline{p}^u_n = \overline{\underline{p}} - \underline{C}^s \underline{x}_n . \tag{7}$$

The ^{s} emphasizes, that the secant matrix is used in order to calculate the internal forces $\underline{p}^i$, see fig. 2.
Since $\underline{C}^s$ is in general not known, $\underline{p}^i$ is calculated as sum of the internal nodal force components.

The left hand side of equation (6) consists of the effective structural matrix

$$\underline{K} = \underline{D}^+ + \underline{C}, \qquad \underline{D}^+ = \frac{1}{\Delta t}\underline{D} \tag{8}$$

and the relaxation increment $\Delta \underline{x}_{n+1}$. $\underline{K}$ rules the size of $\Delta \underline{x}_{n+1}$. Hence $\underline{K}$ can be chosen in a wide range in order to meet different tasks, for example path tracing [6].

Here $\underline{C}$ is chosen as tangent structural matrix $\underline{C}^T$ as in the common incremental approach.

Then $\underline{D}^+ \Delta \underline{x}$ is a vector of artificial damping forces with the damping matrix $\underline{D}^+$. The difference between the relaxation and the incremental approach is, that $\underline{p}^u$ is used instead of a load increment and the effective structural matrix $\underline{K}$ instead of the tangent structural matrix $\underline{C}^T$.

The iteration is numerically stable and convergent for

$$\Delta \underline{x}_{i,n+1} = \lambda_i \Delta \underline{x}_{i,n}, \qquad |\lambda_i| < 1, \tag{9}$$

where n means the iteration step. Equation (9) holds for every component i of $\Delta \underline{x}$. Introducing (9) in (6) and considering the ith component yields

$$(D^{+}_{ij} + C^{T}_{ij})\lambda_j \Delta x_{j,n}, \tag{10}$$

where

$$p^{d}_{i,n} = D^{+}_{ij} \lambda_j \Delta x_{j,n} \tag{11}$$

represents the incremental damping force component i, while

$$p^{i}_{i,n} = C^{T}_{ij} \lambda_j \Delta x_{j,n} \tag{12}$$

means the incremental internal force component with respect to i. Assuming $\underline{D}^{+}$ as "lumped" diagonal matrix, D^{+}_{ii} follows from equation (10) as

$$D^{+} = \frac{p^{u}_{i,n} - p^{i}_{i,n}}{\lambda_j \Delta X_{i,n}} . \tag{13}$$

For $\lambda_1 = 1$ any D^{+}_{ii} larger than D^{+}_{ii} from equation (13) ensures the stability of the iteration process. Hence assuming $p^{i}_{i,n} = 0$ and $\lambda = 1$ D^{+}_{ii} may be chosen as

$$D^{+} = \text{abs}\left[\frac{p^{u}_{i,n}}{\Delta x_{i,n}}\right], \quad \lambda x_{i,n} > \varepsilon \tag{14}$$

In (14) only positive dampings are admitted. The damping reduces always the velocity of the structure. The restriction on $\Delta x_{i,n}$ avoids overdamping for $x_{i,n} \rightarrow 0$. It allows the damping to vanish towards an equilibrium state.

3. The Perturbation Procedure

With the implicit relaxation states of equilibrium can be found in a numerically stable manner. However the physical significance of the obtained solution is still in question. In order to check the stability of an obtained solution point a perturbation is imposed. The perturbation consists of a normalized eigenvector, which is treated as additional imperfection of the structure. The size of the perturbation can be chosen related of the intended intensity of the test. The eigenvector is calculated by a v. Mises iteration of the form

$$(\underline{C}^L + \underline{C}_u^N + \lambda \underline{C}_s^N)\underline{x} = 0, \quad (15)$$

where

$\underline{C}^L$ = linear matrix,

$\underline{C}_u^N$ = nonlinear strain matrix

$\underline{C}_s^N$ = nonlinear stress matrix.

With the perturbation a new state of equilibrium is calculated, which may be a totally different one, if a snap through is initiated by the perturbation. Then the perturbation is taken away and an iteration is performed to an undisturbed state. In all iterations the damping $\underline{D}^+$ is applied in order to ensure convergence. The perturbation procedure is repeated until no lower solution point is found.

4. Numerical Studies

4.1 Relaxation

Examples of cylindrical shells under axial end-shortening are calculated with a curved finite element in a mixed formulation. The nodal unknowns are the tangential and normal deflections u_α, u_3, the axial forces $n^{\alpha\beta}$ and the bending moments $m^{\alpha\beta}$, α, β = 1.2. All unknowns are approximated with the same linear shape functions, see fig. 3. The Flügge shell theorie is used. The nonlinear formulation is based on the Green Strain and the second Piola-Kirchoff stress. The nonlinear solution technique uses equation (6) with the lumped damping of equation (14).

Only the deflection u_3 is damped in the above described manner. The tangent structural matrix $\underline{C}^T$ is not lumped, thus the set of equations has to be solved in every iteration step. In mixed analysis the vectors $\underline{x}$ consist of the nodal displacements and stress resultants, while the load vector $\underline{p}$ contains forces, strains and curvatures. The product $\underline{D\dot{x}}$ represents only damping forces.

The datas of the analysed shell panel are given in fig. 4. The panel was chosen such, that the boundary conditions allow to represent the buckling mode in the deep postbuckling range, which is known from experiments [1].
However, the first aim of the examples is to discuss the stability and effectivity of the iteration process in the situation of multiple snaps in the postbuckling range.

The whole end-shortening ($40^0/00$) was imposed at once, see fig. 5. The relaxation leads to a stable equilibrium state in the deep postbuckling range. The buckling shape coincides very well with the experimental results of [1]. The the cylinder is unloaded by large displacement increments using the damping within each increment. The damping is necessary, because snaps may occur in each increment. One snap occurs at point D. There the small buckle in the corner between edge 1 and edge 4 vanishes, see fig. 5 and 6.

4.2 Relaxation and perturbation

The perturbation is applied to the same cylinder in point A. In the load case with increments the accompanying eigenvalue calculation indicates an unstable point of the solution path. An additional imperfection with the shape of the eigenvector is imposed. The iteration converges to point H. The additional imperfection is removed and the iteration converges to point K. Point K is situated on the branch, which was originally obtained with unloading, see fig. 5.

It should be noted, that the unloading branch snaps back to the prebuckling state in point F underneath of point A. In point F no instability is detected on the prebuckling path. Here the peak load depends on the path even in the elastic case.

With the iteration stabilized by the artificial damping it is possible to change the geometry during the calculation. For example in point D, see fig. 6 instead of a sine-imperfection a local imperfection is imposed. The iteration converges to point W. Then the shell with local imperfection is investigated for unloading, points W, V, R, and for loading from point R. In S a perturbation is applied, T is obtained. After one increment T-U another perturbation is applied and point V on the unloading branch is obtained. It should be noted, that points T and U are locally stable points.

5. Conclusion

The implicit relaxation and the perturbation technique are a tool in order to investigate the behaviour of elastic post-buckling problems in an efficient manner. The proposed strategy is a simplified strategy which does not account for the real behaviour in the snap regions. Only states of equilibrium are sought in a stable manner. In spite of good success with numerous examples further work is necessary in order to find the limites and the application range of the method.

Acknowledgement:
The author acknowledges the work of Dr. D. Dinkler and Mr. J. Hillmann on the examples and the support of the "Deutsche Forschungsgemeinschaft".

References

1. Esslinger, M.; Geier, B.: Postbuckling behaviour of Structures. CISM Courses and lectures No. 236. Springer Verlag 1975.
2. Argyris, J.H.; Hindelang, U.; Menz, W.; Hilpert, O.; Malejannakis, G.A.: Geometrisch nichtlineare Stabilitätsanalyse eines dünnwandigen Schalentragwerks aus Glasfaserbeton. VDI-Forschungsheft Nr. 604 (1981).

3. Kröplin, B.-H.; Dinkler, D.: A Creep Type Strategy for Tracing the Load Path in Elastoplastic Post Buckling Response. Computer methods in applied mechanics and engineering 32 (1982) 365 - 376.
4. Harbord, R.: Berechnungen von Schalen mit endlichen Verschiebungen - Gemischte finite Elemente -. Bericht Nr. 72-7 aus dem Institut für Statik, TU Braunschweig.

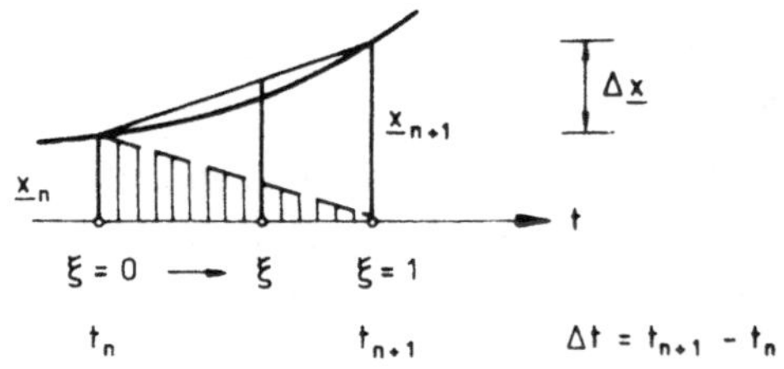

Fig. 1: Shape function in time domain

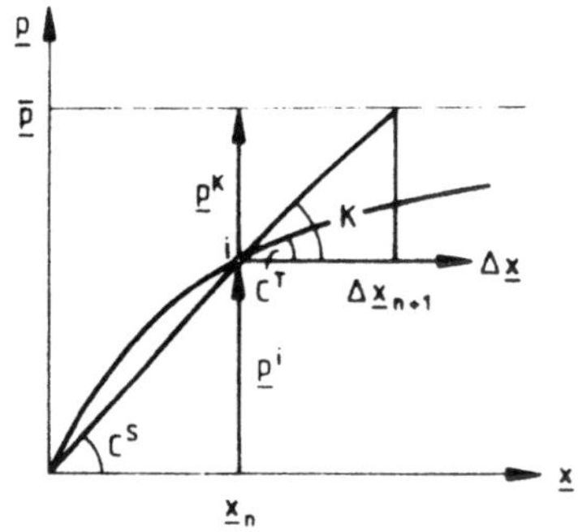

Fig. 2: Load - displacement relationship

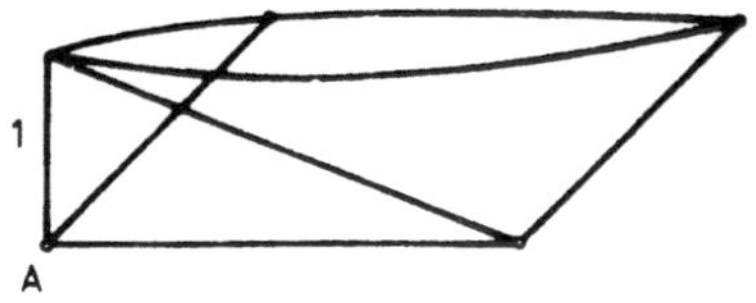

Fig. 3: Shape functions at the node A

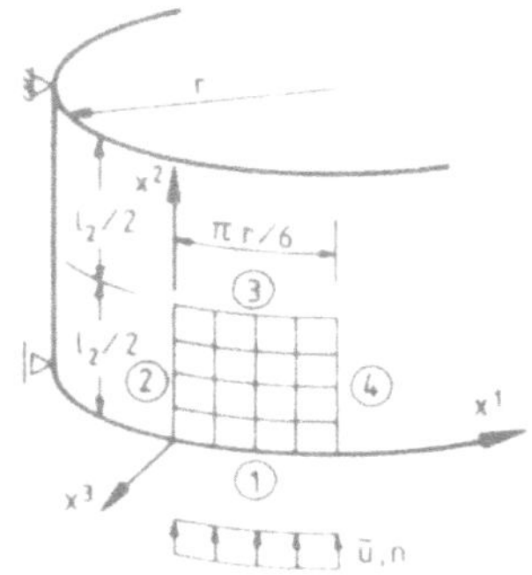

$$r = 100\,\text{mm}$$
$$l_2 = 100\,\text{mm}$$
$$t = 1{,}0\,\text{mm}$$
$$E = 2{,}06 \cdot 10^5\,\text{N/mm}^2$$
$$\nu = 0{,}3$$
$$n = \frac{1}{l_1} \cdot \int n^{22}\, d\vartheta$$
$$p_{cr} = \frac{E \cdot t^2}{r \cdot \sqrt{3 \cdot (1-\nu^2)}}$$
$$\varepsilon = 2\,\bar{u}_2 / l_2$$

boundary conditions

①	②, ③	④
$\begin{pmatrix} u_3 \\ m^{22} \end{pmatrix} = 0$	$\begin{pmatrix} u_1 \\ n^{12} \\ m^{12} \end{pmatrix} = 0$	$\begin{pmatrix} u_2 \\ n^{12} \\ m^{12} \end{pmatrix} = 0$

experimental buckling pattern [1]

Fig 4: Calculated panel

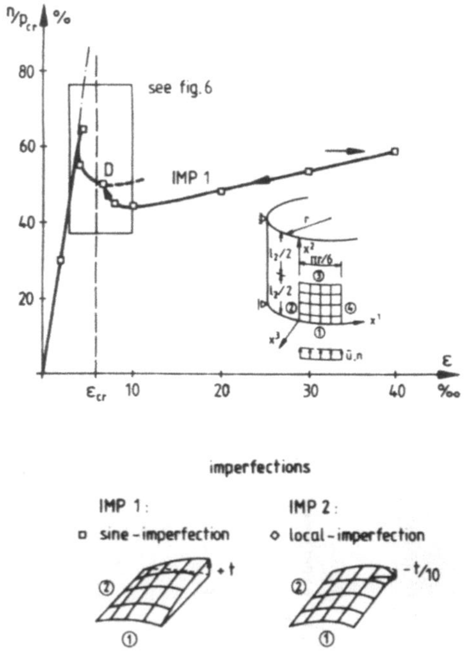

Fig. 5: Cylindrical panel under axial end-shortening - investigation in large steps -

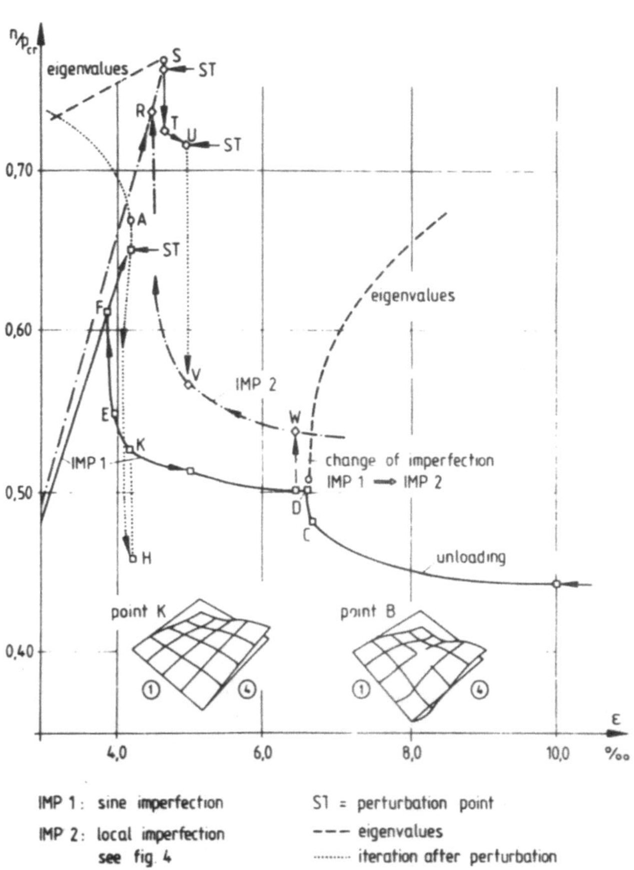

Fig. 6: Pre- and postpuckling states of a cylindrical shell panel with perturbation and artificial damping

Nonlinear Bending of Curved Tubes

F. A. Emmerling

Hochschule der Bundeswehr München
Fachbereich Luft- und Raumfahrttechnik
Institut für Mechanik
Neubiberg, Federal Republic of Germany

Abstracts

The determination of the nonlinear deformation and of the collapse load of elastic curved tubes subjected to bending loads is investigated. The precritical deformation of the tubes is determined on the basis of the semi-membrane theory. The stability analysis is done with the aid of the hypothesis of local buckling. The collapse loads are compared with the results of a bending experiment with Hostaphan-tubes.

1. Infinitely long tubes

A curved tube is an important element of construction in steam power plant and nuclear power station, especially pipework system. The curved tubes mostly are thin-walled. Fig.1 shows a curved tube.

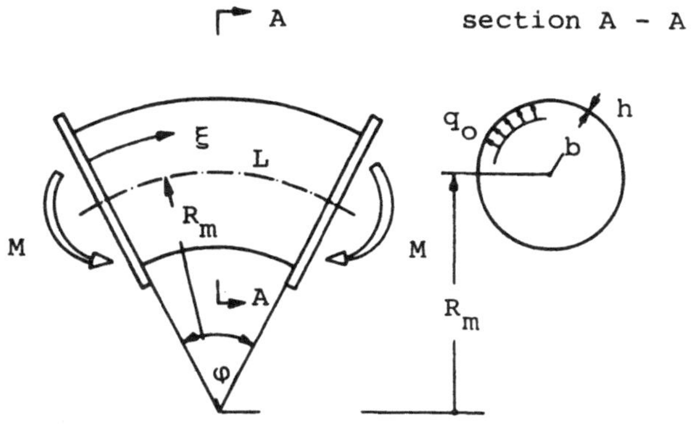

Fig.1. Curved tube, parameters of geometry and load

It is characterized by geometrical parameters, such as wall thickness h, radius of curvature R_m, length $L = R_m\varphi$ of the centre line, radius b of the cross section and by boundary conditions (e.g. flanges) at its ends. The possible boundary conditions are:

- thin flanges (stiff against cross section deformation, flexible in ξ-direction),
- rigid flanges (deformation of cross section is hindered both in the plane and in ξ-direction)
- connection to another tube.

Only two of the possible load configurations are considered (Fig.1):

- bending in plane,
- constant normal pressure q_o.

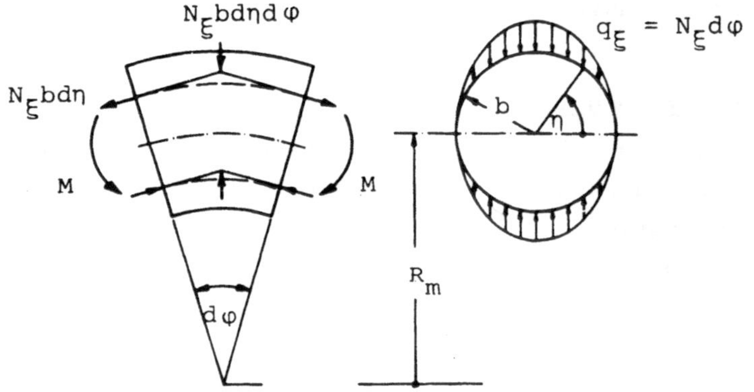

Fig.2. Kármán-effect

The topic of the following investigation is the determination of the nonlinear deformation and of the collapse load of elastic curved tubes subjected to bending loads.

The behavior of a curved tube under bending has been studied by Theodore von Kármán [1](1911). As the following treatments (including nonlinear bending) are based on them, the basic assumptions of Kármán's work are shortly sketched here.

Fig.2 shows an element of a bended curved tube. As can be seen, the resultants of the stress normal forces are directed to the centre of curvature, while the resultant pressure forces are oppositely directed. For thin walled tubes, these forces lead to a flattening of the cross sections. Hence the local stresses change and the bending stiffness of the tube gets smaller. In circumferential direction (η) of the tube, considerable stresses arise from the bending of the wall. This is called Kármán effect.

Kármán made the following assumptions (Fig.3):

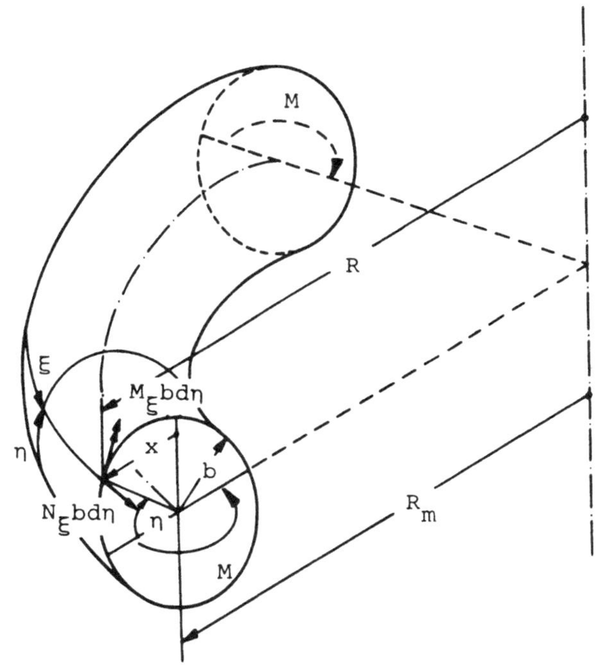

Fig.3. Bending of tubes, Kármán's assumptions

1. Stress and deformation states are independent of the tube's coordinate ξ.
2. The strain ε_η in circumferential direction of the tube is neglected ($\varepsilon_\eta = 0$).
3. In both equations of elasticity for the normal strain ε_ξ in

ξ-direction and for the curvature change $\varkappa_\eta$ in circumferential direction the terms resulting from the orthogonal direction are neglected

$$Eh\varepsilon_\xi = N_\xi - \nu N_\eta \approx N_\xi \ , \quad \frac{Eh^3}{12} \varkappa_\eta = M_\eta - \nu M_\xi \approx M_\eta \ . \qquad (1)$$

4. Only tubes with small initial curvature are considered. The term b/R_m as a small value compared to 1 has been canceled

$$\frac{R}{R_m} = 1 + \frac{b}{R_m} \cos\eta \approx 1 \ . \qquad (2)$$

5. In order to determine the bending moment M the terms resulting from $M_\xi b d\eta$ are neglected

$$M = \int_0^{2\pi} N_\xi \, b d\eta \, x = \int_0^{2\pi} N_\xi \, b d\eta \, b \cos\eta \ . \qquad (3)$$

Because of the first Kármán's assumption, the solution is restricted to the middle region of very long tubes.

Kármán has solved the problem of tube bending by an energetic way and obtains the following formula for the relationship of curvature change vs bending moment

$$\frac{\varphi^* - \varphi}{R_m \varphi} = \frac{M}{KEI} \ . \qquad (4)$$

K is a correcting factor, which takes into account the flattening of the bended tube.

Due to the bending moment, the curvature and the Kármán effect are growing. The bending moment increases sublinear with the curvature change $1/R_m^* - 1/R_m$ and reaches a maximum. This non-linear problem has been formulated by Brazier [2](1927) for the initially straight (cylindrical) tube, and has been solved with the assumptions of Kármán (Fig.4). Due to this approach, the collapse load of the tube is determined by the maximum value of the bending moment [2].

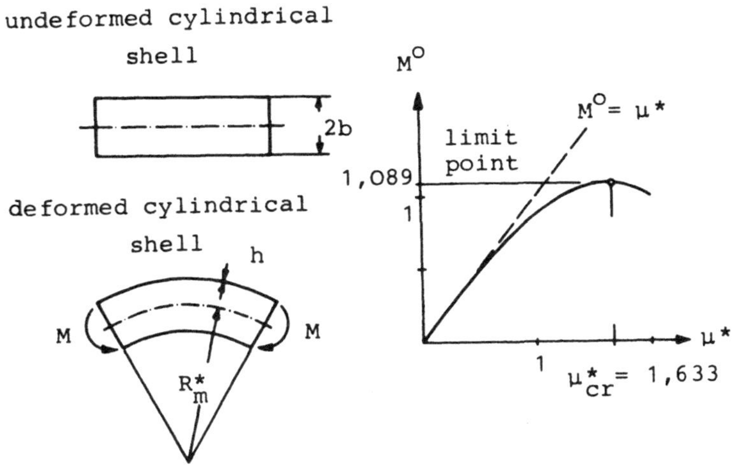

Fig.4. Brazier-problem

$$M = \frac{EIh^o}{b} M^o \ , \quad I = \pi h b^3 \ , \quad h^o = \frac{h}{b\sqrt{12(1-\nu^2)}} \ . \qquad (5)$$

In this presentation M^o is the bending moment acting at the ends of the tube in a nondimensional form. I is the second moment of the cross section and h^o is a small nondimensional shell parameter. In the limit point the maximum stress is $\sigma_{max} = 0.320\ Eh/b$.

There have been dozens of attempts to improve the Brazier-solution. The exact solution has been obtained for the first time in 1963 by E. Reissner and H.J. Weinitschke [6] on the basis of the exact set of differential equations. The energetic methods (dealing the variational problem)- though often applied - did not lead to an improvement. Even in 1979 an English paper has been published, which gives a "limit point" less accurate than the value of Brazier, which had been obtained 50 years ago.

For long tubes, including those with initial curvature, the effective investigation of the Brazier-problem has been made possible in the early 60th by E.L. Axelrad [5], offering simple, nonlinear, generalized Reissner-Meissner-equations:

$$\frac{d^2\psi}{d\eta^2} = \mu \sin\alpha - \mu^* \sin(\alpha+\vartheta) \quad , \quad \frac{d^2\vartheta}{d\eta^2} = \mu^*\psi\cos\alpha \quad , \tag{6}$$

$$N_\xi = Ehh^o \frac{d\psi}{d\eta} \quad , \mu = \frac{b}{h^o R_m} \quad , \quad \mu^* = \frac{b}{h^o R_m^*} \quad ,$$

$$m = \mu^* - \mu \quad , \quad \alpha^* = \alpha + \vartheta \quad . \tag{7}$$

The variables of these equations are: ϑ - angle of tangent change, ψ - stress function ($\dot{\psi} \sim N_\xi$), μ, μ^* - curvature parameters of the undeformed and deformed structure respectively. The equations (6), (7) only apply for tubes with small initial curvature and assume $b/R_m \ll 1$. The asterisk marks the parameters of the deformed tube.

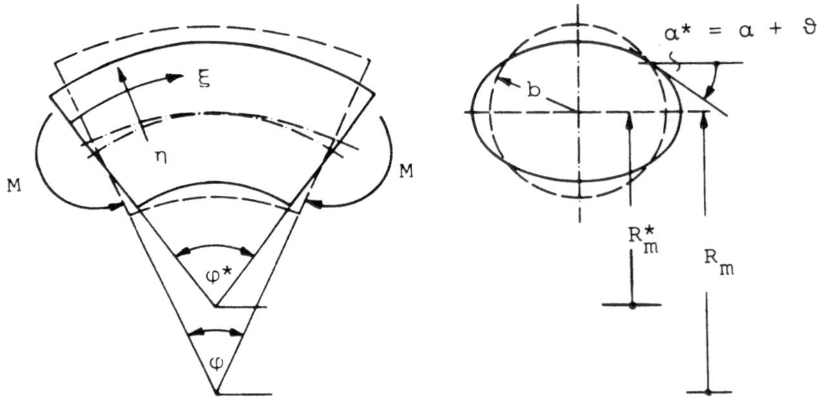

Fig.5. St. Venánt's nonlinear bending of tubes.

The equations are integrated by Fourier series approximation of the solution functions (ϑ,ψ) [9]. The topics of this investigation have been tubes with small initial curvature as well as those with circular and elliptic cross sections. Furthermore, the influence of constant normal pressure has been taken into account. The results of these investigations (Fig.6) show, how much the collapse load of a tube reduces, if there is only a small initial curvature.

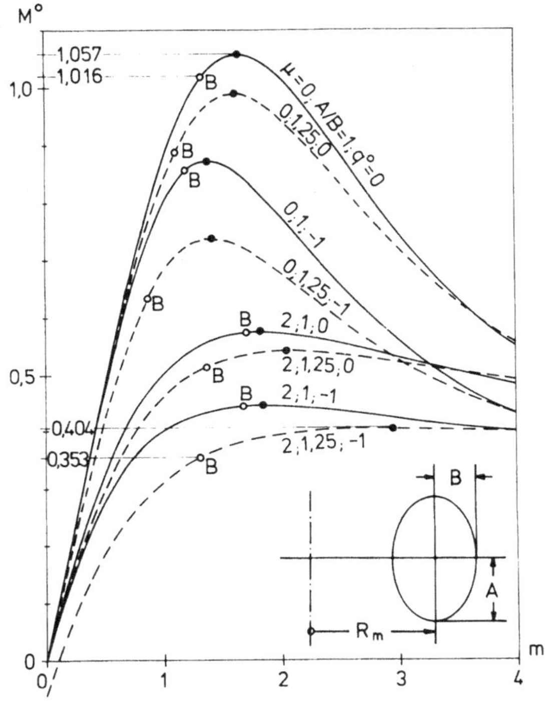

Fig.6. Bending moment M^{O} vs change of curcature $m = \mu^{*} - \mu$. B - Buckling moments according to theory of local buckling

The maximum bending moments (black dots) form an upper limit of the bending load.

However experiments (also made by Brazier) showed, that the tubes don't fail by steadily increasing flattening, but by the beginning of buckling in the domain of pressure. Therefore a loss of stability occurs by bifurcation of the equilibrium state.

Fife years after the basic investigation of Brazier, W. Flügge [3](1932) has treated the buckling of a cylindrical shell under a given axial stress $\sigma_{\xi} = Mx/I$ (Fig.7). He assumed that the cylindrical shell keeps its initial shape until the buckling begins, i.e. the pre-critical deformation was not considered by him. This problem was exactly solved in 1961 by P. Seide and V.I. Weingarten [4]. For the initial pressure stress σ_{cr}, they got a value, which approximately equals the critical pressure stress of a cylinder under axial pressure σ_{cl}

$$\sigma_{cr} \approx \sigma_{cl} = \frac{Eh}{b\sqrt{3(1-\nu^2)}} = 0{,}605\,\frac{Eh}{b} \quad . \tag{8}$$

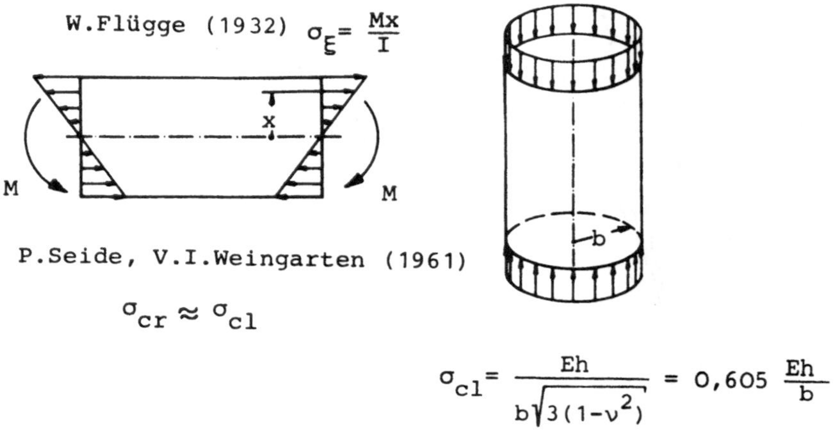

Fig.7. Critical bending stress; classical buckling stress

In the Brazier-problem, considering the bending of cylindrical shell, the maximum stress in the "limit point" is almost half the value of σ_{cr} from the linear stability analysis. Therefore it is closer to the stresses from the experiment than σ_{cr}. Never the less it cannot be used for practical calculations, as the tubes always fail by buckling. Besides this, it still remained unknown, how long a cylindrical shell ought to be, in order to apply the Brazier theory.

The two different approaches to determine the collapse load of curved tubes were combined by E.L. Axelrad 1965 [5]. The collapse load in his approach is determined in two subsequent steps. In the first step, the nonlinear deformation of the tube under bending is determined, taking into account the real boundary conditions. The second step investigates the buckling stability of the deformed tube.

2. Tubes with finite length

To determine the nonlinear deflection of tubes with finite length under bending (i.e. the pre-buckling deformation), E.L. Axelrad used a specialized shell theory, the "semi-membrane-theory". This theory takes into account the knowledge, that at the bending of tubes, the stress state in ξ-direction changes much less than in circumferential (η-)direction. This stress state is typical for a wide category of shells, which are constructed to allow large elastic deformations. They are called in literature "flexible shells".

In the following, the theory of flexible shells shall be sketched briefly. $F(\xi,\eta)$ shall represent the most important forces and strain parameters of a shell. A flexible shell (where curved tubes belong to) is characterized by a stress state with the following condition

$$\left|\frac{\partial^2 F}{a^2 \partial \xi^2}\right| << \left|\frac{\partial^2 F}{b^2 \partial \eta^2}\right| \quad . \tag{9}$$

Hence in the equilibrium conditions, the terms M_ξ, $H_\xi = H_\eta = H$, Q_ξ can be neglected. Analog cancelletion can be done in the compatibility equations . Here the strain parameters ε_η, γ, λ_η (static-geometric analogy to the neglected stress resultants) can be neglected.

These simplifications in the basic equations of shell theory mean, that in ξ-direction only a membrane stress state is assumed, while in η-direction a complete stress state is taken into account. Besides this (as in Kármán's treatment) the terms resulting from the lateral strains are neglected in the elastic equations for the normal strain ε_ξ, and the bending moment M_η, which is in accordance to the condition (9):

$$Eh\varepsilon_\xi = N_\xi - \nu N_\eta \approx N_\xi \;, \quad M_\eta = D\varkappa_\eta + \nu D\varkappa_\xi \approx D\varkappa_\eta \;.$$

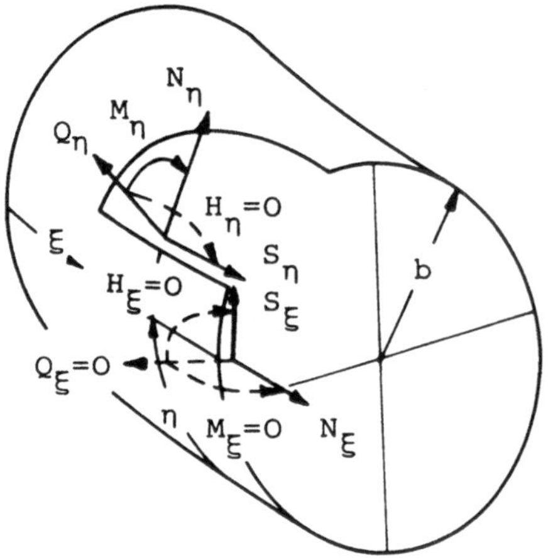

Fig.8. Semi-membrane-theory for tube

For locally axisymmetrical shells (where tubes belong to) it is now possible to express all section forces and strain parameters as functions of two variables - the curvature change $\varkappa_\eta$ and the normal section force N_ξ. By elimination of all the other terms, the following set is obtained [8,10].

$$\left(\frac{a}{b}\frac{\partial^2}{\partial\xi^2} - \frac{\partial}{\partial\eta}\frac{a^3}{b^3}\frac{b}{\rho_\xi} + \frac{\partial}{\partial\eta}\frac{a^2}{b^2}\frac{\partial}{\partial\eta}\frac{a}{b}\frac{R'_\eta}{R_\xi}\right)\varkappa_\eta + \frac{1}{Ehb}W'N_\xi = 0 \;,$$

$$\left(\frac{a}{b}\frac{\partial^2}{\partial\xi^2} - \frac{\partial}{\partial\eta}\frac{a^3}{b^3}\frac{b}{\rho'_\xi} + \frac{\partial}{\partial\eta}\frac{a^2}{b^2}\frac{\partial}{\partial\eta}\frac{a}{b}\frac{R'_\eta}{R'_\xi}\right)N_\xi - \frac{D}{b}W'\varkappa_\eta = bq_\Sigma \;, \tag{10}$$

$$W' = \frac{\partial}{\partial\eta}\frac{a^2}{b^2}\frac{\partial}{\partial\eta}\frac{R'_\eta}{b}\frac{\partial^2}{\partial\eta^2}\frac{a}{b} + \frac{\partial}{\partial\eta}\frac{a^2}{b^2}\frac{b}{R'_\eta}\frac{\partial}{\partial\eta}\frac{a}{b} \;, \quad D = \frac{Eh^3}{12(1-\nu^2)} \;,$$

$$b^3 q_\Sigma = -\,a^2 b\,\frac{\partial q_\xi}{\partial\xi} + \frac{\partial}{\partial\eta}\left(a^3 q_\eta\right) + \frac{1}{b}\frac{\partial}{\partial\eta}\left[a^2\frac{\partial}{\partial\eta}\left(aR'_\eta q\right)\right] . \tag{11}$$

Here, a and b are the Lamé-parameters ($ds_\xi = a d\xi$; $ds_\eta = b d\eta$),

$1/\rho_\xi$ is the geodedic curvature along the ξ-parameter line. Besides this

$$\frac{1}{R'_\xi} = \frac{1+\varepsilon_\xi}{R^*_\xi} = \frac{1}{R_\xi} + \varkappa_\xi \quad , \quad \frac{1}{R'_\eta} = \frac{1+\varepsilon_\eta}{R^*_\eta} = \frac{1}{R_\eta} + \varkappa_\eta \quad , \tag{12}$$

holds, where R^*_ξ, R^*_η are the radia of curvature of the normal sections belonging to the deformed parameter lines. For more details see the papers [8, 10].

The two nonlinear differential equations describe the unlimited, large deformations of flexible shells of revolution with small strains. The integration of these equations leads to the stress state and deformation state in the shell, whereby the first part of stability investigation is done.

In the next step, the deformed state has to be investigated on its stability.

As already mentioned, the tubes under bending, fail by buckling. The buckles arise locally in the pressure domain, therefore the hypothesis of local buckling from E.L. Axelrad [7] can be used to determine the critical stress.

The main statement of the hypothesis is: buckling is only determined by the stress state and the shape of the shell within the zone of the first buckle(s). As a result of this hypothesis can be assumed: If in the domain of the first buckle(s) the section forces (N_ξ, N_η, S), and the curvatures ($1/R^*_\xi$, $1/R^*_\eta$) of the sections normal to the plane coordinates are almost constant these variables can be assumed as constants in the stability analysis. Therefore out of Donnell's stability equations flows a simple relationship [7] between the section forces (N_ξ, N_η, S), which generate the local buckling, and the curvatures in a definite point of the shell. For very thin cylindrical shells and tubes of small curvature, an asymptotic formula can be given to determine the critical stress σ_{cr}, which approximately coincides with the classical buckling stress

$$\sigma_{cr} = \frac{Eh}{R'_\eta \sqrt{3(1-\nu^2)}} = \frac{b}{R'_\eta} \sigma_{cl} \quad . \tag{13}$$

But here is R'_η the actual radius of curvature to the η-parameter line in the deformed tube.

By this asymptotic stability condition, the buckling moments M^o_{cr} have been determined dependent of the nondimensional length

$$l = \frac{L}{\pi b} \sqrt{h^o} \quad . \tag{14}$$

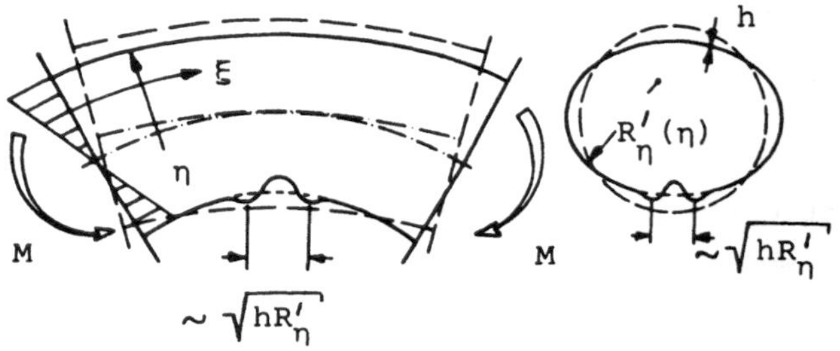

Fig.9. Local buckling of tube under bending

The results are shown in Fig.10 for two boundary conditions (thin and rigid flanges) as function of the initial curvature μ and the nondimensional normal pressure q^o.

The results show, that the tubes can be classified into three classes of length:short,medium and long. In the case of the short tube, the influence of the stiffness at the edges (flanges) is remarkable over the total length of the tube. The precritical deformations are so small, that they can be neglected. In the case of straight tubes the buckling stresses and the buckling moments respectively, coincide with the classical solutions [4]. The collapse loads of the medium tubes essentially depend on the nondimensional length l on the initial curvature μ.

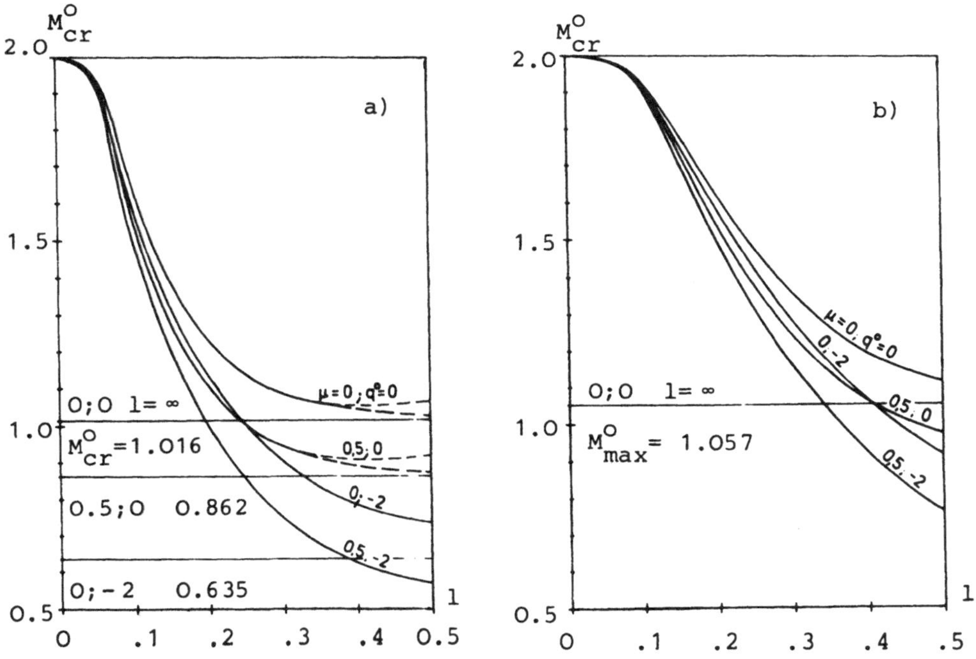

Fig.10. Buckling moments M^o_{cr} [10]
a. thin flanges b. rigid flanges

If the tubes are long enough, the influence of the stiffness at the edges has decreased in the middle domain. In this case, an exact solution exists, which follows from the integration of the nonlinear Reissner-Meissner-equations [9]. This solution is exact for undefinitely long tubes, where the normal force N_ξ and the change of curvature $\varkappa_\eta$ have been obtained without influence of the flanges.

In order to verify the buckling moments obtained by the hypothesis of local buckling, experiments with Hostaphan-tubes have been made at the Institut für Mechanik of the HSBw München. Hostaphan is a synthetic foil, which deforms only elastically up to the buckling moment and even beyond.

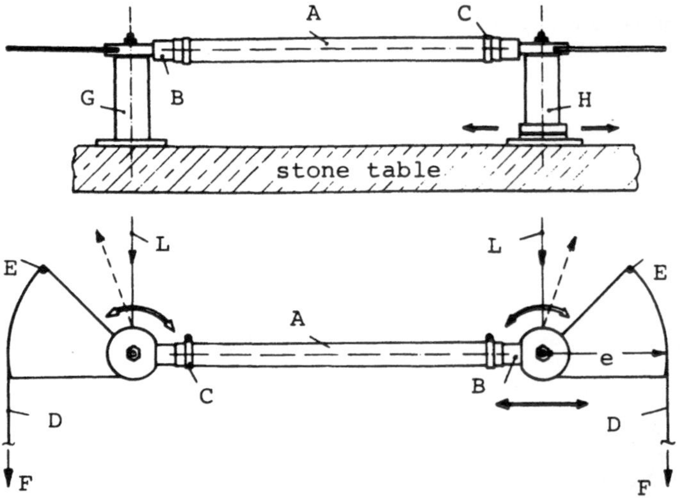

Fig.11. Bending experimental device

Fig.11 shows a schematic view of the bending experimental device, designed at the Institut für Mechanik to determine the buckling moments. The Hostaphan-tube A is fitted on both ends on muffs B and fixed on them by hose clamps C. This fixing corresponds to the boundary case "rigid flanges", as the cross section remains undeformed, and the deflections of the cross sections at the ends in ξ-direction are hindered. The moment $M = Fe$ is applied by two steel ropes D, which are fixed E at the segment discs. The bearings G and H are freely rotatable and in addition the bearing H is freely shiftable in the direction of bearing G. The rotation of the bearings G and H are measured by deflection of the LASER beams L.

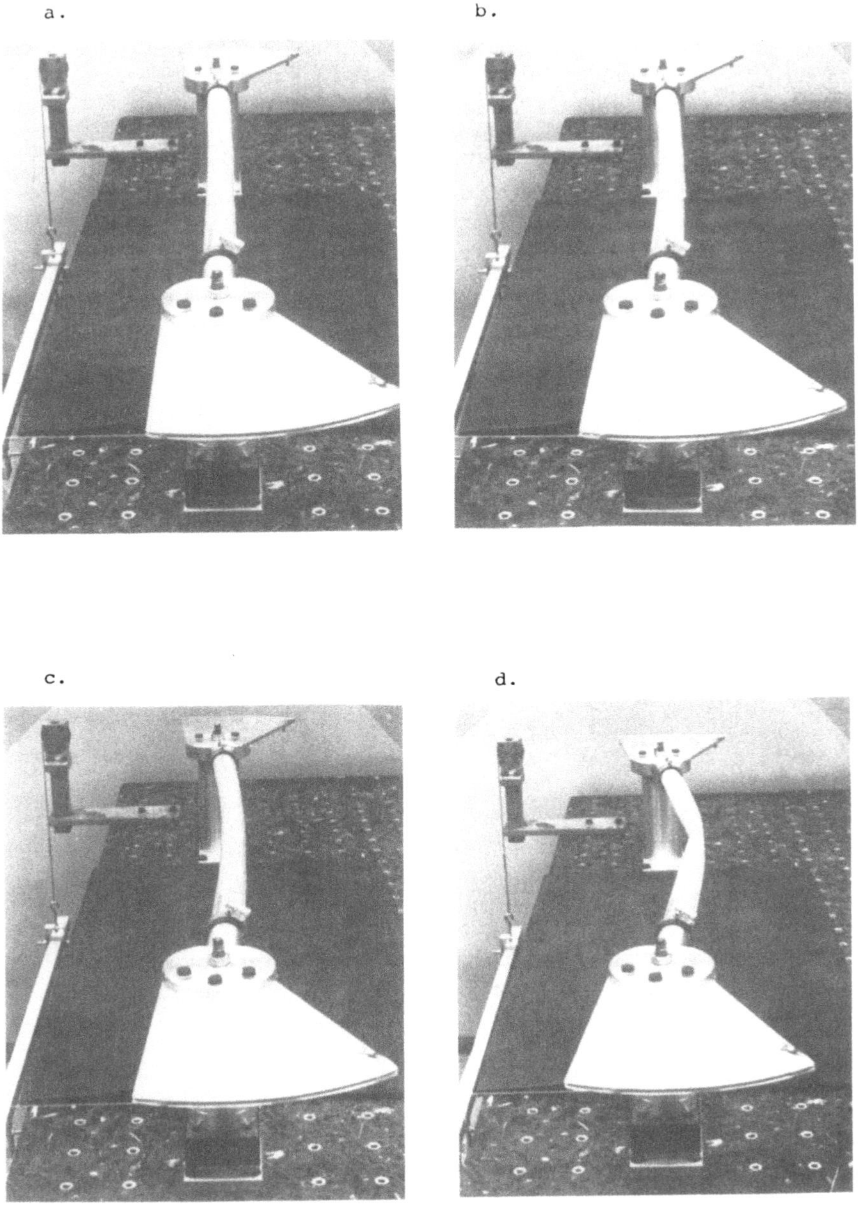

Fig.12: Different phases of the bending experiment

Fig.12 shows the different phases of the bending experiment:

a. Fixed tube at the beginning.

b. The bending of the tube as well as the flattening of the cross sections can be seen clearly.

c. The tube at large deformation. The applied moment is slightly below M_{cr}.

d. The buckled tube. Only one buckle occured, which covers a small domain of the tube.

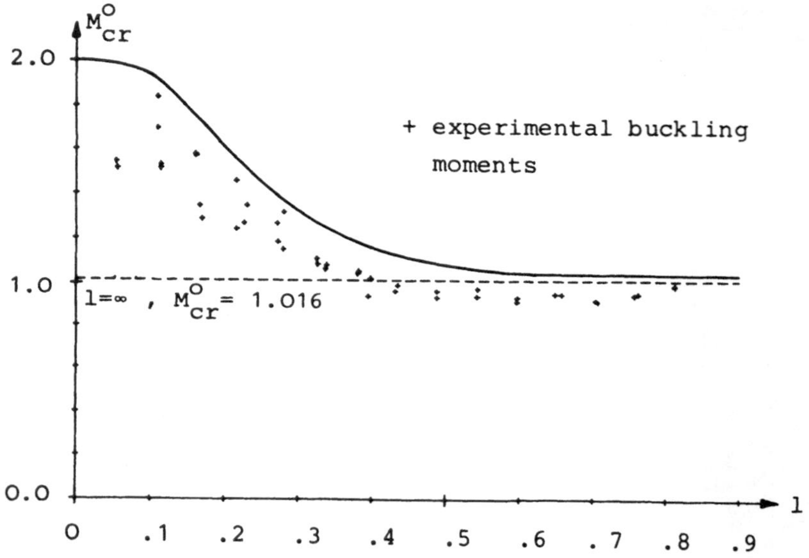

Fig.13. Comparison of the theoretical and experimental buckling moments (+) vs non-dimensional length

The specific bending moments M^o_{cr} of the experiments are shown in Fig.13 as functions of the specific length l. A comparison with the theoretically gained curve of the buckling moment ($\mu = 0$, $q^o = 0$, Fig. 9b) shows a good coincidence. The maximum difference of the experimental and theoretical results is less than 25 %.

Finally, it can be stated, that the buckling moments, calculated by the hypothesis of local buckling, have been proved by the experiments.

References

1. Kármán v., Th.: Über die Formänderung dünnwandiger Rohre, insbesondere federnder Ausgleichsrohre. VDI-Z 55 (1911) 1889-1895.

2. Brazier, L.G.: On the Flexure of Thin Cylindrical Shells and other "Thin" Sections. Proc. Soc. London, Ser. A. 116 (1927) 104-114.

3. Flügge, W.: Die Stabilität der Kreiszylinderschale. Ing.-Archiv 3 (1932) 463-506.

4. Seide, P.; Weingarten, V.I.: On the Buckling of Cylindrical Shells under Pure Bending. J.Appl. Mech. 28 (1961) 112-116.

5. Axelrad, E.L.: Biegung und Stabilitätsverlust dünnwandiger Rohre unter hydrostatischem Druck. Izv. AN SSSR, Mechanika i Maschinostr. 1 (1962) 98-114 (in Russian).

6. Reissner, E.; Weinitschke, H.J.: Finite Pure Bending of Circular Cylindrical Tubes. Q. Appl. Math. 20 (1963) 305-319.

7. Axelrad, E.L.: Präzisierung der oberen kritischen Biegelast eines Rohres unter Berücksichtigung der geometrischen Nichtlinearität, Izv. AN SSSR, Mechanika 4 (1965) 133-139 (in Russian).

8. Axelrad, E.L.: Flexible Shell-Theory and Buckling of Toroidal Shells and Tubes. Ing.-Archiv 47 (1978) 95-104.

9. Emmerling, F.A.: Nichtlineare Biegung und Beulen von Zylindern und krummen Rohren bei Normaldruck. Ing.-Archiv 52 (1982) 1-16.

10. Axelrad, E.L.; Emmerling, F.A.: Große Verformungen und Traglasten elastischer Rohre unter Biegung und Außendruck. Ing.-Archiv 53 (1983) 41-52.

Nonlinear Finite Element Analysis of Shells under Pressure Loads Using Degenerated Elements

Ekkehard Ramm

Institut für Baustatik
Universität Stuttgart
West-Germany

0.Abstract

The paper briefly describes the concept of degeneration for the derivation of shell elements. Then the procedure to describe arbitrarily large displacements and rotations is explained. The problem of displacement dependent pressure loads is discussed in detail. A clear classification of the load definition allows to identify when a load is conservative and when it is not. Some remarks to the solution procedures in nonlinear analyses are added and selected numerical examples are given.

1.Introduction

Shell elements may be derived via a shell theory (classical concept) or directly from the three dimensional theory (degeneration). What the classical elements are concerned mixed and hybrid formulations seem to be more sucessful than displacement based elements because here the rigid body mode and interelement compatibility requirements are difficult to satisfy. Since these problems do not arise if the concept of degeneration is used and because of the simplicity of the formulation today degenerated elements are well established. The present paper focuses on the application of these elements. After a summarizing review of the concept the procedure to describe arbitrarily large displacements and rotations is briefly presented.

After this the paper concentrates on the problem of displacement dependent loads (pressure) and their treatment in nonlinear finite element formulation. This seems to be a controversal topic since often pressure loads are automatically identified as nonconservative forces even in cases where this is not true. The conditions are discussed when the load is conservative or not. For this a clear definition of the kind of loading - either body attached or space attached - is nescessary. Finally the solution procedures used are briefly compiled and some numerical examples are given.

2.Degenerated elements

In the late sixties several finite element users tried for simplicity to apply threedimensional solid elements in thin shell analysis and failed. This is not surprising since not for nothing generations of scientists put a lot of effort into the development of thin shell theories to avoid these difficulties. It was a big step forward when Ahmad [1] 1969 introduced shell-constraints into continuum elements without resorting to a sophisticated shell theory. Essentially these assumptions are a two-dimensional stress state and the kinematic constraint of a straight and inextensional shell normal.

The isoparametric formulation includes properly the rigid body modes. It needs also to be mentioned that degenerated elements use nodal displacements and rotations as independent variables because shear deformations are included. Therefore the compatibility requirements can easily be satisfied. But it was soon recognized [2] that this characteristic has also distinct drawbacks when thin shells have to be analysed: The transverse shear terms in the energy (stiffness) expression increase with decreasing thickness and finally suppress the essential bending terms so that the solution diverges. This is the penalty for not using a Kirchhoff-type of shell theory. The effect known as shear locking phenomenon has been investigated by many papers in the mean time, e.g. [3] . It is especially pronounced when serendipity interpolation schemes (linear, quadratic or cubic) are used, but very low for the bicubic Lagrange 16-node element [4]. A different characteristic often mixed up with shear locking is the membrane locking effect, when low order elements (linear, parabolic) are used as curved models, producing membrane states under pure bending. This typical shell problem can be avoided if the idealization can be based on a facet model [4]. Again the 16-node element does show very little membrane locking.

Several schemes have been proposed in the last years to improve the performance of degenerated elements in the low thickness range, among these are reduced integration, either uniform or selected (only for shear and membrane terms to avoid either kind of locking) [5,6], uniform reduced integration with hourglass control to avoid internal mechanisms [7], penalty methods [8], and discrete Kirchhoff elements (DKT) in which additional constraints based on the Kirchhoff hypothesis are introduced [9,10]. Although not all problems have been solved it can be stated that degenerated elements are very successful in moderately thick and thin shell analysis as well.

The concept of degeneration has been extended to geometrically nonlinear analysis for moderately large rotations in axisymmetric shell analysis in [11] and for arbitrarily large displacements and rotations of general shells in [12,13] (see chapter 3); similar formulations are given in [14],[15],[16],[6]. Furthermore material nonlinearities have been incorporated by a layered model; e.g. an elasto-plastic material law [6],[16] [17] or a concrete model [28].

3.Formulation of arbitrarily large displacements and rotations [12],[13]

The formulation is based on the isoparametric element concept. This is a fundamental feature in nonlinear analysis because original and deformed geometry fit together. Geometry and displacements are defined with respect to a cartesian global system with components ${}^m x_i$ and ${}^m u_i$, respectively. The left superscript indicates a deformed configuration m, corresponding nodal values are denoted by a right superscript k, ${}^m x_i^k$ and ${}^m u_i^k$. Shape functions Φ^k are defined in a local curvilinear (convective) coordinate system r, s of the middle surface which are normalized within the element so that $-1 \leq r$ or $s \leq +1$. The coordinates of any point within the shell are as usual decomposed into a part of the middle surface and a part in thickness direction both interpolated in the same way. The latter is expressed by a normalized coordinate $-1 \leq t \leq +1$ where $t = \pm 1$ defines the upper and lower surface; h^k is the thickness at node k. With this the cartesian component ${}^m x_i$ of an arbitrary configuration m is (figure 1)

$$ {}^m x_i = \sum_{k=1}^{M} \Phi^k(r,s) \cdot {}^m x_i^k + \frac{1}{2} t \sum_{k=1}^{M} \Phi^k(r,s)\, h^k \cdot \cos {}^m \psi_i^k \qquad (i = 1,2,3) \qquad (1) $$

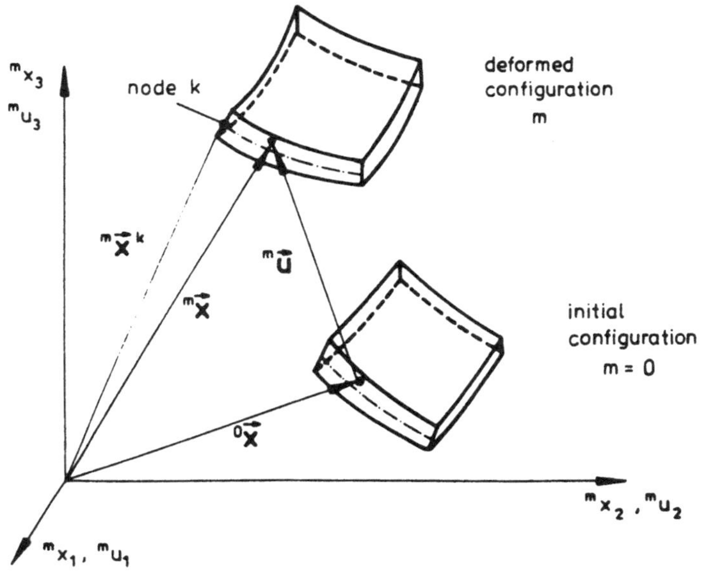

Figure 1: Geometry and Displacements of Shell Element

M is the number of nodes per element, and $\cos {}^m\psi_i^k$ are the direction cosines of the normal at node k with respect to the global coordinate system.
Per definition the displacements are the differences in coordinates of the deformed and undeformed (m=0) configuration:

$$
\begin{aligned}
{}^m u_i &= {}^m x_i - {}^0 x_i \\
&= \sum_{k=1}^{M} \Phi^k(r,s) \cdot {}^m u_i^k + \frac{1}{2}\, t \sum_{k=1}^{M} \Phi^k(r,s)\, h^k \left(\cos {}^m\psi_i^k - \cos {}^0\psi_i^k\right)
\end{aligned} \tag{2}
$$

Eqn (2) implies no restriction in size of displacements or rotations. It reproduces the kinematic constraint of keeping the normal straight and inextensional during deformation. It contains the three nodal displacements ${}^m u_i^k$ of the middle surface and three unknown angles ${}^m\psi_i^k$ of the normal from which only two are independent, so in total there are five unknowns per node.
Essentially eqn (2) is used to compute the internal forces which enter the equilibrium equations. The choice which kinematic variable is introduced into the rotational part of eqn (2) is not unique. For example any two direction cosines or the corresponding two angles ${}^m\psi_i^k$ may be chosen. Usually two rotations are assumed, in most cases the rotations are defined with respect to the tangential r-s-frame; in [12],[13] the rotation ${}^m\alpha^k$ around the global x_1-axis and the variation ${}^m\beta^k$ of ${}^m\psi_1^k$ are taken as unknowns (figure 2). In [6] the cosine terms in eqn (2) are replaced using the displacements at the outer surface instead.

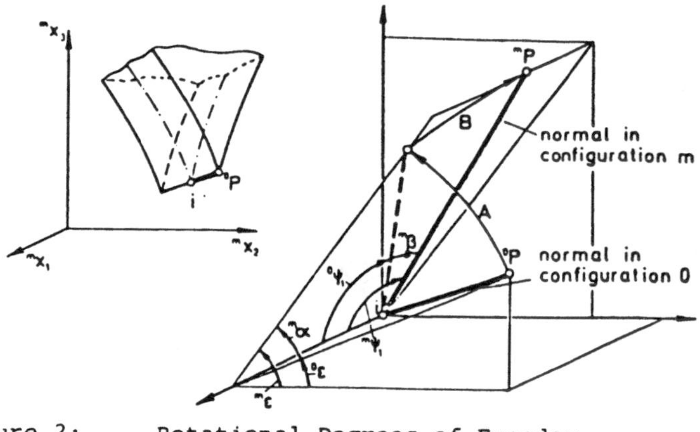

Figure 2: Rotational Degrees of Freedom

The analogous expression for the incremental displacements u_i from configuration m to m+1 (no left superscript) are

$$u_i = {}^{m+1}x_i - {}^{m}x_i$$
$$= \sum_{k=1}^{M} \Phi^k(r,s) \cdot u_i^k + \frac{1}{2} t \sum_{k=1}^{M} \Phi^k(r,s)\, h^k (\cos^{m+1}\psi_i^k - \cos^m \psi_i^k) \qquad (3a)$$

Eqn (3a) is nonlinear in the unknown incremental rotations α^k and β^k because they are arguments of trigonometric functions. In view of the linearization process of the stiffness matrix and a subsequent equilibrium iteration, eqn (3a) is expanded in a series and only the linearized terms in α^k and β^k are considered.

$$u_i \approx \sum_{k=1}^{M} \Phi^k(r,s) \cdot u_i^k + \frac{1}{2} t \sum_{k=1}^{M} \Phi^k(r,s)\, h^k ({}^m f_i^k \cdot \alpha^k + {}^m g_i^k \beta^k) \qquad (3b)$$

Here ${}^m f_i^k$ and ${}^m g_i^k$ are trigonometric functions of the current configuration m. Details of the derivation are given in [12],[13] for a total and updated Lagrangian formulation.

It should be mentioned that all three displacements and the rotational components are interpolated in the same order based on the same nodal mesh. A reduction of the interpolation order by one for the rotations may be considered, see for example the SEMILOOF elements [9],[10]. The element has been derived as cubic, quadratic and linear Lagrange and serendipity, quadrilateral and triangular models (figure 3). Transition elements are given in [18] to connect thin shells with continuum elements.

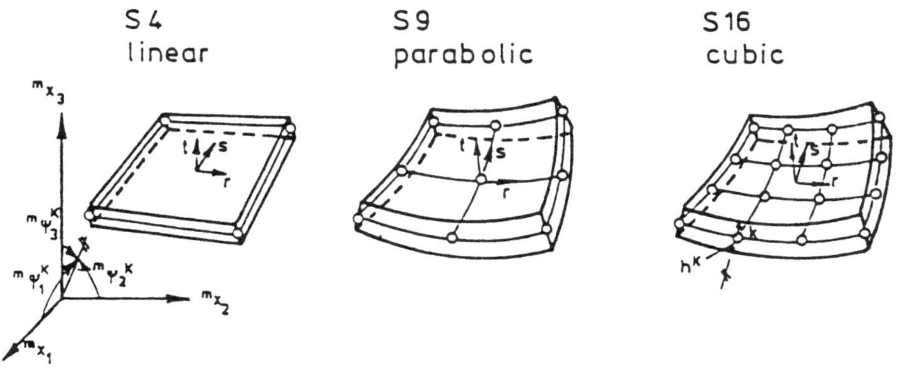

Figure 3: Versions of Quadrilateral Element

The above given expressions for geometry and displacements have been introduced into the principle work expression of the three-dimensional continuum [12],[13] finally (after assembly) resulting in the linearized stiffness expression

$$ {}^{m}\mathbf{K} \cdot \mathbf{u} = {}^{m+1}\mathbf{P} - {}^{m}\mathbf{F} \tag{4} $$

$\mathbf{u}$ is the unknown incremental displacement vector, ${}^{m+1}\mathbf{P} - {}^{m}\mathbf{F}$ are the out - of - balance forces, external load vector ${}^{m+1}\mathbf{P}$ of configuration m+1 minus internal forces ${}^{m}\mathbf{F}$ of position m. By iteration m approaches m+1 and the out - of - balance forces diminish. The tangent stiffness matrix ${}^{m}\mathbf{K}$ is the sum of linear and nonlinear influences among these are load stiffness matrices ${}^{m}\mathbf{K}_L$ for displacement dependent loads (see chapter 4). The stiffness matrix of degenerated elements is derived through numerical integration of the energy terms over the shell volume. Note that in elastic analysis the integration across the shell thickness requires two sampling points. This integration in thickness direction may be performed explicitely utilizing the same approximations adopted in thin shell theory (stress resultant theory). Here the direct relationship between the classical and the degeneration concept is apparent so that any controversal discussion is actually obsolet.

4.Displacement dependent loads - a classification

Pressure load is per definition displacement dependent if the analytical model considers finite deformations. For simplicity this effect is often neglected, thus the loads are assumed to be constant directional. In this context two questions arise:

* What is the influence of this simplification?
* What is the correct formulation if the displacement dependence is included ?

The answer to the second question also allows to classify whether or not a load is conservative. The following discussion is based on the work in [19], [20] where the details for the pressure loads are given. It is exemplified in the sense of the finite element method in which the displacement dependent terms lead to load stiffness matrices. These may be symmetric or not reflecting a conservative or nonconservative load. The discussion is restricted to the pressure load term of the virtual work expression:

$$^{m}\delta W_{ext} = \int_{^{m}A} {^{m}p_i} \; \delta u_i \; d\,^{m}a \tag{5}$$

Here δu_i is the virtual displacement field, and ^{m}A the loaded surface of the current configuration m. The pressure component may be defined as

$$^{m}p_i = {^{m}\lambda} \; {^{l}f} \; {^{k}n_i} \tag{6}$$

with the load multiplier $^{m}\lambda$, the function $^{l}f = {^{l}f}\,(^{l}x_j)$ of the load distribution which may depend on the coordinates of either the initial configuration (l=0) or the current configuration (l=m). The product $^{m}\lambda \cdot {^{l}f}$ is the loading magnitude and $^{k}n_i$ is the component of the surface normal in configuration k=0 or k=m. Different load definitions are classified in figure 4:

	constant directional load (k = 0)		follower normal load (k = m)	
	body attached (l = 0)	space attached (l = m)	body attached (l = 0)	space attached (l = m)
	①	②	③	④
load $^{m}p_i$	$^{m}\lambda \times {^{0}f} \times {^{0}n_i}$	$^{m}\lambda \times {^{m}f} \times {^{0}n_i}$	$^{m}\lambda \times {^{0}f} \times {^{m}n_i}$	$^{m}\lambda \times {^{m}f} \times {^{m}n_i}$
load in configuration m	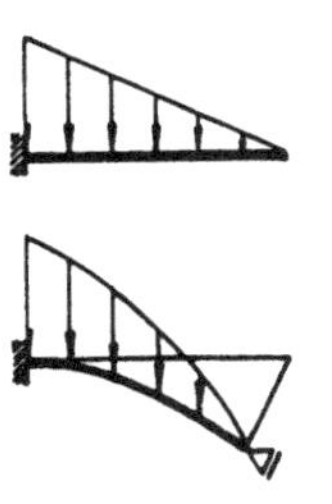	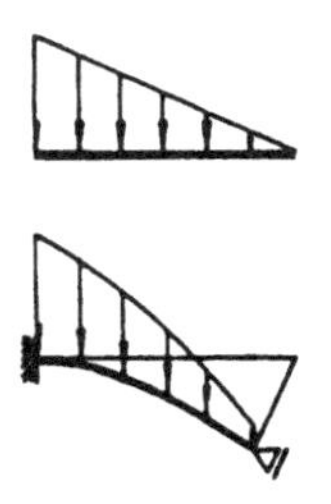	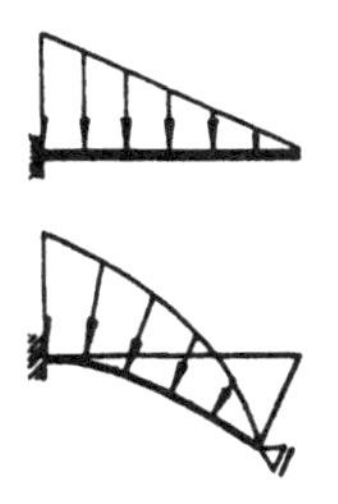	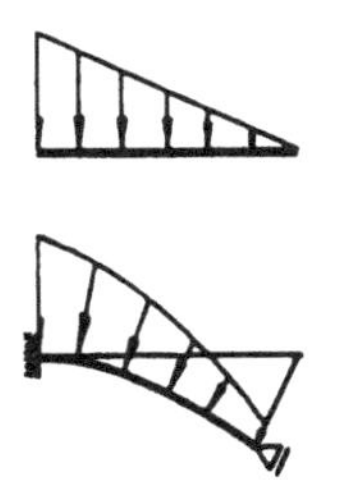

Figure 4: Load Definition

case ① is the usual non-displacement dependent load, case ② is of less practical interest and it is important to distinguish between the body attached load case ③ and the space attached load case ④. These two cases of follower loads (k = m) will be discussed further.

In view of an incremental solution procedure let us identify the position before and after an increment with m=1 and m+1 = 2 so that

$$ {}^{2}x_i\,(r,s) = {}^{1}x_i\,(r,s) + u_i\,(r,s) \qquad (7) $$

Furthermore introducing eqn (6) into eqn (5) and using

$$ {}^{2}n_i\; d\,{}^{2}a = e_{ijk}\, \frac{\partial\, {}^{2}x_j}{\partial r}\, \frac{\partial\, {}^{2}x_k}{\partial s}\; dr\; ds \qquad (8) $$

leads to

$$ {}^{2}\delta W_{ext} = e_{ijk}\, {}^{2}\lambda \iint_{rs} {}^{L}f\; \frac{\partial\, {}^{2}x_j}{\partial r}\; \frac{\partial\, {}^{2}x_k}{\partial s}\; \delta u_i\; dr\; ds \qquad (9) $$

and with eqn (7) omitting nonlinear terms in the displacement increments:

$$ \begin{aligned} {}^{2}\delta W_{ext} &= e_{ijk}\, {}^{2}\lambda \iint_{rs} {}^{L}f\; \frac{\partial\, {}^{1}x_j}{\partial r}\; \frac{\partial\, {}^{1}x_k}{\partial s}\; \delta u_i\; dr\; ds \\ &+ e_{ijk}\, {}^{2}\lambda \iint_{rs} {}^{L}f\; \left(\frac{\partial u_j}{\partial r}\, \frac{\partial\, {}^{1}x_k}{\partial s} + \frac{\partial\, {}^{1}x_j}{\partial r}\, \frac{\partial u_k}{\partial s}\right) \delta u_i\; dr\; ds \end{aligned} \qquad (10) $$

4.1 Space attached loads:

Now eqn (10) is specialized to a pressure field in space and the load magnitude ${}^{l}f$ depending on the current configuration (l=2) is expanded as Taylor series

$$ {}^{2}f = {}^{2}f\,({}^{2}x_n) = {}^{1}f + \frac{\partial\, {}^{1}f}{\partial\, {}^{1}x_n}\, u_n + \cdots \tag{11} $$

The series is introduced in eqn (10) considering only linearized energy terms: $({}^{1}_{1}f_{,n} \equiv \partial\,({}^{1}f)\,/\,\partial\, {}^{1}x_n)$

$$ \begin{aligned} {}^{2}\delta W_{ext} = {} & e_{ijk}\, {}^{2}\lambda \iint_{rs} {}^{1}f\; \frac{\partial\, {}^{1}x_j}{\partial r} \cdot \frac{\partial\, {}^{1}x_k}{\partial s} \cdot \delta u_i \cdot dr \cdot ds \\ & + e_{ijk}\, {}^{2}\lambda \iint_{rs} {}^{1}f \left(\frac{\partial u_j}{\partial r} \cdot \frac{\partial\, {}^{1}x_k}{\partial s} + \frac{\partial\, {}^{1}x_j}{\partial r} \cdot \frac{\partial u_k}{\partial s} \right) \delta u_i \cdot dr \cdot ds \\ & + e_{ijk}\, {}^{2}\lambda \iint_{rs} {}^{1}_{1}f_{,n}\; u_n \cdot \frac{\partial\, {}^{1}x_j}{\partial r} \cdot \frac{\partial\, {}^{1}x_k}{\partial s} \cdot \delta u_i \cdot dr \cdot ds \end{aligned} \tag{12} $$

The first integral is independent of the displacement increments and defines the usual load vector. Only the two other expressions which depend on the incremental displacements are considered below (subscript L). They are modified through integration by parts and the sign change indicates that the terms lead to the load stiffness expression on the left hand side of the linearized incremental equilibrium equations.

$$
\begin{aligned}
{}^2_{-}\delta W_{ext,L} = -\frac{1}{2} e_{ijk}\, {}^2\lambda \Big\{ \iint_{rs} {}^1f \Big[& \frac{\partial\, {}^1x_j}{\partial r} \Big(u_i \cdot \frac{\partial \delta u_k}{\partial s} + \delta u_i \frac{\partial u_k}{\partial s} \Big) \\
& + \frac{\partial\, {}^1x_k}{\partial s} \Big(u_i \cdot \frac{\partial \delta u_j}{\partial r} + \delta u_i \frac{\partial u_j}{\partial r} \Big) \Big] dr \cdot ds \qquad \text{(I)} \\
& + 2 \iint_{rs} {}^1_1f_{,n} \frac{\partial\, {}^1x_j}{\partial r} \cdot \frac{\partial\, {}^1x_k}{\partial s} \cdot u_n \cdot \delta u_i \cdot dr \cdot ds \\
& + \iint_{rs} {}^1_1f_{,n} \Big(\frac{\partial\, {}^1x_n}{\partial r} \cdot \frac{\partial\, {}^1x_j}{\partial s} - \frac{\partial\, {}^1x_n}{\partial s} \cdot \frac{\partial\, {}^1x_j}{\partial r} \Big) u_k \cdot \delta u_i \cdot dr \cdot ds \qquad \text{(II)} \\
& + \int_b {}^1f\, u_j \frac{\partial\, {}^1x_k}{\partial s} \delta u_i \cdot ds \qquad \text{(III)} \\
& - \int_b {}^1f\, u_j \frac{\partial\, {}^1x_k}{\partial r} \delta u_i \, dr \Big\} \qquad \text{(IV)}
\end{aligned}
\tag{13}
$$

The second domain term (II) is slightly reordered

$$
\begin{aligned}
{}^2_{-}\delta W^{(II)}_{ext,L} = & -\frac{1}{2} e_{njk}\, {}^2\lambda \iint_{rs} {}^1_1f_{,n} \Big(\frac{\partial\, {}^1x_n}{\partial r} \cdot \frac{\partial\, {}^1x_j}{\partial s} - \frac{\partial\, {}^1x_n}{\partial s} \cdot \frac{\partial\, {}^1x_j}{\partial r} \Big) \\
& \qquad (u_k \cdot \delta u_n + u_n \cdot \delta u_k)\, dr \cdot ds \\
& - e_{njk}\, {}^2\lambda \iint_{rs} {}^1_1f_{,n} \frac{\partial\, {}^1x_j}{\partial r} \cdot \frac{\partial\, {}^1x_k}{\partial s}\, u_n \cdot \delta u_n \cdot dr \cdot ds
\end{aligned}
\tag{14}
$$

Interchanging the indices it can be recognized that both domain parts (I) and (II) are symmetric, and the boundary terms (III) and (IV) are skew-symmetric. In [19], [20] the above equations are cast into a matrix notation using operator matrices so that the structure of the equations is even more obvious. It can be shown that the boundary terms disappear if one of the following symmetry conditions hold.

1. The loading magnitude ${}^1f$ vanishes on the entire boundary in all deformed configurations.

2. At least two displacement components on the boundary are prescribed.
3. The shell is always restrained perpendicular to the deformed surface (fixed boundary) or to any continuous surface, e.g. the deformed surface.

The question whether arbitrary space attached loads are conservative or not depends only on the boundary conditions.

4.2 Body attached loads

Here the load function is defined through coordinates ${}^0x_n$ of the initial configuration (l=0): ${}^0f = {}^0f\,({}^0x_n)$. Now equation (10) is

$$
{}^2\delta W_{ext} = e_{ijk}\,{}^2\lambda \iint\limits_{rs} {}^0f\, \frac{\partial\, {}^1x_j}{\partial r}\, \frac{\partial\, {}^1x_k}{\partial s}\, \delta u_i\, dr\cdot ds
$$

$$
+ e_{ijk}\,{}^2\lambda \iint\limits_{rs} {}^0f \left(\frac{\partial u_j}{\partial r}\, \frac{\partial\, {}^1x_k}{\partial s} + \frac{\partial\, {}^1x_j}{\partial r}\, \frac{\partial u_k}{\partial s} \right) \delta u_i\cdot dr\cdot ds \qquad (15)
$$

Again the first term defines the usual load vector. A series expansion is not required but as above integration by parts of the second integral - the load stiffness term - leads to $({}^0_0f_{,n} \equiv \partial\, ({}^0f)\, /\, \partial\, {}^0x_n)$

$$
{}^2_-\delta W_{ext,L} = -\frac{1}{2}\, e_{ijk}\,{}^2\lambda \Bigg\{ \iint\limits_{rs} {}^0f \Bigg[\frac{\partial\, {}^1x_j}{\partial r} \left(u_i\, \frac{\partial \delta u_k}{\partial s} + \delta u_i\, \frac{\partial u_k}{\partial s} \right) \qquad \text{(I)}
$$

$$
+ \frac{\partial\, {}^1x_k}{\partial s} \left(u_i\, \frac{\partial \delta u_j}{\partial r} + \delta u_i\, \frac{\partial u_j}{\partial r} \right) \Bigg]\, dr\, ds
$$

$$
+ \iint\limits_{rs} {}^0_0f_{,n} \left(\frac{\partial\, {}^0x_n}{\partial r}\, \frac{\partial\, {}^1x_j}{\partial s} - \frac{\partial\, {}^0x_n}{\partial s}\, \frac{\partial\, {}^1x_j}{\partial r} \right) u_k\, \delta u_i\, dr\, ds \qquad \text{(II)}
$$

$$
+ \int\limits_b {}^0f\, u_j\, \frac{\partial\, {}^1x_k}{\partial s}\, \delta u_i \cdot ds \qquad + \int\limits_b {}^0f\, \frac{\partial\, {}^1x_j}{\partial r}\, u_k\, \delta u_i\, dr \Bigg\}
$$

$$
\text{(III)} \qquad\qquad \text{(IV)} \qquad\qquad (16)
$$

The first domain part (I) is similar to that of equation (13) except ^1f is replaced by ^0f and is symmetric. But the second domain term (II) is skew-symmetric as well as the two boundary terms so both domain and boundary parts influence the question whether the problem is conservative. Term (II) vanishes if ^0f is constant ($^0_0f,n = 0$), i.e. for uniform pressure. In this case both kinds of loading - space or body attached loads - coincide. But the first condition at the boundary given above is violated so that the others must be satisfied in order to get a symmetric expression, i.e. a conservative problem.

4.3 Discussion

Summarizing the results given above (figure 5) it can be stated:

origin of load stiffness terms			pressure load	
			body attached	field in space
domain	load on surface	nonuniform	N	S
		uniform	S identical	S
boundary	arbitrary b.c.		N	N
	specific b.c.		S	S

Figure 5: Symmetry (S) and Non-Symmetry (N) of Structural Load Stiffness Matrix

If the boundary terms disappear non-uniform body attached loads lead to non-symmetrical matrices for the domain and therefore are nonconservative. Then space attached loads have always symmetric domain terms and therefore are always conservative. But the boundary terms vanish only for certain boundary conditions which have to be satisfied during the entire deformation process.

The subject of the existence of a potential for displacement dependent loads in nonlinear analysis has been investigated by several "classical" papers, e.g. [21], [22], [23]. In a finite element formulation the discretization process has to be introduced into the above equations leading to load stiffness matrices. For the degenerated shell element the matrices have been derived in [19], [20]. Note that these matrices on the element level are always unsymmetric because of the boundary terms for either kind of loading. During the assembly procedure the non-symmetric boundary terms of two neighboured elements cancel each other provided the load function is continuous. Thus the discussion of symmetry is related to the remaining terms at the boundary of the structure and the domain terms (for non-uniform body attached loads). Or in other words the linearized structural tangent stiffness matrix - or to be more specific - the load stiffness matrix gives the answer whether the load is conservative or not.

In the finite element literature the topic is often discussed in a controversial way. For example nonsymmetrical matrices are derived for problems where a potential must exist [24],[25]. This may be due to the fact that pressure loads defined as a field in space are assumed to be body attached using a simplified mathematical definition. So unsymmetries come in because of a false load definition. It should be mentioned that most pressure loads in reality are space attached fields, e.g. hyrostatic pressure.

It is well known that nonconservative systems may fail by flutter instead of divergence. This requires the application of a kinetic stability criterion using the complete unsymmetric stiffness matrix in the corresponding eigenvalue analyses. This means a lot of numerical effort which can be avoided in nonlinear analyses based on incremental and iterative solution procedures. Then the unsymmetrical terms or even the entire load stiffness matrix can be omitted and the effect of displacement dependent loads is gradually included through iteration via the right hand side. But note that then flutter instabilities are not indicated. Various proposals of using symmetrized matrices are discussed in [19],[20],[26],[27] where also further references to the present subject are given.

Summarizing this part it is obligatory to use a poper load definition. What the influence of the deformation dependence of loads on the structural response is concerned further numerical studies are nescessary. From cylindrical shells it is known that the influence is decreasing when the deformation and failure mode has an increasing wave number (e.g. thin and short cylinders). In other words for many real structures the influence is minor. It seems that this tendency may be generalized.

5.Solution procedures

Additional solution schemes used in this study are only briefly summarized.

Material Nonlinearities: An integrating model based on a layered approach is applied. Simpson integration rule is used in thickness direction. Classical metal plasticity is incorporated (v. Mises yield condition, Prandtl - Reuss flow rule, isotropic hardening) [17] . Hypoelastic material model (modified Darwin/ Pecknold model) is implemented for reinforced concrete shells [28]. Within a load step subincrement technique is introduced to follow the yield progress.

Solution strategies: The nonlinear response is analysed by an incremental solution procedure using the linearized stiffness expression, equation (4). The iteration scheme within the load step is based on the modified or standard Newton - Raphson method in combination with a constant - arc - length method (modified Riks - Wempner) or displacement control approach, i.e. the iteration takes place in the displacement and load space allowing also to trace the response beyond the limit point[29].

Stability analyses: Stability analyses can be performed in the sense of either initial linear stability investigations or a supplementary eigenvalue analysis started in a deformed equilibrium configuration [30]. The latter procedure allows to judge the current status of the shell on the nonlinear prebuckling path. In addition the determinant of the stiffness matrix according to the static stability criterion may be monitored. Eigenvalue analyses are based on the subspace iteration scheme.

Program: The elements and solution procedures are incorporated in the nonlinear computer program NISA80 [31]. The code has a multi - purpose function and includes other kinds of elements as well. It has an out - of - core solver and various plot facilities. The following examples are run on CDC 6600 or Cyber 174 computers.

6. Numerical Examples

6.1 Axially loaded cylindrical shells [32]

This example shows the influence of a circumferential weld seam on the stability of an axially loaded cylindrical shell. The structure given in figure 6 represents a typical silo shell with a radius to thickness ratio of 500. Classical boundary conditions (SS3) are assumed and one sector representing 13 waves in the circumferential direction has been idealized by 4 × 21 cubic elements (S16). At the weld symmetry conditions are assumed. Near the weld geometric deviations of the ideal shape are defined in figure 7 with a maximum amplitude of half of the wall thickness. In addition to the axisymmetric imperfection residual stresses given also in figure 7 are assumed. Figure 8a shows a load - deflection diagram for the imperfect shell without residual stresses. The structure bifurcates into a non axisymmetric mode before yielding. The elastic stability load is 42% of the classical buckling load. If residual stresses are assumed (figure 8b) yielding starts earlier. Nevertheless this has a minor influence on the failure load. Figure 9 shows the displacement pattern of the sector before and after buckling.

6.2 Nonlinear collapse analysis of stringer stiffened cylindrical shells under axial load[4]

The analysis of this example follows an experimental study of axially compressed stringer-stiffened cylindrical shells investigated in [34] . The geometrical and material properties given in figure 10 correspond to a typical offshore component with a relatively small number of stringers compared to aeronautical structures. The example analysed was the specimen UC 8 in [34]. The shell has a radius of 291.6 mm, a wall thickness of 0.81 mm and a height of 323.7 mm. In the analysis only a panel with one stiffener and symmetry boundary conditions along the longitudinal edges was idealized, corresponding to the buckling mode

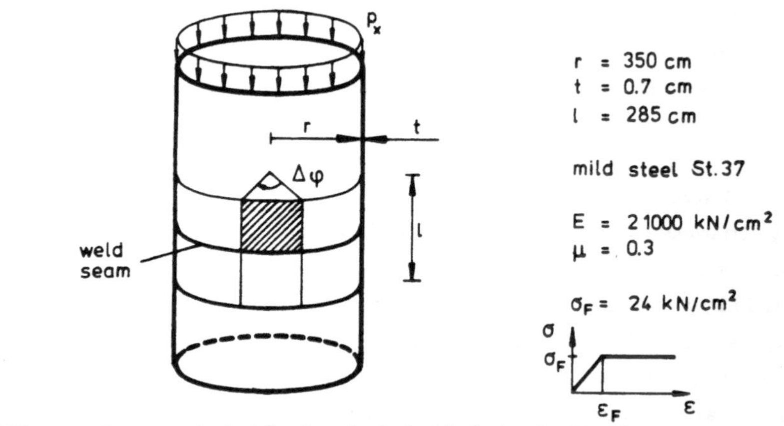

Figure 6: Axially Loaded Cylindrical Shell

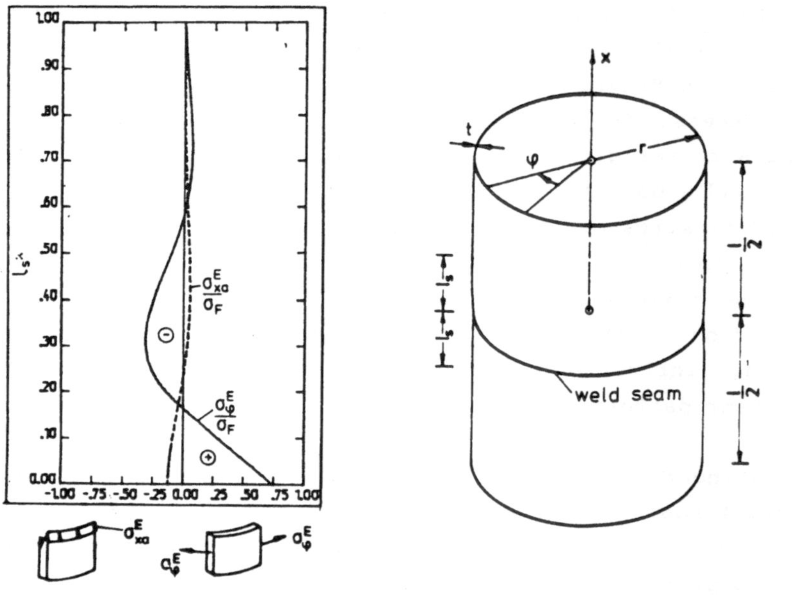

	i	$w_0 = w_0(x)$		l_i
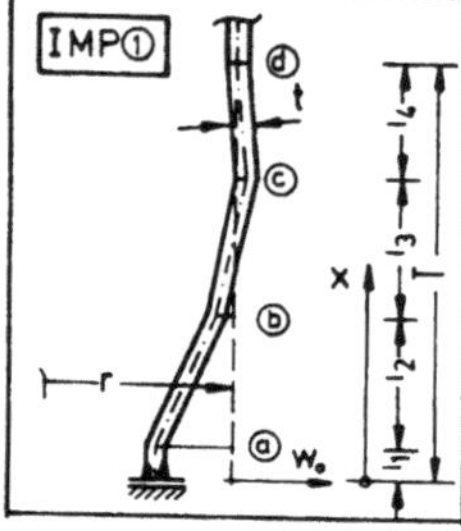	1	$w_0 = w_0^{(a)}$	$0 \le x \le l_1$	4,28t
	2	$w_0 = \frac{w_0^{(b)} - w_0^{(a)}}{l_2}[x - l_1] + w_0^{(a)}$	$l_1 \le x \le (l_1 + l_2)$	15,72t
	3	$w_0 = \frac{w_0^{(c)} - w_0^{(b)}}{l_3}[x - (l_1 + l_2)] + w_0^{(b)}$	$(l_1 + l_2) \le x \le (l_1 + l_2 + l_3)$	15,71t
	4	$w_0 = -\frac{w_0^{(c)}}{l_4}[x - (l_1 + l_2 + l_3)] + w_0^{(c)}$	$(l_1 + l_2 + l_3) \le x \le \bar{l}$	14,29t

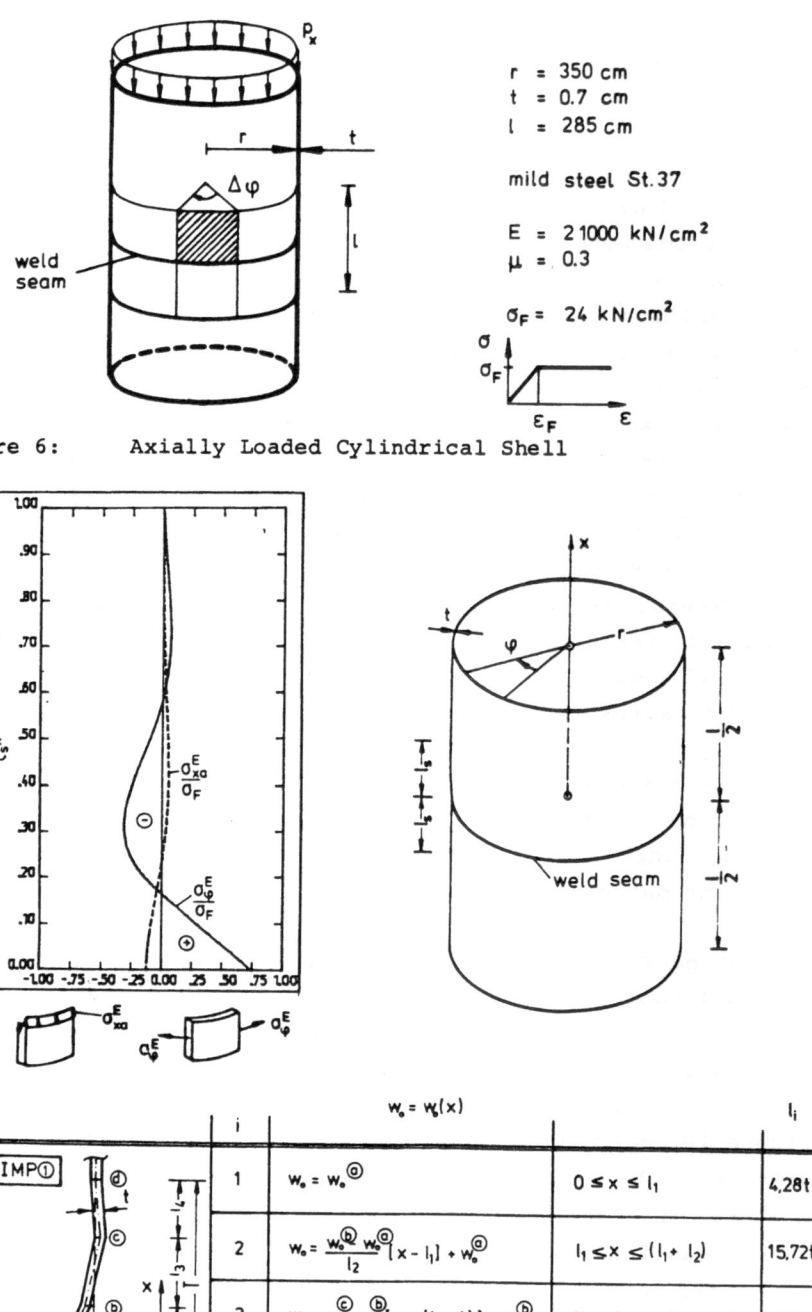

Figure 7: Definition of Residual Stresses ($l_s = 30 \cdot t$) and Geometrical Imperfection ($\bar{l} = 50 \cdot t$)

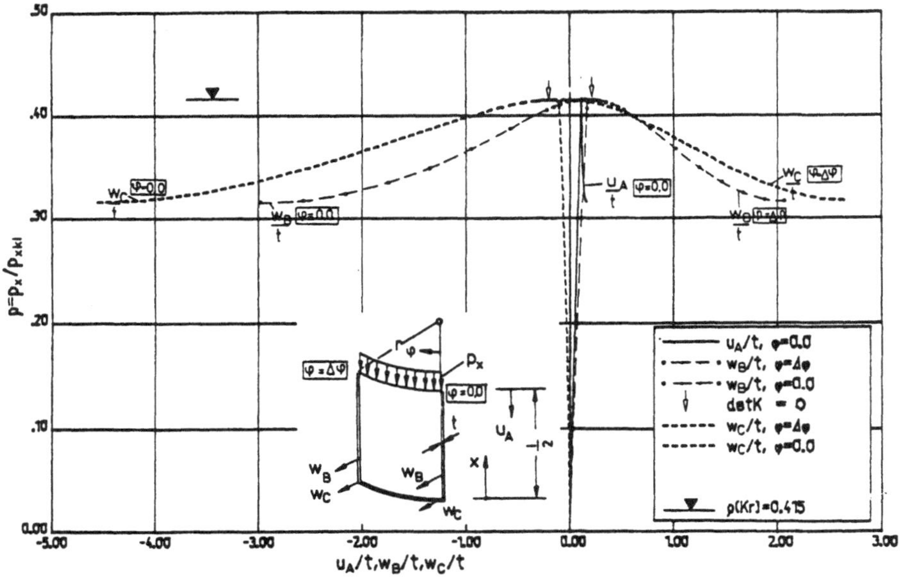

a) Without Residual Stresses

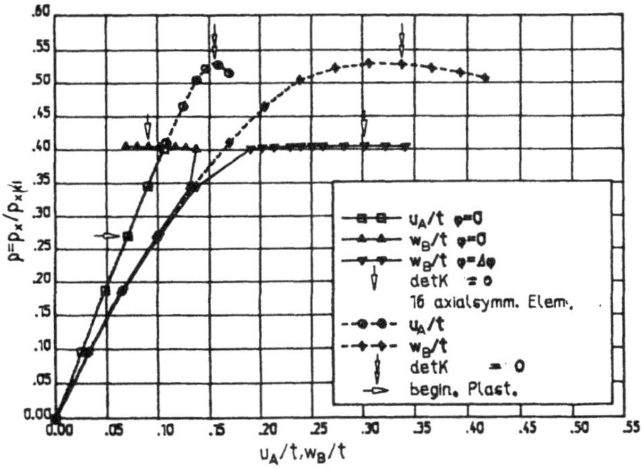

b) With Residual Stresses

Figure 8: Load - Deflection Diagrams of Axially Loaded Shell

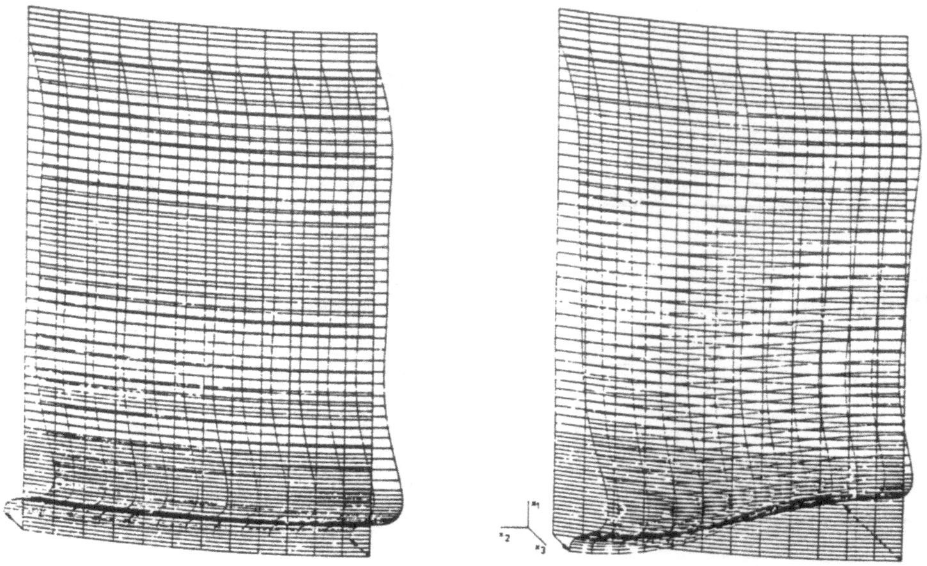

a) before buckling   b) after buckling

Figure 9: Displacement Pattern

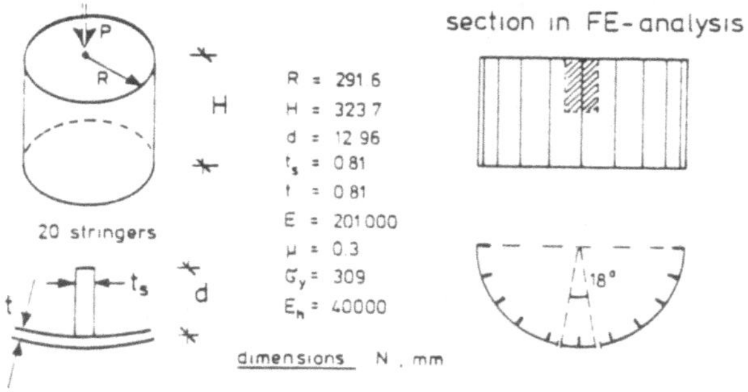

Figure 10:
Stringer Stiffened Cylindrical Shell under Axial Load

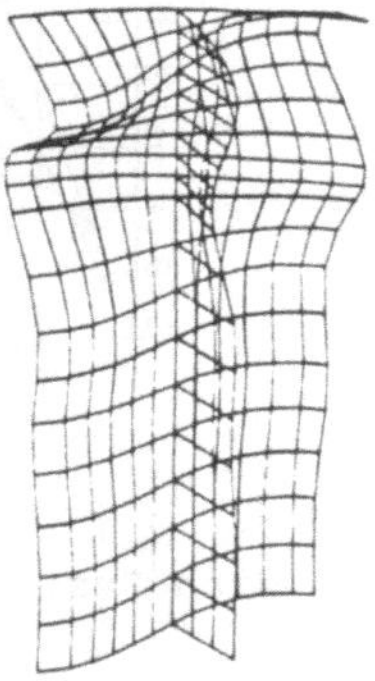

Figure 11:
Linear Buckling Mode

which is one half wave between the stiffeners in circumferential direction. The failure mechanism - local buckling of shell and stringer while the junction between both remains straight - is typical for a broad-panelled cylinder. For the imperfect shell the linear buckling mode, shown in figure 11 was superimposed on the perfect geometry with a maximum initial imperfection amplitude equal to the wall thickness. The loads were applied by uniform axial shortening suppressing any warping of the loaded edge. It should be noted here that in the case of constant edge loads there is a remarkable influence on the results. Two different material models were used in the analysis. First the computation was carried out with a linearly elastic ideally plastic material model without strain-hardening. In the second run strain-hardening was introduced to the material law. The results of analyses compared to those of the experiment are shown in figure 12. The average axial stress $\sigma = P/A$ is normalized to the yield stress σ_y and the end shortening ε is normalized to ε_y which corresponds to the linear elastic axial shortening due to the load $P = \sigma_y * A$. The difference in the maximum load carrying capacity may be explained by two facts: The analysis did not consider residual stresses and no local imperfections in addition to the global mode described above were applied. Both effects reduce the ultimate load of the shell.

6.3 Closed, wind loaded cylindrical shell [33]

The elastic buckling analysis of the closed cylindrical shell under wind load studied in [30] has been extended to the post-buckling range. The simply supported structure (figure 13) is extremely thin with a radius to thickness ratio of 2095. The wind load defined in figure 13 for the circumference is not varied in axial direction. It is assumed to be constant directional. The maximal load p at the stagnation point is normalized to the linear buckling load of the shell under uniform pressure

$$p_{cl} = \frac{0.918 \cdot E \left(\frac{h}{R}\right)^2}{\frac{L}{R}\sqrt{\frac{R}{h}} - 0.657}$$

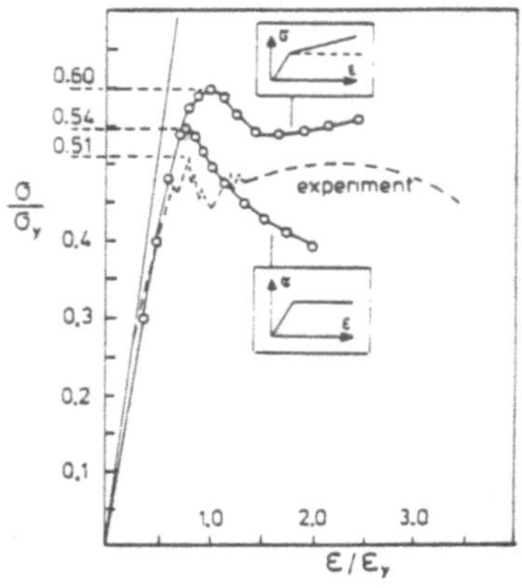

Figure 12: Load - Deflection Diagram of Stringer Stiffened Cylinder

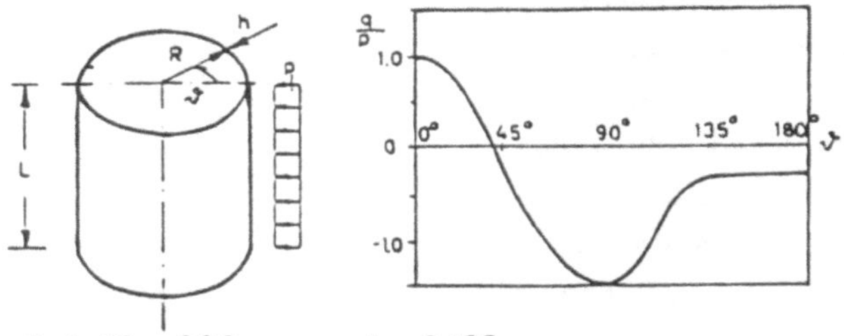

$R = L/2 = 220$ mm , $h = 0.105$ mm
$E = 6.87 \cdot 10^4$ N/mm², $\mu = 0.3$

Figure 13: Geometry and Load Function of Closed Cylindrical Shell under Wind Load

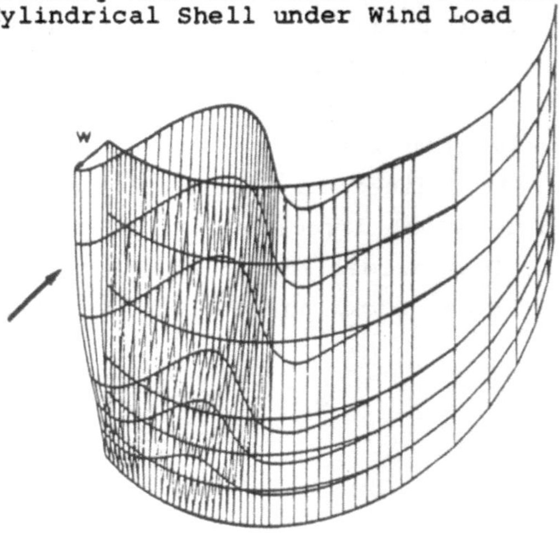

Figure 14: Displacement Pattern

One quarter of the shell is idealized by 2x18 cubic elements S 16. Two elements of unequal length are arranged in the axial direction. 18 elements are used in the circumferential direction assuming a mesh refinement in the stagnation zone. The perfect and two imperfect shells were analysed. Figure 14 shows the displacement pattern of one quarter of the perfect shell near the limit point. The failure mode has one half wave in the axial direction a few circumferential waves in the compression zone. The postbuckling minimum of the load -deflection diagram (Figure 15) is about 60% of the limit load . The imperfections assumed for the remaining analyses correspond to the failure mode of the perfect shell. The maximum amplitude η of the imperfection is 2.5 and 5.0 times the wall thickness. The load deflection path for $\eta = 2.5$ indicates a reduction of the limit load to 68% of that for the perfect shell. The postbuckling minima nearly coincide. For $\eta = 5.0$ no limit point is indicated The imperfection sensitivity of this structure corresponds to the knock-down factors obtained from buckling experiments. It should be noted that the example is numerically very sensitive because of the extreme slenderness ratio and the local nature of the failure mode. The analysis was repeated using the bilinear S 4 element with one point integration with the same number of d. o. f. Nearly identical results were obtained for much less computer time.

6.4 Open cantilever shell under wind load [19]

The open cantilever shell shown in figure 16 together with the wind load definition has been analysed for two kinds of loads: constant directional and body attached load (nonconservative). Because the shell is open an additional suction (first term in a_0) is included. One half of the shell is idealized by 32 cubic elements, 4 in longitudinal direction and 8 in circumferential direction with a mesh refinement in the compression zone. Only elastic analyses are performed. Figure 18 shows a load - deflection diagram for both kinds of loading, an experimental value [35] is added. The differences are minor. Figure 17 shows a displacement plot of the shell. In [19] additional details are given. Among these are linear buckling analyses using different kinds of symmetrization of the stiffness matrix. It is shown

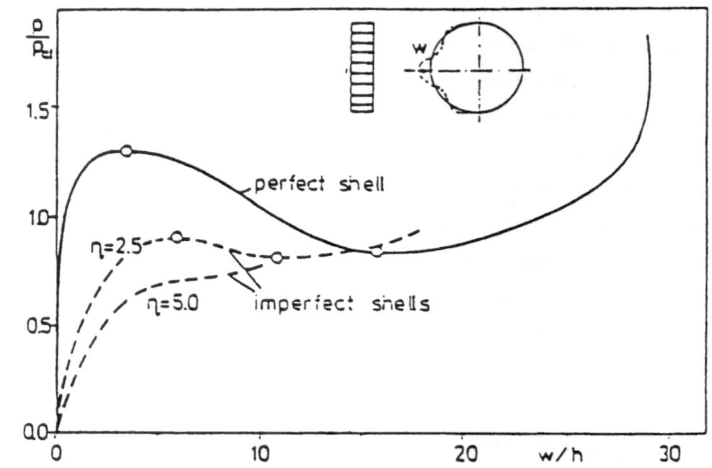

Figure 15: Load - Deflection Diagram of Closed Wind - Loaded Shell

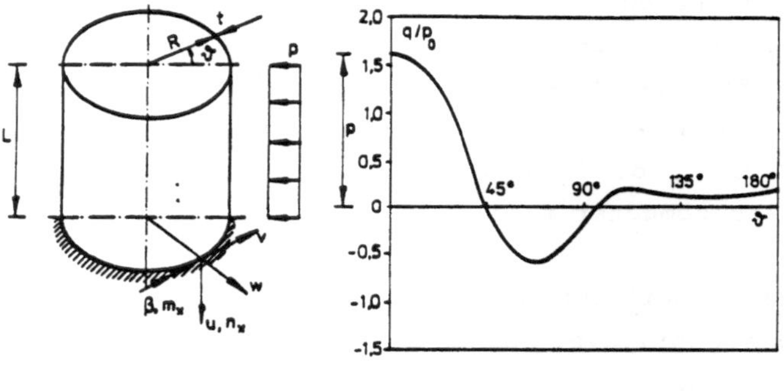

boundary conditions

↑ $m_x = n_x = q_x = n_{x\vartheta} = 0$

↓ $\beta = u = w = v = 0$

geometry

R = 40 in , L = 120 in , t = 0,1064 in

$E = 3 \cdot 10^7$ psi , $\gamma = 0{,}3$

Figure 16: Geometry and Load Function of Open Cantilever Shell under Wind Load

$$q = p_0 \cdot \sum_{n=0}^{6} a_n \cdot \cos n \cdot \vartheta$$

$a_0 = 0.607 - 0.387$ $a_3 = 0.338$ $a_6 = 0.166$

$a_1 = 0.533$ $a_4 = 0.471$

$a_2 = -0.066$ $a_5 = -0.055$

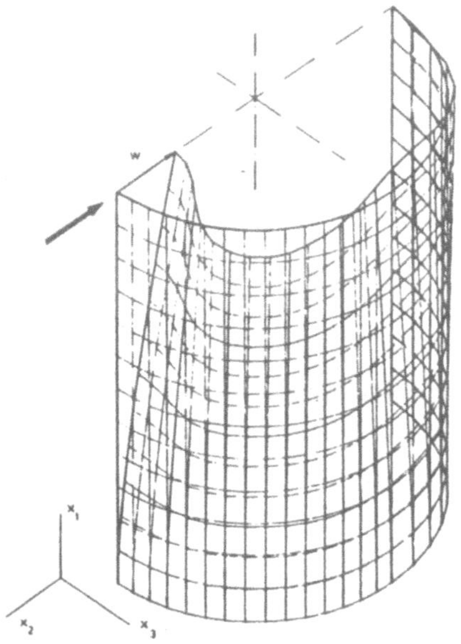

Figure 17: Displacement Pattern

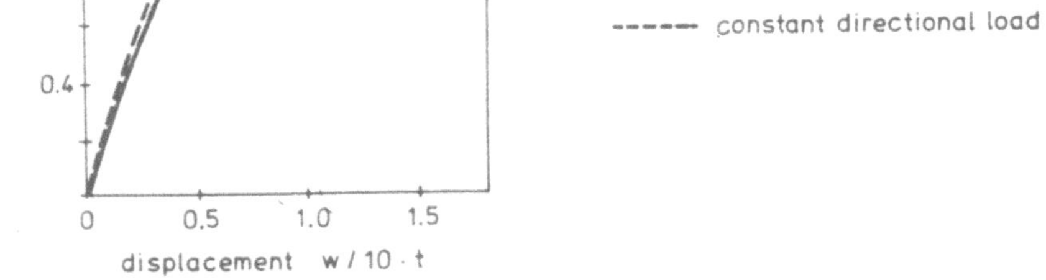

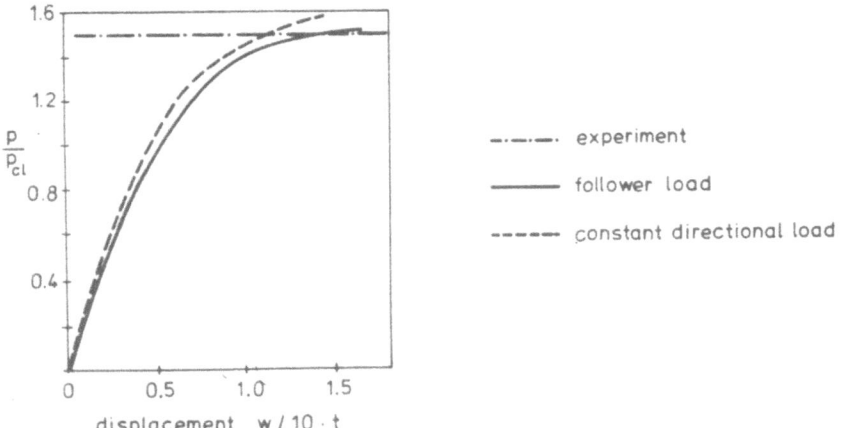

Figure 18: Load - Deflection Diagram of Stagnation Point

that the influence is neglegible.

7. Conclusions

This study on the application of degenerated elements in nonlinear shell analysis allows the following conclusions:

* Even extremely thin shells may be analysed by degenerated elements.
* The concept allows to incorporate arbitrarily large displacements and rotations as well as material nonlinearities via a layered model.
* For displacement dependent pressure loads a clear distinction in body and space attached load fields is nescessary.
* On the basis of this classification domain and boundary pressure load terms can be derived leading to load stiffness matrices.
* The structure of the structural load stiffness matrix (symmetric or not) allows to identify when the problem is conservative or when it is not.

Further numerical studies are nescessary to judge the influence of the deformation dependence of loads. In particular this is nescessary when flutter instabilities may occur.

References

[1] Ahmad, S. — Curved Finite Elements in the Analysis of Shell and Plate Structures. PhD. thesis, University of Wales, Swansea, 1969.

[2] Zienkiewicz, O.C, Too, J., Taylor, R.L. — Reduced Integration Technique in General Analysis of Plates and Shells. Int. J. Num. Meth. Eng. 3(1979) 275/290.

[3] Hughes, T.J.R., Cohen, M., Haroun, M. — Reduced and Selective Integration Techniques in the Finite Element Analysis of Plates. Nucl. Eng. Design. 46(1978) 203/222.

[4] Ramm, E., Stegmüller, H. — The Displacement Finite Element Method in Nonlinear Buckling Analysis of Shells. in "Buckling of shells" (Ed. E.Ramm) Springer, 1982

[5] Parisch, H. — A Critical Survey of the Nine-Node Degenerated Shell Element with Special Emphasis on Thin Shell Application and Reduced Integration. Comp. Meth. Appl. Mech. Eng. 20(1979) 323/350

[6] Hughes, T.J.R., Liu, W.V. — Nonlinear Finite Element Analysis of Shells. Part I: Three-Dimensional Shells. Comp. Meth. Appl. Mech. Eng. 26 (1981) 331/362

[7] Belytschko, T., Tsay, C.S. — Explicit Algorithms for Nonlinear Dynamics of Shells. in "Nonlinear Finite Element Analysis of Plates and Shells" Proc. ASME-WAM, Washington 1981, AMD-Vol.48, 209/232

[8] Brockman, R.A. A Penalty Function Approach for the Nonlinear Finite Element Analysis of Thin Shells
Dissertation, University of Dayton
Dayton, Ohio 1979

[9] Javaherian, H.
Dowling, P.J.
Lyons, L.P.R.
Nonlinear Finite Element Analysis of Shell Structures using the SEMILOOF Element.
Comp. & Struc. 12(1980) 147/159

[10] Martins, R.A.F.
Owen, D.R.J.
Elastic PLastic and Geometrically Nonlinear Thin Shell Analysis by the SEMILOOF Element
Comp. & Struc. 13(1981) 505/513

[11] Larsen,P.K. Large Displacement Analysis of Shells of Revolution, Including Creep, Plasticity and Viscoelasticity.
SESM-Report No. 71-22, Dep. of Civil Engineering, University of California, Berkeley, 1971

[12] Ramm, E. Geometrisch nichtlineare Elastostatik und finite Elemente.
Habilitation thesis, Universität Stuttgart 1975

[13] Ramm, E. A Plate/Shell Element for Large Deflections and Rotations.
in Proc. US-Germany Symp. on "Formulations and Computational Algorithms in Finite Element Analysis" MIT,1976, MIT-Press,1977

[14] Parisch, H. Geometrical Nonlinear Analysis of Shells.
Comp. Meth. Appl. Mech. Eng. 14(1978) 159/178

[15] Bathe, K.J.
Bolourchi,S.
A Geometric and Material Nonlinear Plate and Shell Element.
Comp. & Struc. 11(1980), 23-48

[16] Krakeland, B. Large Displacement Analysis of Shells Considering Elastic-Plastic and Elastic-Viscoelastic Materials.
Report 776, Norwegian Institute of Technology; The University of Trondheim, Norway, Dec.1977

[17] Sättele,J.M. Ein finites Elementkonzept zur Berechnung von Platten und Schalen bei stofflicher und geometrischer Nichtlinearität.
Dissertation, Universität Stuttgart, 1980

[18] Bathe, K.-J.
Ho, L.W. Some Results in the Analysis of Thin Shell Structures
in "Nonlinear Finite Element Analysis in Structural Mechanics" (eds. Wunderlich, W. Stein, E., Bathe, K.J.) Springer, 1981

[19] Schweizerhof,K. Nichtlineare Berechnung von Tragwerken unter verformungsabhängiger Belastung mit finiten Elementen.
Dissertation, Universität Stuttgart, 1982

[20] Schweizerhof,K.
Ramm, E. Displacement Dependent Pressure Loads in Nonlinear Finite Element Analyses.
Comp. & Struct., 1984

[21] Pearson, G.E. General Theory of Elastic Stability.
Quart. Appl. Math. 14 (1956), 133-144.

[22] Sewell, M.J. On Configuration - Dependent Loading.
Arch. Rational Mech. Anal. 23 (1967), 327-351.

[23] Romano, G. Potential Operators and Conservative Systems.
Meccchanica 7 (1972), 141-146.

[24] Argyris, J.H. Symeonitis, S. — Nonlinear Finite Element Analysis of Elastic Systems under Nonconservative Loading - Natural Formulation. Part I. Quasistatic Problems Comp. Meth. Appl. Mech. Eng. 26 (1981) 75-123.

[25] Frey, F. Cescotto, S. — Some New Aspects of the Incremental Total Lagrangian Description in Nonlinear Analysis. Int. Conf. "Finite Elements in Nonlinear Mechanics" Vol. 1, Geilo, Norway (1977) Tapir, 1977.

[26] Mang, H.A. — Symmetricability of Pressure Stiffness Matrices for Shells with Loaded Free Edges Int. J. Num. Meth. Eng. 15 (1980) 981-990

[27] Mang, H.A. Gallagher, R.H. — On the Unsymmetric Eigenproblem for the the Buckling of Shells under Pressure Loading. Trans. ASME, J. Appl. Mech. 50 (1983), 95 -100

[28] Kompfner, T.A. — Ein finites Elementmodell für die geometrisch und physikalisch nichtlineare Berechnung von Stahlbetonschalen. Dissertation (submitted), Universität Stuttgart 1983

[29] Ramm, E. — Strategies for Tracing the Nonlinear Response Near Limit Points. Proc. Europe - U.S. workshop on "Nonlinear Finite Element Analysis in Structural Mechanics" Bochum 80, Springer - Verlag 1981

[30] Brendel, B. Ramm, E. — Linear and Nonlinear Stability Analysis of Cylindrical Shells. Comp. & Struc. 12 (1980) 549-558.

[31] Häfner, L.
Ramm, E.
Sättele, J.M.
Stegmüller, H.
Programmdokumentation - Programmsystem NISA80.
report, Institut für Baustatik, Universität Stuttgart, 1980.

[32] Häfner, L.
Einfluß einer Rundschweißnaht auf die Stabilität und Traglast des axialbelasteten Kreiszylinders.
Dissertation, Universität Stuttgart 1982

[33] Ramm, E.
Sättele, J.M.
Elasto - Plastic Large Deformation Shell Analysis using Degenerated Elements.
in "Nonlinear Finite Element Analysis of Shells", ASME - WAM, Washington 1981, AMD-Vol. 48, 265-282

[34] Walker, A.C.
Sridharan, S.
Analysis of the Behaviour of Axially Compressed Stringer - Stiffened Shells.
Proc. Inst. Civ. Engrs., Part 2, 69 (1980) 447-472

[35] Prabhu, S.K.
Gopalacharyulu, S.
Johns, D.J.
Stability of Cantilever Shells under Wind Loads.
Proc. ASCE, J. Eng. Mech. Div. 101 (1975) 517/530

Contribution on the Numerical Analysis of Thin Shell Problems

M. BERNADOU

INRIA - Domaine de Voluceau
B.P. 105 - Rocquencourt
78153 Le Chesnay - France

Summary

This contribution reviews our results on the numerical analysis of thin shell problems. Essentially, it includes theorems of existence (and uniqueness) for a solution in the linear and non linear static case as well as the study of the free vibration problem ; next some results from the mathematical study of the approximation are given ; the implementation of these methods is illustrated by numerical simulation of arch dam problems. Finally some open problems are pointed out.

1. Introduction

There are numerous general models of thin shells, specially that of Koiter [15] which is the last improvement of the "classical" theories and that of Naghdi [19] which originates from the surface theory of Cosserat [9]. This paper is essentially based upon our works on numerical analysis of thin shell problems according to Koiter's equations. In the second paragraph we study the existence of solutions for the following continuous thin shell problems :

i) the displacement formulation of the general linear equations ;
ii) the corresponding free vibration equations ;
iii) the displacement formulation of the general nonlinear equations for shallow shells.

Next, in the third paragraph we examine the approximation of the linear and free vibration problems by conforming finite element methods in conjunction with numerical integration techniques. We record the error estimate results as well as sufficient conditions on the numerical integration schemes according to Bernadou [2]. In a fourth paragraph, we consider the application of previous results to the computation of an arch dam. Corresponding numerical experiments are not expensive and in agreement with experimental results for similar arch dams. In the fith paragraph we conclude by giving some open problems.

2. The continuous problems

2.1. Notations

Let Ω be a bounded open subset in a plane $\mathscr{E}^2$, with boundary Γ. The middle surface S of the shell is the image of the set $\bar{\Omega}$ by the mapping $\vec{\phi} : \bar{\Omega} \subset \mathscr{E}^2 \to \mathscr{E}^3$, where $\mathscr{E}^3$ is the usual Euclidean space. Subsequently, we shall assume that $\vec{\phi} \in (\mathscr{C}^3(\bar{\Omega}))^3$ and that all points of $S = \vec{\phi}(\bar{\Omega})$ are regular, in the sense that the vectors $\vec{a}_\alpha = \vec{\phi}_{,\alpha}$, $\alpha = 1,2$, are linearly independent for all point $\xi = (\xi^1,\xi^2) \in \bar{\Omega}$. With the covariant basis $(\vec{a}_\alpha)$ of the tangent plane, we associate the contravariant basis $(\vec{a}^\alpha)$ through the relations $\vec{a}^\alpha \cdot \vec{a}_\beta = \delta^\alpha_\beta$. The normal vector is $\vec{a}_3 = \vec{a}^3 = \vec{a}_1 \times \vec{a}_2 \,/\, |\vec{a}_1 \times \vec{a}_2|$.

The unknowns are the components $u_i(\xi)$ of the displacement $\vec{u} = \vec{u}(\xi)$ of the point $\vec{\phi}(\xi)$, i.e.,

$$\vec{u} = u_i \vec{a}^i \tag{2.1}$$

In [15], Koiter shows that the evaluation of the strain tensor of the shell can be approximated by using the middle surface strain tensor $\gamma_{\alpha\beta}(\vec{u})$, and the tensor of changes of curvature $\bar{\rho}_{\alpha\beta}(\vec{u})$.

We assume that (*) the material of the shell is elastic, homogeneous and isotropic ; (**) the strains are small everywhere in the shell ; (***) the shell is in a state of stress in which all nonzero stress components are developed on surfaces parallel to the middle surface. Then, the symmetric tensors of tangential (membrane) stress resultants $n^{\alpha\beta}(\vec{u})$ and stress couples $m^{\alpha\beta}(\vec{u})$, are given by

$$n^{\alpha\beta}(\vec{u}) = eE^{\alpha\beta\lambda\mu}\gamma_{\lambda\mu}(\vec{u}) \quad , \quad m^{\alpha\beta}(\vec{u}) = \frac{e^3}{12}E^{\alpha\beta\lambda\mu}\bar{\rho}_{\lambda\mu}(\vec{u}) \tag{2.2}$$

where e denotes the thickness of the shell and $E^{\alpha\beta\lambda\mu}$ denotes the tensor of elastic moduli for plane stress at the middle surface.

In the following, we consider for simplicity the case of a clamped shell. Let $\vec{p} = p^i\vec{a}_i$ be the resultant of external applied forces per unit surface area defined on the middle surface of the shell. Then, according to [15], the equations of equilibrium are given in section 2.2 for linear case and in section 2.4 for non linear case. The corresponding linear free vibration equations are given in section 2.3.

2.2. The system of linear equations

$$\left.\begin{aligned} & n^{\alpha\beta}(\vec{u})|_{\beta} + p^{\alpha} = 0 \,, & (2.3)\\ & m^{\alpha\beta}(\vec{u})|_{\alpha\beta} - b_{\alpha\beta}n^{\alpha\beta}(\vec{u}) - p^{3} = 0 \,, & (2.4) \end{aligned}\right\} \text{on } \Omega,$$

$$\vec{u} = \vec{0} \,, \quad \partial u_3/\partial n = 0 \quad \text{on } \partial\Omega, \qquad (2.5)$$

where $b_{\alpha\beta}$ denotes the second fundamental form of S and " | " denotes the covariant partial derivatives.

The variational formulation of the problem is : Find $\vec{u} \in \vec{V}$ such that

$$a(\vec{u},\vec{v}) = \int_{\Omega} \vec{p}\vec{v} \sqrt{a}d\xi^1 d\xi^2 \,, \quad \forall\vec{v} \in \vec{V} \,, \text{ where} \qquad (2.6)$$

$$a(\vec{u},\vec{v}) = \int_{\Omega} eE^{\alpha\beta\lambda\mu}[\gamma_{\alpha\beta}(\vec{u})\gamma_{\lambda\mu}(\vec{v}) + \frac{e^2}{12}\bar{\rho}_{\alpha\beta}(\vec{u})\bar{\rho}_{\lambda\mu}(\vec{v})]\sqrt{a}d\xi^1 d\xi^2 \,, \qquad (2.7)$$

$$\gamma_{\alpha\beta}(\vec{u}) = \frac{1}{2}(u_{\alpha|\beta} + u_{\beta|\alpha}) - b_{\alpha\beta}u_3 \,, \qquad (2.8)$$

$$\bar{\rho}_{\alpha\beta}(\vec{u}) = u_{3|\alpha\beta} - b^{\lambda}_{\alpha}b_{\lambda\beta}u_3 + b^{\lambda}_{\beta}|_{\alpha}u_{\lambda} + b^{\lambda}_{\beta}u_{\lambda|\alpha} + b^{\lambda}_{\alpha}u_{\lambda|\beta} \qquad (2.9)$$

$$\vec{V} = H^1_0(\Omega) \times H^1_0(\Omega) \times H^2_0(\Omega) \,, \qquad (2.10)$$

$H^k_0(\Omega) = \{v \in H^k(\Omega) \,, \; v_{|\partial\Omega} = 0\}$, $H^k(\Omega)$ = Sobolev spaces, k = 1 or 2.

Theorem 2.1 : The bilinear form (2.7) is $\vec{V}$-elliptic and the problem (2.6) has one and only one solution. The proof is given in [4]. Essentially, we show that

$$\Phi(\vec{v}) = \left\{ \sum_{\alpha,\beta=1}^{2} |\gamma_{\alpha\beta}(\vec{v})|^2_{L^2(\Omega)} + |\bar{\rho}_{\alpha\beta}(\vec{v})|^2_{L^2(\Omega)} \right\}^{1/2}$$

is an equivalent norm to the usual norm of space $\vec{V}$, and hence $a(.,.)$ is $\vec{V}$-elliptic. We conclude by using Lax-Milgram Lemma. □

For more general boundary conditions, see [4], and for extension to Naghdi's equations see Coutris [10].

2.3. The linear free vibration equations

We start from the three-dimensional elasticity equations applied to the shell $\mathscr{C}$ in the absence of loads, i.e.,

$$\left.\begin{aligned}&\int_{\mathscr{C}} \rho \frac{\partial^2 \vec{U}}{\partial t^2} \vec{V}\, d\mathscr{C} + \int_{\mathscr{C}} E^{*ijkl} \gamma^*_{ij}(\vec{U}) \gamma^*_{kl}(\vec{V})\, d\mathscr{C} = 0 \\ &\qquad\qquad\qquad \forall \vec{V} \in \vec{\mathscr{V}}\end{aligned}\right\} \quad (2.11)$$

where $\vec{U}$ = displacement field of the shell $\mathscr{C}$, ρ = mass density of the material, $\vec{\mathscr{V}} = \{\vec{V} \in (H^1(\mathscr{C}))^3 ,\ \vec{V}_{|\partial\mathscr{C}} = \vec{0}\}$, γ^*_{ij} = covariant component of spatial strain tensor and E^{*ijkl} = contravariant tensor of elastic moduli in $\mathscr{E}^3$.

By integrating through the thickness, we get the approximations :

$$\vec{U} = \vec{u} - \xi^3 (u_{3|\alpha} + b^\lambda_\alpha u_\lambda) \vec{a}^\alpha , \quad (2.12)$$

$$\int_{\mathscr{C}} E^{*ijkl} \gamma^*_{ij}(\vec{U}) \gamma^*_{kl}(\vec{V})\, d\mathscr{C} \simeq a(\vec{u},\vec{v}) . \quad (2.13)$$

Then free vibrations can be represented by

$$u_j(\xi^1,\xi^2,t) = \tilde{u}_j(\xi^1,\xi^2) \cos\omega t \quad , \ j = 1,2,3 \quad (2.14)$$

where ω is the frequency of the vibration. From equations (2.11) to (2.14), we finally obtain the free vibration equations (for simplicity, we denote $\vec{u}$ instead of $\tilde{\vec{u}}$ and $\lambda = \omega^2$) :

$$\left.\begin{aligned}&\text{Find couples } (\lambda,\vec{u}) \in \mathbb{R}^+ \times \vec{V} \text{ such that} \\ &a(\vec{u},\vec{v}) = \lambda \tilde{b}(\vec{u},\vec{v}) ,\quad \forall \vec{v} \in \vec{V} ,\end{aligned}\right\} \quad (2.15)$$

with

$$\tilde{b}(\vec{u},\vec{v}) = \int_\Omega \rho e\{[1 + \frac{e^2}{12}(b_1^1 b_2^2 - b_1^2 b_2^1)][a^{\alpha\beta}u_\alpha v_\beta + u_3 v_3]$$

$$+ \frac{e^2}{12} a^{\alpha\beta}[(u_{3|\alpha} + b_\alpha^\lambda u_\lambda)(v_{3|\beta} + b_\beta^\mu v_\mu) \tag{2.16}$$

$$+ (u_\alpha v_{3|\beta} + v_\beta u_{3|\alpha} + 2b_\alpha^\lambda u_\lambda v_\beta) b_\eta^\eta]\}\sqrt{a}\,d\xi^1 d\xi^2 \quad .$$

This problem enters in the abstract setting of Riesz-Nagy [22] :

Theorem 2.2 : The eigenvalues of (2.15) form an increasing sequence

$$0 < \lambda_1 \leq \lambda_2 \leq \ldots \leq \lambda_m \leq \ldots, \tag{2.17}$$

growing to $+\infty$ when the space $\vec{V}$ is of infinite dimension, each of these eigenvalues having a finite multiplicity. Moreover, the associated eigenvectors can be chosen such that

$$a(\vec{u}_j,\vec{v}) = \lambda_j \tilde{b}(\vec{u}_j,\vec{v}) \quad , \quad \forall\vec{v} \in \vec{V} \text{ and } \tilde{b}(\vec{u}_j,\vec{u}_i) = \delta_{ij} \tag{2.18}$$

□

2.4. The system of nonlinear shallow shell equations

Koiter [15,§11] proposes the following equations :

$$n^{\alpha\beta}(\vec{u})|_\beta + p^\alpha = 0 \ , \tag{2.19}$$

$$m^{\alpha\beta}(\vec{u})|_{\alpha\beta} - b_{\alpha\beta}\, n^{\alpha\beta}(\vec{u}) - (u_{3|\alpha} n^{\alpha\beta}(\vec{u}))|_\beta - p^3 = 0 \ , \tag{2.20}$$

with boundary conditions (2.5). Relations (2.2) are still available and, instead of relations (2.8)(2.9), we have

$$\left.\begin{aligned} \gamma_{\alpha\beta}(\vec{u}) &= \frac{1}{2}(u_{\alpha|\beta} + u_{\beta|\alpha}) - b_{\alpha\beta}u_3 + \frac{1}{2}u_{3,\alpha}u_{3,\beta} \ , \\ \bar{\rho}_{\alpha\beta}(\vec{u}) &= u_{3|\alpha\beta} \ . \end{aligned}\right\} \tag{2.21}$$

In [6], we have given a variational formulation stated in the space $\vec{V}$ and we have proved :

Theorem 2.3 : For a large class of nonlinear shallow shell equations, the associated variational formulation has at least one solution whenever tangential components p^α of the loads are sufficiently small.

The proof requires two steps. First, we fix u_3 and we solve a linear equation with respect to u_1 and u_2 by using Lax-Milgram lemma. Then, it remains a nonlinear equation with respect to u_3. We prove that the corresponding operator is pseudo-monotone and coercive. We conclude according to Lions [17, chapter 2, theorem 2.7]. □

In addition, we have proved that solutions are unique whenever the loads are sufficiently small. Also, note that Kubrusly [16] has used the results of [6] to study the existence of post-buckling solutions of shallow shells under a certain unilateral constraint. □

3. The approximated problem

3.1. The discrete space $\vec{V}_h$

From now on, we restrict our attention to linear problems and we assume that the set $\bar{\Omega}$ is a polygon. Then, with the terminology of Ciarlet [8], we may cover the set $\bar{\Omega}$ by an affine regular family of triangulations $\mathcal{T}_h$. To every triangle $K \in \mathcal{T}_h$, we associate two finite elements so that we define two spaces of finite elements V_{h1} and V_{h2} such that $V_{h1} \subset H_0^1(\Omega)$, $V_{h2} \subset H_0^2(\Omega)$ and

$$\vec{V}_h \subset \vec{V} \quad . \tag{3.1}$$

3.2. The discrete problems

We could define an approximated solution of problem (2.6) by restricting $\vec{u}$ and $\vec{v}$ to belong to the subspace $\vec{V}_h$ of $\vec{V}$ and by using numerical quadrature scheme over the set K

$$\int_K \psi(x)\,dx \sim \sum_{\ell=1}^{L} \omega_{\ell,K}\psi(b_{\ell,K}) \quad . \tag{3.2}$$

So, in the expressions (2.6)(2.7), we write $\int_\Omega (\ ) = \sum_{K\in\mathcal{T}_h} \int_K (\ )$

and we approximate every integral upon K with the help of (3.2). That amounts to substitute to a(.,.) a new bilinear form $a_h(.,.)$. The corresponding discrete problem is : Find $\vec{u}_h \in \vec{V}_h$ such that

$$a_h(\vec{u}_h,\vec{v}_h) = \sum_{K\in\mathcal{T}_h} \sum_{\ell=1}^{L} \omega_{\ell,K}(\vec{p}\vec{v}_h \sqrt{a})(b_{\ell,K}) \tag{3.3}$$

In [2] we have (i) shown that the problem (3.3) has a unique solution. This has been achieved by proving that, under mild assumptions, the bilinear form $a_h(.,.)$ is $\vec{V}_h$-elliptic, uniformly with respect to h ; (ii) proved the convergence ; (iii) obtained sufficient conditions on the quadrature schemes in order to preserve the error estimate of the exact integration case. Some examples of these results are given on Figure 3.1 upon which we have used the notations

$$\|\vec{u}-\vec{u}_h\| = (\sum_{\alpha=1}^{2} \|u_\alpha-u_{\alpha h}\|^2_{1,\Omega} + \|u_3-u_{3h}\|^2_{2,\Omega})^{1/2} = O(h^k) ,$$

$$E_K(\psi) = \int_K \psi(\xi^1,\xi^2)\, d\xi^1 d\xi^2 - \sum_{\ell=1}^{L} \omega_{\ell,K}\, \psi(b_{\ell,K}) .$$

Other examples are given in Bernadou [2]. Moreover, note that
i) it is possible to extend these results to the case of a curved boundary Γ ;
ii) in this analysis, the geometry of the shell appears only through variable coefficients which are not approximated, but only evaluated at the nodes of the numerical integration scheme. For an approximation of the geometry of the shell we refer to [8] in case of conforming methods and to [5] in case of nonconforming methods applied to general arch problems. In the last case, the arch is approximated by straight beam elements.

3.3. Approximation of free vibration modes

The first free vibration discrete problem

Inclusion (3.1) allows us to associate to problem (2.15) the first discrete problem :

$$\left.\begin{aligned} &\text{Find couples } (\tilde{\lambda}_h,\tilde{\vec{u}}_h) \in \mathbb{R}^+ \times \vec{V}_h \text{ such that} \\ &a(\tilde{\vec{u}}_h,\vec{v}_h) = \tilde{\lambda}_h\, \tilde{b}(\tilde{\vec{u}}_h,\vec{v}_h) \quad , \forall \vec{v}_h \in \vec{V}_h \quad . \end{aligned}\right\} \tag{3.4}$$

Since $\vec{V}_h \subset \vec{V}$, Theorem 2.2 can be applied to the problem (3.4).

V_{h1} \ V_{h2}	ARGYRIS 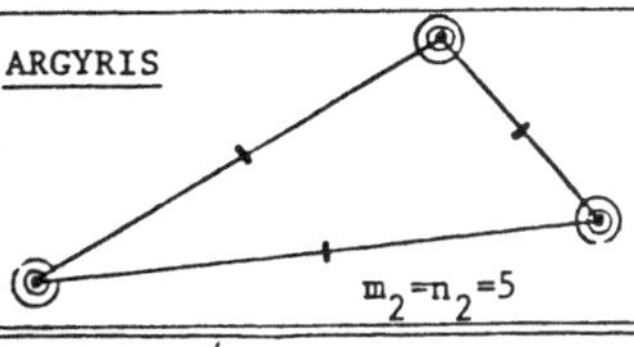 $m_2=n_2=5$	BELL 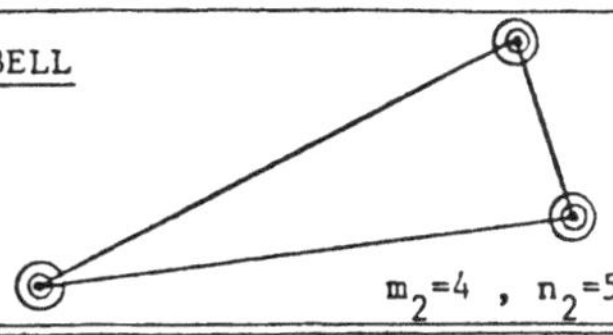 $m_2=4$, $n_2=5$
ARGYRIS $m_1=n_1=5$	$m=3$; $O(h^4)$ $\{\forall\phi\in P_8, E_K(\phi)=0\}$ Scheme at 16 points $\vec{u}\in(H^5(\Omega))^2\times H^6(\Omega)$ $A_{IJ}\in W^{4,\infty}(\Omega)$, $p^i\in W^{4,q}(\Omega)$	
BELL $m_1=4$, $n_1=5$	$m=3$; $O(h^4)$ $\{\forall\phi\in P_8$, $E_K(\phi)=0\}$ Scheme at 16 points $\vec{u}\in(H^5)^2\times H^6$; $A_{IJ}\in W^{4,\infty}$;$p^i\in W^{4,q}$	$m=2$; $O(h^3)$ $\{\forall\phi\in P_8$, $E_K(\phi)=0\}$ Scheme at 16 points $\vec{u}\in(H^4)^2\times H^5$; $A_{IJ}\in W^{3,\infty}$;$p^i\in W^{3,q}$
HERMITE of type 3 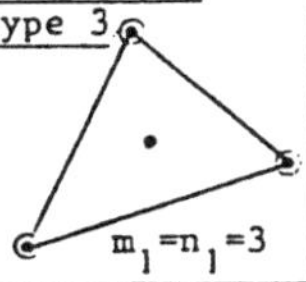 $m_1=n_1=3$	$m=2$; $O(h^3)$ $\{\forall\phi\in P_6$, $E_K(\phi)=0\}$ Scheme at 12 points or Scheme at 13 points $\vec{u}\in(H^4)^2\times H^5$; $A_{IJ}\in W^{3,\infty}$; $p^i\in W^{3,q}$	$m=2$; $O(h^3)$ $\{\forall\phi\in P_6$, $E_K(\phi)=0\}$ Scheme at 12 points or Scheme at 13 points $\vec{u}\in(H^4)^2\times H^5$; $A_{IJ}\in W^{3,\infty}$; $p^i\in W^{3,q}$
HERMITE of type 3 (reduc.) 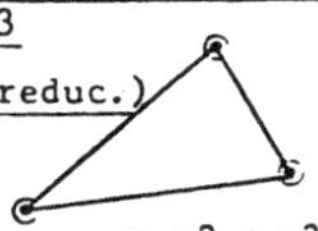 $m_1=2, n_1=3$		$m=1$; $O(h^2)$ $\{\forall\phi\in P_6$, $E_K(\phi)=0\}$ Scheme at 12 points or Scheme at 13 points $\vec{u}\in(H^3)^2\times H^4$; $A_{IJ}\in W^{2,\infty}$; $p^i\in W^{2,q}$
LAGRANGE of type 2 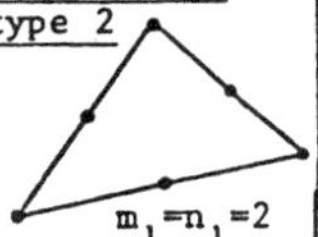 $m_1=n_1=2$		$m=1$; $O(h^2)$ $\{\forall\phi\in P_6$, $E_K(\phi)=0\}$ Scheme at 12 points or Scheme at 13 points $\vec{u}\in(H^3)^2\times H^4$; $A_{IJ}\in W^{2,\infty}$;$p^i\in W^{2,q}$

Figure 3.1 : Triangular finite elements

(Successively, in each case : i) $m=-1+\min(m_1,m_2-1)$; ii) the error estimate $o(h^{m+1})$; iii) the assumptions of the integration schemes ; iv) the number of nodes of suitables schemes ; v) the regularity assumptions on $\vec{u}$, on the variable coefficients A_{IJ} and on the loads p^i).

The second free vibration discrete problem

But, in order to compute integrals which appear in (2.7) and (2.16) we need to use numerical integration schemes of type (3.2). Thus, we define the approximate bilinear forms $a_h(.,.)$ and $b_h(.,.)$. Hence the second free vibration discrete problem is stated as follows :

$$\left.\begin{array}{l}\text{Find couples } (\lambda_h, \vec{u}_h) \in \mathbb{R}^+ \times \vec{V}_h \text{ such that} \\[2ex] a_h(\vec{u}_h, \vec{v}_h) = \lambda_h b_h(\vec{u}_h, \vec{v}_h) \quad , \forall \vec{v}_h \in \vec{V}_h \,.\end{array}\right\} \qquad (3.5)$$

Convergence and error estimates

By using Chatelin [7], Rappaz [21] or Strang and Fix [25], we can prove that

(i) the discrete problem has an increasing finite sequence of eigenvalues $\tilde{\lambda}_{hj}$, $j = 1,\ldots,M_h = \dim \vec{V}_h$ such that $0 < \lambda_j \leq \lambda_{hj}$ where λ_j are the first M_h eigenvalues of Problem (3.4) ;

(ii) for j fixed, $j = 1,\ldots,M_h$, we have $|\tilde{\lambda}_{hj} - \lambda_h| = 0(h^{2k})$, where k is the order of the interpolation error $\|\vec{v} - \overrightarrow{\pi_h v}\|_{\vec{V}}$ and π_h is the $\vec{V}_h$-interpolation operator in $\vec{V}$;

(iii) when the eigenvalue $\tilde{\lambda}_j$ is distinct, then the corresponding eigenvector $\vec{\tilde{u}}_{hj}$ satisfies the estimate $a(\vec{u}_j - \vec{\tilde{u}}_{hj}, \vec{u}_j - \vec{\tilde{u}}_{hj}) = 0(h^{2k})$;

(iv) when the eigenvalue $\tilde{\lambda}_j$ is repeated, then we get a similar estimate on the associated eigensubspace.

□

The extension of these results to the second discrete problem (3.5) is possible : see [7] or [21].

Examples : Similarly to Fig. 3.1, we illustrate the results of approximation of free vibration modes on Fig. 3.2.

4. Application to the computation of an arch dam

4.1. Geometrical definition of the dam

In this paragraph, we consider the project of Grand'Maison arch dam studied by Coyne et Bellier [11] (see Fig. 4.1. and 4.2.) :

Step 1 : Definition of the middle surface of the dam. The reference system and the reference domain are defined on Fig. 4.1.. Then, with notations of Fig. 4.2., the coordinates x^i of

V_{h1}, V_{h2} } ARGYRIS	V_{h1}, V_{h2} } Complete H.C.T.	V_{h1}: Tr. type (2); V_{h2}: Complete H.C.T.	V_{h1}, V_{h2} } Reduced H.C.T.	V_{h1}: Tr. type (1); V_{h2}: Reduced H.C.T.
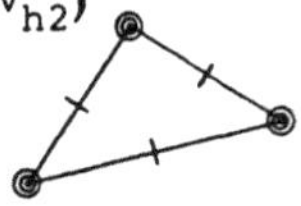	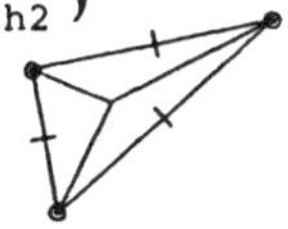	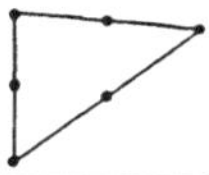	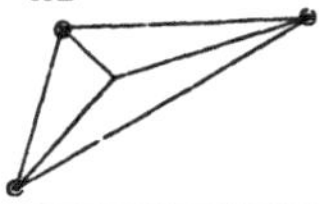	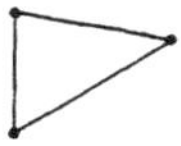
$\lvert\lambda_{hj}-\lambda_j\rvert = O(h^8)$ $\lVert u_j - u_{hj}\rVert = O(h^4)$	$\lvert\lambda_{hj}-\lambda_j\rvert = O(h^4)$ $\lVert u_j - u_{hj}\rVert = O(h^2)$	$\lvert\lambda_{hj}-\lambda_j\rvert = O(h^4)$ $\lVert u_j - u_{hj}\rVert = O(h^2)$	$\lvert\lambda_{hj}-\lambda_j\rvert = O(h^2)$ $\lVert u_j - u_{hj}\rVert = O(h)$	$\lvert\lambda_{hj}-\lambda_j\rvert = O(h^2)$ $\lVert u_j - u_{hj}\rVert = O(h)$

Figure 3.2 :
Error estimates for free vibration mode approximations
(case of distinct eigenvalue)

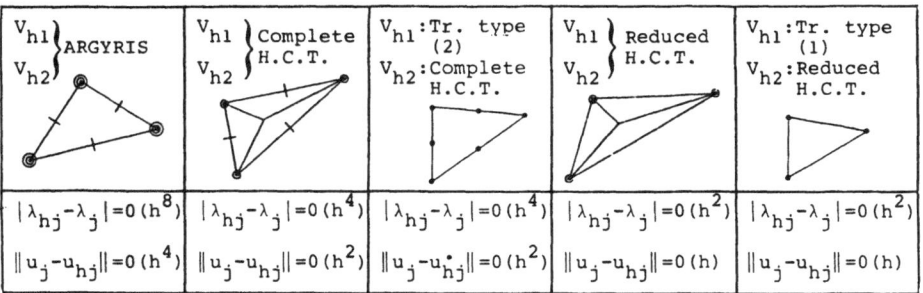

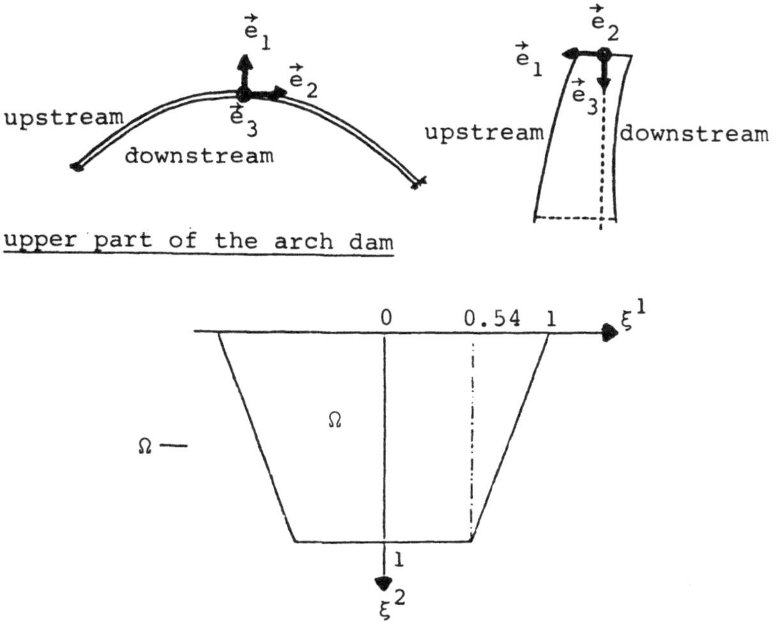

Figure 4.1 :
Reference system and reference domain Ω

any point M of the middle surface S of the dam are given as the components of the mapping

$$\vec{\phi} : (\xi^1,\xi^2) \in \bar{\Omega} \to \vec{OM} = \vec{\phi}(\xi^1,\xi^2) = x^i(\xi^1,\xi^2)\vec{e}_i \ ,$$

by the relations

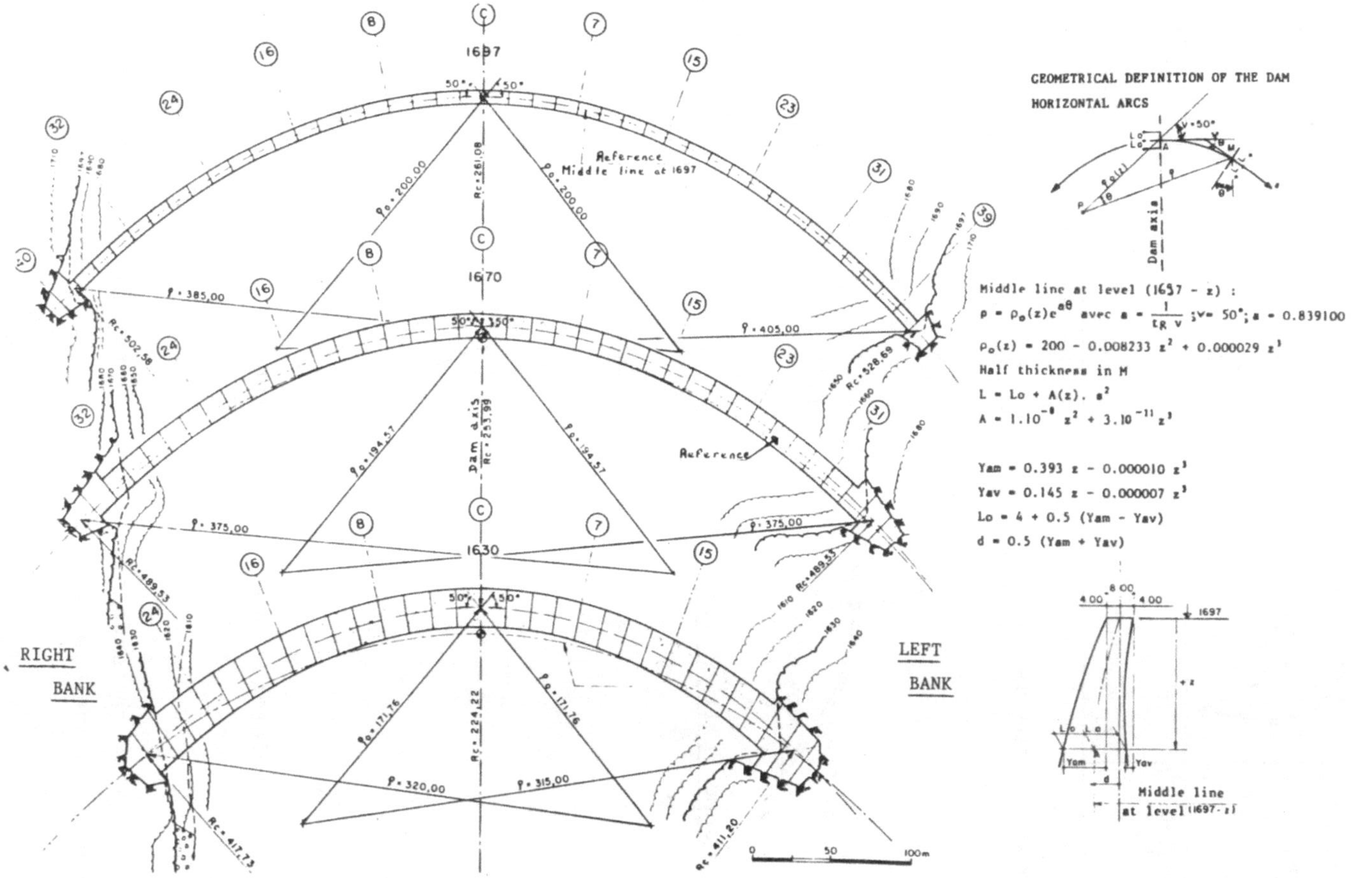

Figure 4.2 : Geometric definition of the dam - Upper horizontal sections

$$\begin{cases} x^1(\xi^1,\xi^2) = \rho_o(\xi^2)\left[e^{\alpha\theta_o|\xi^1|}\cos(\theta_o|\xi^1|+40°) - \cos 40°\right] \\ \qquad + 0.269\ z_o\xi^2 - 0.0000085\ z_o^3(\xi^2)^3 \\ x^2(\xi^1,\xi^2) = \frac{|\xi^1|}{\xi^1}\rho_o(\xi^2)\left[e^{\alpha\theta_o|\xi^1|}\sin(\theta_o|\xi^1|+40°) - \sin 40°\right] \\ x^3(\xi^1,\xi^2) = z_o\xi^2 \end{cases}$$

where

$$\alpha = \mathrm{tg}\ 40°\ ,\ \theta_o = 48°\ 178\ ,\ z_o = 157\ \mathrm{m}$$

$$\rho_o(\xi^2) = 200 - 0.008233(z_o)^2(\xi^2)^2 + 0.000029(z_o)^3\ (\xi^2)^3$$

Step 2 : Definition of the thickness of the dam

$$e(\xi^1,\xi^2) = 8 + 0.248\ z_o\xi^2 - 0.000003\ (z_o\xi^2)^3$$
$$+ 2.10^{-8}(z_o\xi^2)^2\ [1 + 0.003\ z_o\xi^2]\left[\frac{e^{\alpha\theta_o|\xi^1|}-1}{\sin 40°}\rho_o(\xi^2)\right]^2$$

4.2. Variational formulation of the problem

The potential energy of the external loads is approximated by (see [3]) :

$$f(\vec{v}) = \frac{E\bar{\alpha}}{2(1-\nu)}\int_\Omega [eT_1\gamma_\lambda^\lambda + \frac{e^3}{12}T_2(2b_\eta^\lambda\gamma_\lambda^\eta - b_\lambda^\lambda\gamma_\eta^\eta - \bar{\rho}_\lambda^\lambda)]\sqrt{a}d\xi^1 d\xi^2$$
$$+ \int_\Omega p\ [\frac{1}{2}e_{,\beta}v_\lambda a^{\lambda\beta} - (1-\frac{1}{2}eb_\beta^\beta)v_3]\sqrt{a}d\xi^1 d\xi^2$$
$$+ \int_\Omega \rho_1 g_o\ e[(a^{12}v_1 + a^{22}v_2)\ z_o + (\vec{a}_3.\vec{e}_3)v_3]\sqrt{a}d\xi^1 d\xi^2\ ,$$

where the three integrals take respectively into account the effects of

(i) the thermal loads : On the middle surface S, we denote T_1 and T_2 the moments of order 0 and 1 through the thickness (in a mathematical sense) of the steady-state temperature distribution; here $E = 2\ 10^6$ ton/m^2 ; $\nu = 0.2$; $\bar{\alpha} = 10^{-5}$ by centigrade degree ;

(ii) the hydrostatic pressure p : If $\xi^2 = \bar{\xi}^2$ refers to the level of water in the reservoir, then

$$p = 0 \text{ if } 0 \le \xi^2 \le \bar{\xi}^2 \ , \ p = \rho_2 g_o z_o(\xi^2 - \bar{\xi}^2) \text{ if } \bar{\xi}^2 \le \xi^2 \le 1 \ ,$$

where $\rho_2 = 10^3$ Kg/m^3 , $g_o = 9.81$ m/s^2 , $z_o = 157$ m ;

(iii) the self weight. We take $\rho_1 = 2500$ Kg/m^3.

Boundary conditions : Assume that the middle surface S is free on the upper part of its boundary and clamped everywhere else, that is on Γ_o, so that the space of admissible displacement is

$$\vec{V} = \{\vec{v} \mid \vec{v} \in (H^1(\Omega))^2 \times H^2(\Omega) \ , \ \vec{v}|\Gamma_o = \vec{0} \ , \ \frac{\partial v_3}{\partial n}|\Gamma_o = 0\} \ .$$

Then, the problem can be stated : For any $T_\alpha \in L^2(\Omega)$, find $\vec{u} \in \vec{V}$ such that $a(\vec{u},\vec{v}) = f(\vec{v})$, $\forall\vec{v} \in \vec{V}$. An extension of theorem 2.1 gives the existence and the uniqueness for a solution.

4.3. Computation of displacements and stresses in the arch dam problem

For simplicity, we have assumed that the temperature factors T_1 and T_2 are symmetrical in ξ^1. Then, introducing "pseudo" boundary conditions on $\xi^1 = 0$, we can formulate the problem on the half domain $\Omega_1 = \{(\xi^1,\xi^2) \in \Omega \mid \xi^1 \ge 0\}$. This implementation is detailed in Bernadou-Boisserie [3]. In this section, we present some results obtained by using Argyris'triangle to approximate the three components of the displacement. The system is solved by a Choleski method using the sky-line bandwidth factorization - see Parlett [20]. From the solution of the system, we obtain approximations of the displacement, strain tensor and change of curvature tensor at any point of the middle surface S. Then, applying the basic hypotheses of [15], we derive an approximation of the displacement and mixed stress tensor σ^{*i}_{j} everywhere in the dam. In order to get components having natural physical dimensions, we introduce the so-called right physical components of the stress tensors, as in Truesdell [27].

Figures 4.3 and 4.4 show the distribution of stresses on the downstream and upstream faces of the arch dam subjected to the combined effect of weight, temperature and water pressure. The level of water in the reservoir is assumed to be 152 meters so that $\bar{\xi}^2 = 0.032$ meanwhile $E = 2.10^6$ ton/m^2 , $\nu = 0.2$. These figures are obtained from a triangulation with 32 triangles.

Combined effect of weight, temperature and
water pressure :

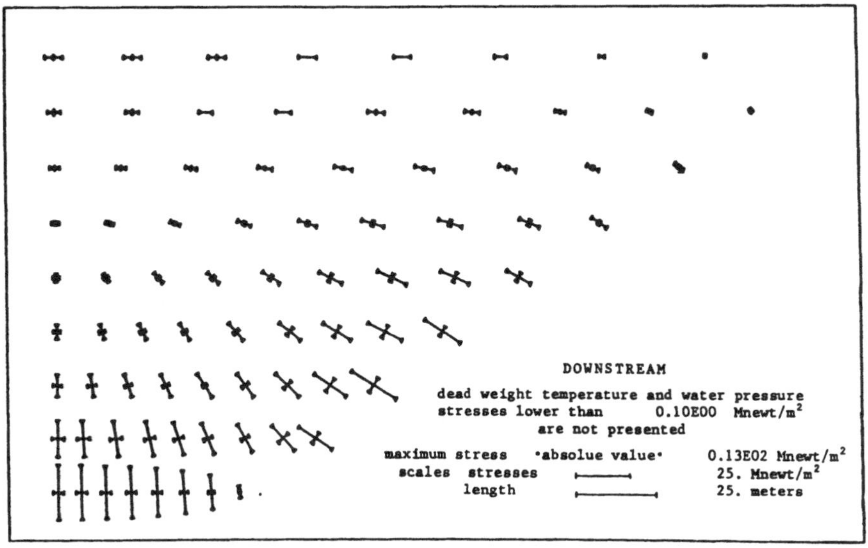

Figure 4.3 :

Stress distribution on downstream face (32 triangles)

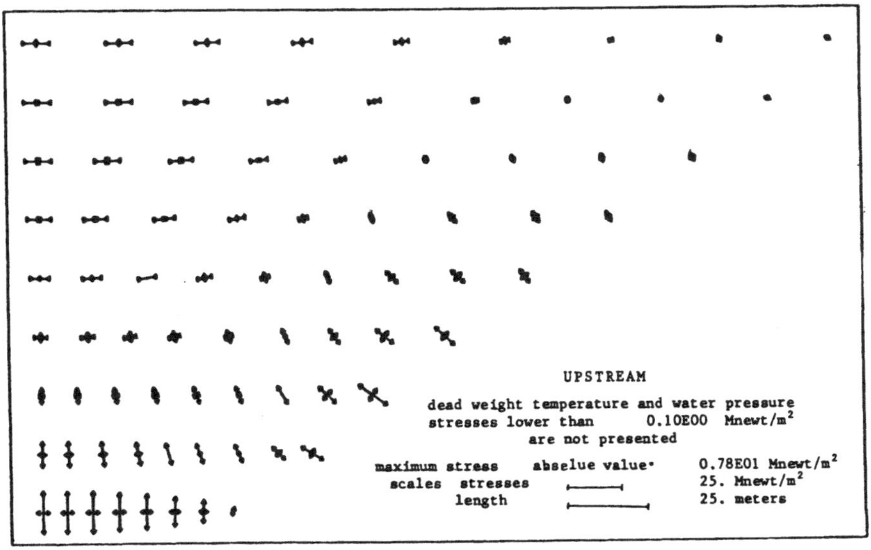

Figure 4.4 :

Stress distribution on upstream face (32 triangles)

There are 618 degrees of freedom, 463 unknows and the computing time on CRAY 1 is 11.2s. Other computations using triangulations with 8, 18, 50, 72, 98 and 128 triangles have been performed : the corresponding results show the excellent approximation properties of the Argyris element. Particularly, the results obtained from a coarse triangulation with only 8 triangles are really acceptable and non expensive.

Let us observe that these numerical experiments can be qualitatively compared with the experimental results given in Rydzewski [23 p. 639] for a similar arch dam.

4.4. Computation of the free vibration modes are done by combined use of

(i) a space $\vec{V}_h$ constructed from Argyris or reduced-H.C.T. triangles ;

(ii) the simultaneous iteration method - see Parlett [20].

In Table 4.1 and 4.2, we list the results concerning the approximation of the first five frequencies. In both cases we observe an excellent convergence for the first frequency. As usual we observe that the higher eigenvalues are progressively more difficult to approximate. Finally, in Table 4.3 we check that these results are in good agreement with experimental results obtained on similar arch dams by Medvedev and Sinitzyn [18] and Takahashi [26] (the frequencies are decreasing when the dimensions of the structure increase).

5. Some open problems

Hereafter we list some open shell problems with respect to the numerical analysis wiewpoint.

5.1. Continuous linear problems

5.1.1. - Mathematical justification of the different linear models by asymptotic methods. An important work in this direction is due to Destuynder [12].

5.1.2. - Coupling between two and three dimensional models. For instance, this is the case of a thin shell lying on elastic foundations.

Frequencies \ Number of triangles	8 ($h = \frac{1}{2}$)	18 ($h = \frac{1}{3}$)	32 ($h = \frac{1}{4}$)	50 ($h = \frac{1}{5}$)
$1^{\underline{st}}$ frequency	2.491	2.485	2.483	2.483
$2^{\underline{nd}}$ frequency	3.688	3.674	3.670	3.670
$3^{\underline{rd}}$ frequency	5.691	5.503	5.494	
$4^{\underline{th}}$ frequency	5.729	5.668	5.662	
$5^{\underline{th}}$ frequency	8.570	7.871	7.581	

Table 4.1 : Approximation of the first five frequencies by using Argyris triangle (global computing time : 183s. on CRAY 1)

Frequencies \ Number of triangles	32 ($h = \frac{1}{4}$)	72 ($h = \frac{1}{6}$)	128 ($h = \frac{1}{8}$)	200 ($h = \frac{1}{10}$)
$1^{\underline{st}}$ frequency	2.609	2.545	2.521	2.508
$2^{\underline{nd}}$ frequency	4.366	4.038	3.900	3.823
$3^{\underline{rd}}$ frequency	6.121	5.863	5.777	5.733
$4^{\underline{th}}$ frequency	6.812	6.735		
$5^{\underline{th}}$ frequency	8.969	8.916		

Table 4.2 : Approximation of the first five frequencies by using reduced H.C.T. element (global computing time : 426 s. on CRAY 1)

5.1.3. - Singular connections of middle surface. In the classical models of shells, it is implicitely assumed that the middle surface is defined through a mapping of class $\mathscr{C}^3$ (or, at least $\mathscr{C}^2$). What happens in the neighborhood of a connection of class $\mathscr{C}^0$ (case of discontinuous tangent plane) or of class $\mathscr{C}^1$ (case of curvature discontinuities) of two different middle surfaces ?

5.2. Discrete linear problems

5.2.1. - Numerical analysis of mixed, hybrid, equilibrium methods... There are some contributions on these subjects : for instance see Destuynder-Lutoborski [13], Stephan-Weisgerber [24].

5.2.2. - Approximation by flat plate elements : the middle surface is approximated by plane triangular facets. On every facet,

Arch dam / Characteristics	Saranami-gawa dam	Russian arch dam	Kamishiiba dam	Project of Grand ' Maison arch dam
Max. height	67.4 m	103 m	110 m	157 m
Crest length	127 m	340 m	310 m	670 m
Crest width	2.4 m	5 m	7 m	8 m
Base width	8.8 m	31 m	27.7 m	36 m
Radius of crest arch	74.74 m	150 m	142.4 m	261 m
First frequency	5.5 hertz (cycle par s.	3.57 hz.	3.83 hz.	2.48 hz.
Second frequency	6.83 hz.		5.83 hz.	3.67 hz.
First frequency			8.67 hz.	5.49 hz.
Fourth frequency				5.66 hz.
Fifth frequency				≃ 6.8 hz.

Figure 4.3 :
Comparison with experimental results issued from existing arch dam

we define a plate energy ; next we have to suitably assemble these elementary different energies in order to obtain the energy of the entire shell. These methods are intensively used by engineers but it seems that the corresponding mathematical studies remain open. We just mention convergence results for general arches [5].

5.2.3. - Numerical analysis of D.K.T. (Discrete Kirchhoff Triangle) methods for shells (these methods were introduced by Wempner-Oden-Kross [28] and next improved by Batoz-Bathe-Ho [1] among others). For plate problems let us mention the convergence results of Kikuchi [14].

5.2.4. - Discrete approximation of coupling of two and three dimensional models.

5.2.5. - Discrete approximation of connections of class $\mathscr{C}^0$ or $\mathscr{C}^1$ of two different middle surfaces.

5.3. <u>Linear dynamical problems and approximations</u>
For intance, vibration and earthquake problems...

5.4. <u>Steady-state or dynamical nonlinear problems - Approximation of these problems</u>. Particularly, buckling, bifurcation, post-buckling, crack, fracture... which are very open with respect to the numerical analysis viewpoint.

5.5. Fluid-Structure coupling. Some typical cases are the fuel tanks, the flows along a structure.

6. References

1. Batoz J.L., Bathe K.J., Ho L.W., [1980] : A study of three node triangular plate bending elements, Internat. J. Numer. Methods Engng., 15, n° 12, 1771-1812.

2. Bernadou M., [1980] : Convergence of conforming finite element methods for general shell problems. Int. J. Engng. Sc., 18, 249-276.

3. Bernadou M., Boisserie J.M., [1982] : The Finite Element Method in Thin Shell Theory ; Application to an Arch Dam, Birkhäuser, Boston Inc.

4. Bernadou M., Ciarlet P.G., [1976] : Sur l'ellipticité du modèle linéaire de coques de W.T. Koiter. "Computing Methods in Applied Sciences and Engineering". Lectures Notes in Economics and Mathematical Systems, Vol. 134, pp. 89-136, Springer-Verlag, Berlin.

5. Bernadou M., Ducatel Y., [1982] : Approximation of general arch problems by straight beam elements, Numer. Math.,40, 1-29.

6. Bernadou M., Oden J.T., [1981] : Existence theorem for general nonlinear shallow shell problems, J. Math. Pures Appl., 60, 285-308.

7. Chatelin F., [1982] : Spectral Approximation of Linear Operator, Academic Press, New York.

8. Ciarlet P.G. [1978] : The Finite Element Method for Elliptic Problems. North-Holland.

9. Cosserat E. et F., [1909] : Théorie des Corps Déformables. Hermann, Paris.

10. Coutris N., [1978] : Théorème d'existence et d'unicité pour un problème de coque élastique dans le cas d'un modèle linéaire de P.M. Naghdi. R.A.I.R.O., Analyse Numérique, 12, n° 1, 51-58.

11. Coyne & Bellier [1977] : Barrage de Grand'Maison, Dossier préliminaire.

12. Destuynder P., [1980] : Sur une Justification des Modèles de Plaques et de Coques par les Méthodes Asymptotiques, Thèse d'Etat, Paris VI.

13. Destuynder P., Lutoborski [1982] : A penalty method for the Budiansky-Sanders shell model, Comput. Methods Appl. Mech. Engng., 35, 127-151.

14. Kikuchi F., [1981] : On the discrete Kirchhoff approach for plate bending problems - Theoretical and Applied Mechanics, Vol. 31, pp. 3-21, University of Tokyo Press.

15. Koiter W.T., [1966] : On the nonlinear theory of thin elastic shells. Proc. Kon. Ned. Akad. Wetensch, B69, 1-54.

16. Kubrusly R., [1982] : On the existence of post-buckling solutions of shallow shells under a certain unilateral constraint, Int. J. Engng. Sci., 20, n° 1, 93-99.

17. Lions J.L., [1969] : Quelques Méthodes de Résolution des Problèmes aux Limites Non Linéaires, Dunod, Gauthier-Villars, Paris.

18. Medvedev S.V., Sinitzyn A.P., [1964] : Tests and theoretical studies on the earthquake resistant properties of the arch dams, Proc. 8e Congrès International des Grands Barrages, Vol. II, Mai 1964, pp. 899-907.

19. Naghdi P.M., [1972] : The Theory of Shells and Plates. Handbuch der Physik, Vol. VI a-2, pp. 425-640, Springer Verlag, Berlin.

20. Parlett B.N., [1980] : Symmetric eigenvalue problem, Prentice Hall, Englewood Cliffs.

21. Rappaz J., [1979] : Analyse numérique de certains problèmes aux valeurs propres. Application à la magnétohydrodynamique. Conférence donnée à l'Ecole Supérieure d'Electricité, à la Journée d'Etude du 25.4.1979.

22. Riesz F., Nagy B.Sz., [1952] : Leçons d'Analyse Fonctionnelle, Budapest, Akadémiai Kiado.

23. Rydzewski J.R., [1965] : Theory of Arch Dams, Pergamon Press, Oxford.

24. Stephan E., Weisgerber, [1978] : Zur approximation von schalen mit hybriden elementen, Computing, 20, n° 1, 75-94.

25. Strang G., FIX G.J., [1973] : An Analysis of the Finite Element Method, Prentice-Hall, Englewood Cliffs.

26. Takahashi T., [1964] : Results of vibration tests and earthquake observations on concrete dams and their considerations, Proc. 8e Congrès International des Grands Barrages, Vol. II, Mai 1964, pp. 239-250.

27. Truesdell C., [1953] : The physical components of vectors and tensors, Z. Angew. Math. Mech., 33, n° 10-11, 345-356.

28. Wempner G.R., Oden J.T., Kross D., [1968] : Finite element anal. of thin shells, J. Engng. Mech. Div.ASCE, 94, n° EM6, 1273-1294.

A Total Lagrangian Finite Element Formulation for the Geometrically Nonlinear Analysis of Shells

J. Oliver and E. Oñate

E.T.S. Ingenieros de Caminos, Canales y Puertos
Jordi Girona Salgado, 31
Barcelona (34)
SPAIN

1. INTRODUCTION

The analysis of structures subjected to large displacements by means of the finite element method has attracted the attention of many researchers in recent years and different publications have been reported in the literature [1]-[8].

In this work the authors suggest an alternative finite element formulation for the analysis of shells which allows for large displacements together with finite rotations. The deformation process of the structure is defined via a total Lagrangian approach. Stresses and strains over the shell surface are defined using a local set of cartesian axes based on the principal curvature directions of the shell middle surface. This allows to obtain useful explicit expressions of the finite element matrices in a simple manner. Additionaly, normals to the midsurface before deformation are assumed to remain straight but not necessarily normal to the midsurface after deformation, thus allowing for shear deformation effects. Finally, it is worth pointing out that no restrictions are made on the magnitude of the curvatures. This is of special interest for the analysis of non-shallow shells using an small number of elements.

The formulation uses two dimensional finite elements for the analysis of 3-D shells. The discretization over the shell thickness is eliminated using what is usually known as "degenerated element technique" [2,4]. This allows for a substantial reduction in the number of variables and eliminates the possibility of ill-conditioning of the element matrices which takes place when using 3-D elements and the thickness of the shell is small.

In the first part of this work the formulation for 3-D shells is presented. Then a series of examples of shells undergoing large displacements are presented.

2. GEOMETRIC DESCRIPTION

The middle surface of the shell can be expressed in parametric form as (see. Fig.1) [9]

$$\underset{\sim}{r_0} = [\, x_0(\mu_1,\mu_2),\, y_0(\mu_1,\mu_2),\, z_0(\mu_1,\mu_2)\,]^T \tag{1}$$

where μ_1 and μ_2 are the principal curvature lines at point 0 of the shell middle surface.

Let $\vec{a}$ and $\vec{b}$ be unit vectors tangent to the curvature lines μ_1 and μ_2 in 0, respectively, and $\vec{n}$ the normal vector to the middle surface in 0. Parameters r,s and t are defined as the lenghts measured along the lines μ_1, μ_2 and along the normal $\vec{n}$, respectively.

A second set of orthogonal vectors $\vec{l},\vec{m}$ and $\vec{n}$ is defined at 0, as shown in Fig. 2. Vector $\vec{l}$ is taken as parallel to the global plane xz and tangent to the shell middle surface, $\vec{n}$ is the normal vector, previously defined, and $\vec{m}$ is orthogonal to the plane

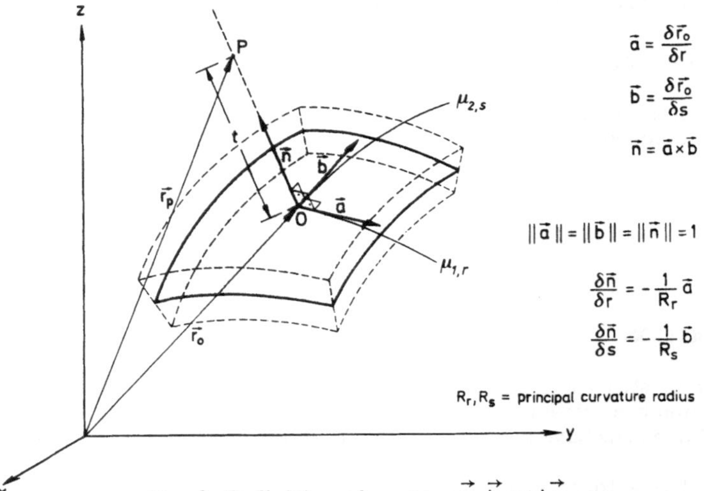

Fig. 1. Definition of vectors $\vec{a}$, $\vec{b}$ and $\vec{n}$

$\vec{n}$-$\vec{l}$ and also tangent to the shell middle surface at point O. Note that $\vec{l}$ and $\vec{m}$ are not unit vectors and their modulus is $e=(1-n_y^2)^{1/2}$

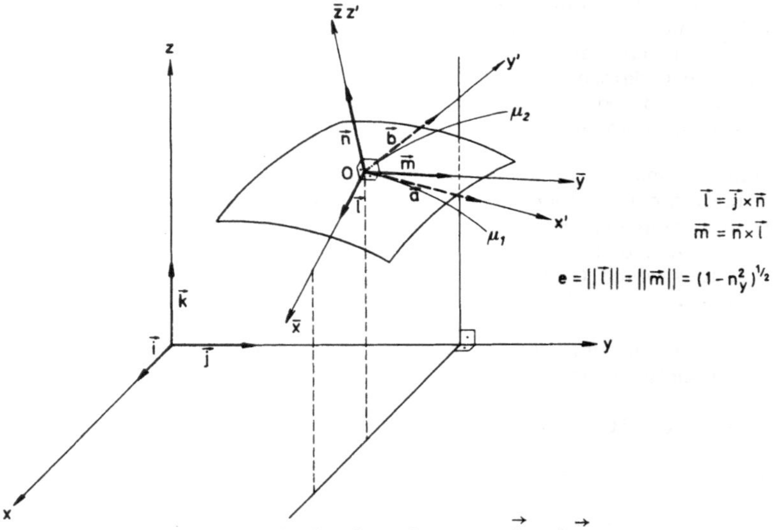

Fig. 2. Definition of vectors $\vec{l}$ and $\vec{m}$

The components of vectors $\vec{a},\vec{b},\vec{n},\vec{l}$ and $\vec{m}$ in the global reference system x,y,z (with associated unit vectors $\vec{i},\vec{j},\vec{k}$, respectively) can be written in matrix form as

$$\underset{\sim}{a} = \begin{bmatrix} a_x \\ a_y \\ a_z \end{bmatrix} \; ; \; \underset{\sim}{b} = \begin{bmatrix} b_x \\ b_y \\ b_z \end{bmatrix} \; ; \; \underset{\sim}{n} = \begin{bmatrix} n_x \\ n_y \\ n_z \end{bmatrix} \; ; \; \underset{\sim}{l} = \begin{bmatrix} l_x \\ l_y \\ l_z \end{bmatrix} \; ; \; \underset{\sim}{m} = \begin{bmatrix} m_x \\ m_y \\ m_z \end{bmatrix} \quad (2)$$

Vectors $\vec{a},\vec{b}$ and $\vec{n}$ define a set of local axes x',y' and z', associated with the shell principal curvature lines. On the other hand, vectors $\vec{l},\vec{m}$ and $\vec{n}$ define a second set of local axes $\bar{x},\bar{y}$ and $\bar{z}$ which is easily identified within the structure ($\bar{x}$ is parallel to plane $\bar{x}\bar{z}$, etc... seè Fig. 2).

A point P over the shell middle surface can be defined by a vector $\vec{r}$ (see Fig. 1) such that we can write, in matrix form

$$\underset{\sim}{r} = \underset{\sim}{r}_p = \underset{\sim}{r}_0 + t\underset{\sim}{n} \tag{3}$$

where t is the distance measured along the normal.

Using eq.(3) and the relationships shown in Fig. 1 the following expressions can be obtained

$$\left[\frac{\partial(x,y,z)}{\partial(r,s,t)}\right] = \underset{\sim}{T}^T \underset{\sim}{R} \tag{4}$$

where $\underset{\sim}{T}$ is the Jacobian matrix of the transformation x'y'z' → xyz given by

$$\underset{\sim}{T} = [\underset{\sim}{a}, \underset{\sim}{b}, \underset{\sim}{n}]^T \tag{5}$$

and matrix $\underset{\sim}{R}$ is the Jacobian matrix of the transformation x'y'z' → rst given by

$$\underset{\sim}{R} = \begin{bmatrix} 1-\frac{t}{R_r} & 0 & 0 \\ 0 & 1-\frac{t}{R_s} & 0 \\ 0 & 0 & 1 \end{bmatrix} \tag{6}$$

Additionaly, vectors $\vec{l},\vec{m}$ and $\vec{n}$ can be defined in the system $\vec{a},\vec{b}$ and $\vec{n}$ using the following expression

$$[\underset{\sim}{l},\underset{\sim}{m},\underset{\sim}{n}] = [\underset{\sim}{a},\underset{\sim}{b},\underset{\sim}{n}]\,\bar{\underset{\sim}{T}} \tag{7}$$

where

$$\bar{\underset{\sim}{T}} = \begin{bmatrix} b_y & a_y & 0 \\ -a_y & b_y & 0 \\ 0 & 0 & 1 \end{bmatrix} \tag{8}$$

For the finite element analysis we discretize the middle surface of the shell into a mesh of curved cuadrangular elements as shown in Fig. 3. The geometry of the shell is defined, within each element, in the standard isoparametric form [10] as

$$\underset{\sim}{r} = \sum_i^{n_e} N_i(\xi,\eta)\underset{\sim}{r}_{0_i} + \tau\frac{h}{2}\underset{\sim}{n} \tag{9}$$

where $\underset{\sim}{r}_{0_i}$ is the vector of each node of the finite element mesh, n_e the number of nodes per element, $N_i(\xi,\eta)$ is the shape function of node i, and ξ and η the normalized isoparametric coordinates. The third normalized coordinate is defined as

$$\tau = \frac{2t}{h} \tag{10}$$

where h is the thickness of the element. Finally, the normal vector $\vec{n}$ is obtained from the approximate middle surface given in (9) and the expressions (1)-(9). More details about the choice of element will be given in section 7.

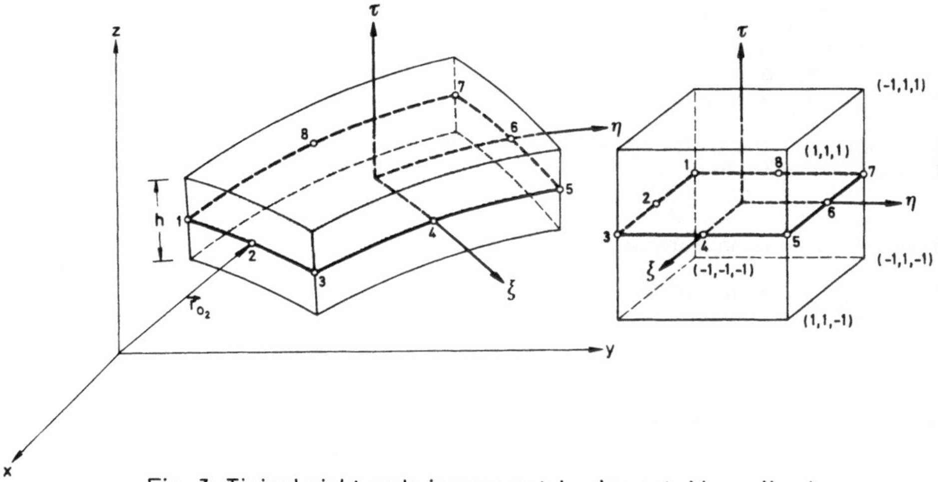

Fig. 3 Tipical eight node isoparametric element. Normalized coordinate system and associated normalized volume.

Eq.(9) transforms the element volume into a cube of unit side defined in the coordinate system ξ, η, τ [10](see Fig. 3.).

3. KINEMATIC DESCRIPTION

The deformation of the structure is based in the following two main assumptions:

a) Normals to the middle surface before deformation remain straight but not necessarily normal to the middle surface after deformation.

b) The lenght of the normal vector does not change during the deformation.

Assumption a) allows to express the displacement vector $\vec{u}$, of a point P (laying over the normal, $\vec{n}$, at a distance t from the corresponding point 0 over the middle surface (see Fig. 4), in terms of the displacement vector of point 0, $\vec{u}_0$, and the relative displacement of the end of the normal vector in 0, with respect to this point, $\vec{u}_1$, i.e.

$$\vec{u} = \vec{u}_0 + t\vec{u}_1 \tag{11}$$

We define

$$\underset{\sim}{u}_0 = [u_0, v_0, w_0]^T \quad \text{and} \quad \underset{\sim}{\bar{u}}_1 = [\bar{u}_1, \bar{v}_1, \bar{w}_1]^T \tag{12}$$

where u_0, v_0, w_0 refer to components of $\vec{u}_0$ in the global system $\vec{i}, \vec{j}, \vec{k}$, respectively, and $\bar{u}_1, \bar{v}_1, \bar{w}_1$ refer to components of $\vec{u}_1$ on the local system $\vec{l}, \vec{m}, \vec{n}$ (see Fig. 4).

The vector of "fundamental displacements" $\underset{\sim}{p}$ is defined now as

$$\underset{\sim}{p} = [\underset{\sim}{u}_0^T, \underset{\sim}{u}_1^T]^T \tag{13}$$

The components of $\underset{\sim}{p}$ define generically the displacement of any point of the shell. Using the finite element interpolation, it can be written

$$\underset{\sim}{p} = \sum_{1}^{n_e} \underset{\sim}{N}_i \underset{\sim}{p}_i \quad ; \quad \underset{\sim}{N}_i = N_i \underset{\sim}{I}_6 \tag{14}$$

where p_i is the value of p at node i.

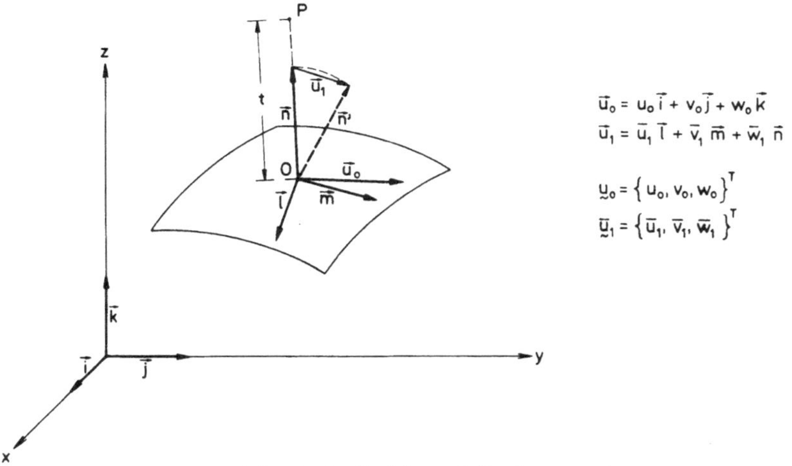

Fig. 4. Definition of displacements.

Assumption b) allows to express the components of $\bar{\underset{\sim}{u}}_1$, at node i, as

$$\bar{\underset{\sim}{u}}_{1_i} = \left[\frac{-\sin\alpha_i \cos\beta_i}{e_i}, \frac{-\sin\alpha_i \sin\beta_i}{e_i}, \cos\alpha_i - 1 \right]^T \tag{15}$$

Figs. 5 and 6 show the relationships between angles α_i and β_i and the two components of the rotation vector $\vec{\theta}_i$, which expresses the anti-clockwise rotation of the normal.

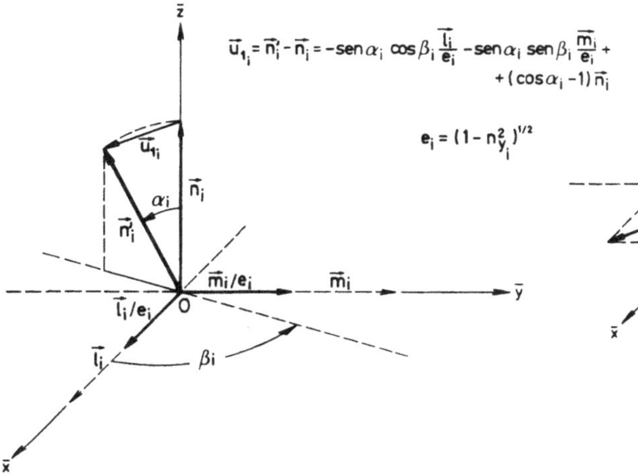

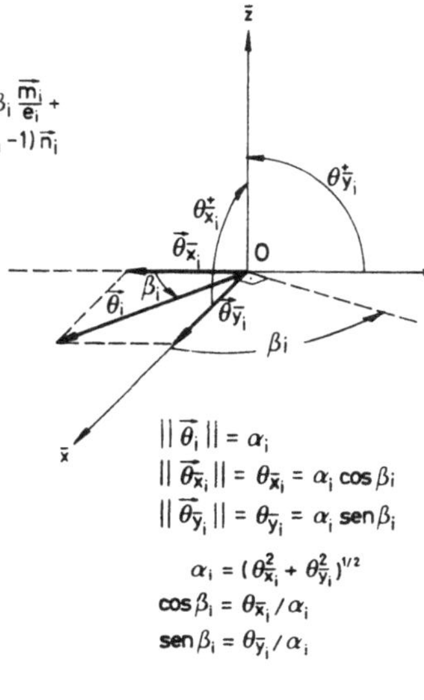

Fig. 5. Definition of angles α_i and β_i

Fig. 6. Sign convention for rotations

We define the vector of displacements of node i as

$$\underset{\sim}{a} = [\underset{\sim}{u}_{0_i}^T , \theta_{\bar{x}_i} , \theta_{\bar{y}_i}]^T \tag{16}$$

where vector $\underset{\sim}{u}_{0_i}$ is obtained from eq.(12). On the other hand, the relationships linking the rotations and the components of vector $\underset{\sim}{\bar{u}}_1$ can be deduced from eq.(15) and the expressions of Fig. 6.

4. STRAIN FIELD

Let's define $\underset{\sim}{u}' = [u',v',w']$, as the vector containing the components of the displacement vector $\underset{\sim}{u}$ of **any** point of the shell measured in the coordinate system defined by vectors $\vec{a},\vec{b}$ and $\vec{n}$, corresponding at a **particular** point, O, laying over the shell middle surface.

The vector of displacement gradients at a point, P, laying over the normal vector in O (see Fig. 4), can be defined now as

$$\underset{\sim}{g} = [\underset{\sim}{g}_1^T , \underset{\sim}{g}_2^T , \underset{\sim}{g}_3^T]^T \tag{17}$$

with

$$\underset{\sim}{g}_1 = \begin{bmatrix} g_1 \\ g_2 \\ g_3 \end{bmatrix} = \begin{bmatrix} \frac{\partial u'}{\partial x'} \\ \frac{\partial v'}{\partial x'} \\ \frac{\partial w'}{\partial x'} \end{bmatrix}_P \qquad \underset{\sim}{g}_2 = \begin{bmatrix} g_4 \\ g_5 \\ g_6 \end{bmatrix} = \begin{bmatrix} \frac{\partial u'}{\partial y'} \\ \frac{\partial v'}{\partial y'} \\ \frac{\partial w'}{\partial y'} \end{bmatrix}_P \qquad \underset{\sim}{g}_3 = \begin{bmatrix} g_7 \\ g_8 \\ g_9 \end{bmatrix} = \begin{bmatrix} \frac{\partial u'}{\partial z'} \\ \frac{\partial v'}{\partial z'} \\ \frac{\partial w'}{\partial z'} \end{bmatrix}_P \tag{18}$$

where subscrit P denotes values in point P.

The Green strain vector at point P, (associated to the local directions x',y',z' of Fig. 2) can be written, using eqs.(18), as

$$\underset{\sim}{\varepsilon} = \begin{bmatrix} \varepsilon_{x'} \\ \varepsilon_{y'} \\ \gamma_{x'y'} \\ \gamma_{x'z'} \\ \gamma_{y'z'} \end{bmatrix} = \begin{bmatrix} g_1+\frac{1}{2}(g_1^2+g_2^2+g_3^2) \\ g_5+\frac{1}{2}(g_4^2+g_5^2+g_6^2) \\ g_2+g_4+g_1g_4+g_2g_5+g_3g_6 \\ g_2+g_7+g_1g_7+g_2g_8+g_3g_9 \\ g_6+g_8+g_4g_7+g_5g_8+g_6g_9 \end{bmatrix} \tag{19}$$

On the other hand, vector $\underset{\sim}{g}$ can be explicitely obtained in terms of the fundamental displacement vector $\underset{\sim}{p}$, of eq.(13) as [11,12]

$$\underset{\sim}{g} = \begin{bmatrix} C_r \, \underset{\sim}{T}\frac{\partial}{\partial r} & tC_r(\underset{\sim}{\Delta}_1+\bar{\underset{\sim}{T}}\frac{\partial}{\partial r}) \\ C_s \, \underset{\sim}{T}\frac{\partial}{\partial s} & tC_s(\underset{\sim}{\Delta}_2+\bar{\underset{\sim}{T}}\frac{\partial}{\partial s}) \\ \underset{\sim}{0} & \underset{\sim}{T} \end{bmatrix} \underset{\sim}{p} = \underset{\sim}{L} \, \underset{\sim}{p} \tag{20}$$

with

$$C_r = \frac{1}{1-\frac{t}{R_r}} \qquad ; \qquad C_s = \frac{1}{1-\frac{t}{R_s}} \tag{21}$$

$$\underset{\sim}{\Delta}_1 = \frac{1}{R_r} \begin{bmatrix} 0 & n_y & -1 \\ -n_y & 0 & 0 \\ b_y & a_y & 0 \end{bmatrix} \qquad \underset{\sim}{\Delta}_2 = \frac{1}{R_s} \begin{bmatrix} n_y & 0 & 0 \\ 0 & n_y & -1 \\ -a_y & b_y & 0 \end{bmatrix} \tag{22}$$

Substituting eq.(14) in (20) it can be finally written

$$\underset{\sim}{g} = \underset{\sim}{L} \sum_1^{n_e} \underset{\sim i}{N} \underset{\sim i}{p} = \sum_1^{n_e} \underset{\sim i}{M} \underset{\sim i}{p} \tag{23}$$

where

$$\underset{\sim i}{M} = \begin{bmatrix} C_r \underset{\sim}{T} \frac{\partial N_i}{\partial r} & t\, C_r (N_i \underset{\sim}{\Delta}_1 + \frac{\partial N_i}{\partial r} \underset{\sim}{\overline{T}}) \\[2mm] C_s \underset{\sim}{T} \frac{\partial N_i}{\partial s} & t\, C_s (N_i \underset{\sim}{\Delta}_2 + \frac{\partial N_i}{\partial s} \underset{\sim}{\overline{T}}) \\[2mm] \underset{\sim}{0} & N_i \cdot \underset{\sim}{\overline{T}} \end{bmatrix} = [\underset{\sim i}{\overline{M}}, \underset{\sim i}{\overline{\overline{M}}}] \tag{24}$$

$$\underbrace{\qquad\qquad}_{\overline{M}_i} \quad \underbrace{\qquad\qquad\qquad}_{\overline{\overline{M}}_i}$$

Taking into account the linearity of matrix $\underset{\sim i}{M}$ (independent of the nodal unknowns) we can write

$$\delta \underset{\sim}{g} = \sum_1^{n_e} \underset{\sim i}{M} \, \delta \underset{\sim i}{p} \tag{25}$$

It can be shown that a relationship between fundamental displacements and nodal displacement increments can be obtained in the form [11]

$$\delta \underset{\sim i}{p} = \underset{\sim i}{C} \, \delta \underset{\sim i}{a} \tag{26}$$

where

$$\underset{\sim i}{C} = \begin{bmatrix} \underset{\sim}{I}_3 & 0 \\[2mm] \underset{\sim}{0} & \underset{\sim}{V}_i \end{bmatrix} \qquad \underset{\sim i}{V} = \begin{bmatrix} V_{1i} & V_{2i} \\ V_{2i} & V_{3i} \\ V_{4i} & V_{5i} \end{bmatrix} \tag{27}$$

$$V_{1i} = -\frac{1}{e_i}(\cos\alpha_i \cos^2\beta_i + \frac{\sin\alpha_i}{\alpha_i} \sin^2\beta_i)$$

$$V_{2i} = -\frac{1}{e_i} \sin\beta_i \cos\beta_i \, (\cos\alpha_i - \frac{\sin\alpha_i}{\alpha_i}) \qquad \alpha_i = (\theta_{\bar{x}_i}^{\,2}+\theta_{\bar{y}_i}^{\,2})^{\frac{1}{2}}$$

$$V_{3i} = -\frac{1}{e_i}(\cos\alpha_i \sin^2\beta_i + \frac{\sin\alpha_i}{\alpha_i} \cos^2\beta_i \qquad \cos\beta_i = \frac{\theta_{\bar{x}_i}}{\alpha_i} \tag{28}$$

$$V_{4i} = -\sin\alpha_i \cos\beta_i \qquad \sin\beta_i = \frac{\theta_{\bar{y}_i}}{\alpha_i}.$$

$$V_{5i} = -\sin\alpha_i \sin\beta_i$$

Thus, after substituting eq.(26) in eq.(25), it is finally found

$$\delta \underset{\sim}{g} = \sum_1^{n_e} \underset{\sim}{M}_i \underset{\sim}{C}_i \, \delta \underset{\sim}{a}_i = \sum_1^{n_e} \underset{\sim}{G}_i \, \delta \underset{\sim}{a}_i \tag{29}$$

where matrix $\underset{\sim}{G}_i = \underset{\sim}{M}_i \underset{\sim}{C}_i$ can be explicitely obtained using eqs.(24), (27) and (28).

The relationship between Green strains and nodal displacements is simply obtained from eqs.(19) and (29) as

$$\delta \underset{\sim}{\epsilon} = \underset{\sim}{A} \, \delta \underset{\sim}{g} = \underset{\sim}{A} \sum_1^{n_e} \underset{\sim}{G}_i \, \delta \underset{\sim}{a}_i = \sum_1^{n_e} \underset{\sim}{B}_i \, \delta \underset{\sim}{a}_i \tag{30}$$

Explicit expressions of $\underset{\sim}{A}$ and $\underset{\sim}{B}$ matrices can be found in references [11] and [12].

5. CONSTITUTIVE EQUATIONS

The constitutive relationship must be incorporated to the general formulation in an incremental or rate form. This can be written generically as

$$\delta \underset{\sim}{\sigma} = \underset{\sim}{D}^* \, \delta \underset{\sim}{\varepsilon} \tag{31}$$

where $\underset{\sim}{\sigma} = [\sigma x', \sigma y', \tau x'y', \tau x'y', \tau y'z']^T$ is the second Piola-Kirchoff stress vector.

We will not go here into details of the different alternative forms of matrix $\underset{\sim}{D}^*$ for the various types of non-linear material behaviour. A comprehensive study in this subject can be found in reference [13].

6. DISCRETIZED EQUILIBRIUM EQUATIONS

The simplest procedure to obtain the discretized form of the equilibrium equations in a general way is to make use of the virtual work expression which in the total Lagrangian formulation can be written as

$$\int_V \delta \underset{\sim}{\varepsilon}^T \cdot \underset{\sim}{\sigma} dv - \int_V \delta \underset{\sim}{u}^T \underset{\sim}{b} dV - \int_\Gamma \delta \underset{\sim}{u}^T \underset{\sim}{t} d\Gamma = 0 \tag{32}$$

where $\underset{\sim}{\varepsilon}$, $\underset{\sim}{u}$ and $\underset{\sim}{\sigma}$ have been defined in previous sections. V and Γ are the undeformed volume and surface of the structure, over which the body forces $\underset{\sim}{b}$ and surface loads $\underset{\sim}{t}$ respectively act. Vectors $\underset{\sim}{b}$ and $\underset{\sim}{t}$ are defined by

$$\underset{\sim}{b} = [\,b_x, b_y, b_z]^T \quad \text{and} \quad \underset{\sim}{t} = [\,t_x, t_y, t_z]^T \tag{33}$$

where x,y and z refer to components in the global systems of Fig. 2.

Using eqs.(30) and (11)-(14) we can write the discretized finite element form of eq. (32) in the following standard manner [10].

$$\underset{\sim}{\Psi}(\underset{\sim}{a}) = \underset{\sim}{P}(\underset{\sim}{a}) - \underset{\sim}{R} = \underset{\sim}{0} \tag{34}$$

which for the ith node gives

$$\underset{\sim}{\Psi}_i(\underset{\sim}{a}) = \int_V \underset{\sim}{B}_i^T \underset{\sim}{\sigma} dv - \underset{\sim}{R}_i = \underset{\sim}{0} \tag{35}$$

In eq.(34), $\underset{\sim}{\Psi}$ (a) is the residual force vector and $\underset{\sim}{R}$ is the equivalent nodal force vector due to exterior loads given by

Surface loads

$$\underset{\sim}{R}_i = \int_A [\,N_i t_x, N_i t_y, N_i t_z, N_i M_{\bar{x}_i}, N_i M_{\bar{y}_i}\,]^T dA \tag{36}$$

where $M_{\bar{x}_i}$ and $M_{\bar{y}_i}$ are the components of the surface moment in the local axes $\bar{x}$ and $\bar{y}$ of Fig. 2.

Body forces

$$\underset{\sim}{R}_i = \int_V [\,N_i b_x, N_i b_y, N_i b_z, 0, 0]^T \, dV \tag{37}$$

All the integrals which appear in eqs.(32)-(37) are evaluated assembling the contributions from the different elements following standard finite element procedures [10].

Eq.(34) is a system of non linear equations which must be solved using an iterative numerical technique. From the many existing procedures [14] an standard Newton-Raphson algorithm [10] has been chosen here. Thus, the corresponding displacement increment vector for the nth iteration is calculated by

$$\Delta \underset{\sim}{a}^n = -\underset{\sim}{K}_T(\underset{\sim}{a}^n)\ \underset{\sim}{\psi}(\underset{\sim}{a}^n) \tag{38}$$

from which the updated displacement field can be computed as $\underset{\sim}{a}^{n+1} = \underset{\sim}{a}^n + \Delta \underset{\sim}{a}^n$. Iterations stop when the values of the residual forces are suficiently small.

In eq.(38) $\underset{\sim}{K}_T$ is the, so called, tangent matrix, obtained as $K_T(a) = \dfrac{\partial \underset{\sim}{\Psi}(\underset{\sim}{a})}{\partial \underset{\sim}{a}}$.

A typical submatrix relating nodes i and j can be obtained as follows

$$\underset{\sim}{K}_{T_{ij}}(\underset{\sim}{a}) = \frac{\partial \Psi_i(\underset{\sim}{a})}{\partial \underset{\sim}{a}_j} \tag{39}$$

Taking the first variation of eq.(35) we obtain

$$\delta \underset{\sim}{\Psi}_i (\underset{\sim}{a}) = \sum_{j=1}^{n} \underset{\sim}{K}_{T_{ij}} \delta \underset{\sim}{a}_j = \int_v \delta \underset{\sim}{B}_i^T \underset{\sim}{\sigma} dv + \int_v \underset{\sim}{B}_i^T \delta \underset{\sim}{\sigma} dv \tag{40}$$

It can be shown that [12] .

$$\int_v \underset{\sim}{B}_i^T \delta \underset{\sim}{\sigma} dv = \sum_{j=1}^{n} \underset{\sim}{K}_{ij}^L \delta \underset{\sim}{a}_j \tag{41}$$

$$\int_v \delta \underset{\sim}{B}_i^T \underset{\sim}{\sigma} dv = \sum_{j=1}^{n} (\underset{\sim}{K}_{ij}^{\sigma_I} + \underset{\sim}{K}_{ij}^{\sigma_{II}}) \delta \underset{\sim}{a}_j \tag{42}$$

where the different components of the tangent matrix can be evaluated as follows

a) $\underset{\sim}{K}_{ij}^L$

$$\underset{\sim}{K}_{ij}^L = \int_v \underset{\sim}{B}_i^T \underset{\sim}{D}^* \underset{\sim}{B}_j \, dv \tag{43}$$

where matrix $\underset{\sim}{B}$ can be obtained from eqs.(30)

b) $\underset{\sim}{K}_{ij}^{\sigma_I}$

$$\underset{\sim}{K}_{ij}^{\sigma_I} = \int_v \underset{\sim}{G}_i^T \underset{\sim}{S} \underset{\sim}{G}_j \, dv \tag{44}$$

where matrix $\underset{\sim}{G}$ is obtained from eqs.(29) and matrix $\underset{\sim}{S}$ is given by

$$\underset{\sim}{S} = \begin{bmatrix} \sigma x' \underset{\sim}{I}_3 & \tau x'y' \underset{\sim}{I}_3 & \tau x'z' \underset{\sim}{I}_3 \\ \tau x'y' \underset{\sim}{I}_3 & \sigma y' \underset{\sim}{I}_3 & \tau y'z' \underset{\sim}{I}_3 \\ \tau x'z' \underset{\sim}{I}_3 & \tau y'z' \underset{\sim}{I}_3 & 0 \end{bmatrix} \tag{45}$$

c) $\underset{\sim}{K}_{ij}^{\sigma_{II}}$

$$\underset{\sim}{K}_{ij}^{\sigma_{II}} = 0 \qquad \text{for} \qquad i \neq j \tag{46}$$

$$\underset{\sim}{K}_{ij}^{\sigma_{II}} = \begin{bmatrix} \underset{\sim}{0} & \underset{\sim}{0} \\ \underset{\sim}{0} & \underset{\sim}{H}_i \end{bmatrix} \tag{47}$$

where $\underset{\sim}{H}$ is obtained as:

$$\underset{\sim}{H}_i \, (2\times 2) = \int_v \left[\frac{\partial \underset{\sim}{V}_i^T}{\partial \theta_{\bar{x}_i}} \cdot \underset{\sim}{F} \, , \, \frac{\partial \underset{\sim}{V}_i^T}{\partial \theta_{\bar{y}_i}} \cdot \underset{\sim}{F} \right] dv \, , \; \underset{\sim}{F} = \bar{\bar{\underset{\sim}{M}}}_i^T \underset{\sim}{A}^T \underset{\sim}{\sigma} \tag{48}$$

Explicit expressions of matrices $\frac{\partial \underset{\sim}{V}_i^T}{\partial \theta_{\bar{x}_i}}$ and $\frac{\partial \underset{\sim}{V}_i^T}{\partial \theta_{\bar{y}_i}}$ can be found in references [11] and [12].

7. NUMERICAL COMPUTATION OF THE INTEGRALS AND FINITE ELEMENT CHOSEN

All the different integrals are numerically evaluated using a Gauss-Legendre quadrature 10 which gives

$$\int_V f(x',y',z')\,dv = \int_{-1}^{+1}\int_{-1}^{+1}\int_{-1}^{+1} g(\xi,\eta,\zeta)\,d\xi d\eta d\zeta \qquad (49)$$

It can be shown [11] that the following relationships exists between differentials of volume:

$$dV = dx'dy'dz' = \left(1-\frac{t}{R_r}\right)\left(1-\frac{t}{R_s}\right)\frac{h}{2}\,\frac{\frac{\partial x_o}{\partial \xi}\frac{\partial y_o}{\partial \eta}-\frac{\partial x_o}{\partial \eta}\frac{\partial y_o}{\partial \xi}}{\frac{\partial x_o}{\partial r}\frac{\partial y_o}{\partial s}-\frac{\partial x_o}{\partial s}\frac{\partial y_o}{\partial r}}\,d\xi\, d\eta\, d\zeta \qquad (50)$$

where x_0 and y_0 are components of vector $\underline{r}_0$ of eq.(3).

Eq.(50) allows to use the normalized volume in ξ , η and ζ coordinates (see Fig. 3) for the numerical integration of all volume integrals. Same procedure applies for the surface integrals and will not be given here.

With respect to the type of element chosen we have to note that the formulation here presented is, in fact, a "thick" shell formulation, as the terms of shear deformation are included in the analysis. We, therefore, have to be aware that, when using this formulation in the context of very thin shell analysis, some precautions must be taken. It is well known that, for very thin shells or plates, unrealistic overstiff numerical results can be found due to the overestimation of the shear terms which tend to dominate the solution. This numerical error has been studied by many researchers and different alternatives to overcome the problem have been suggested [10]. Here we use one of the simplest procedures which consists in underintegrating the terms due to shear in the numerical integration of the stiffness matrix. This method, commonly known as "reduced integration technique", has been extensively used with success by many authors in the context of thick and thin shell and plate analysis [15]. For the examples presented in the paper, the eight node isoparametric element (Fig. 3) has been chosen with the following integrating rule:

a) A 2x2 Gauss-Legendre rule along directions ξ and η over the middle surface of the shell.

b) A 2 point integration rule over the thickness (τ direction).

8. NUMERICAL EXAMPLES

Example 1. Parabolic Shell

The geometry of the shell and finite element mesh used can be seen in Fig. 7. The problem has been analyzed by a process of incrementing the central deflection from a zero initial value to a value of 50".

Numerical results for the central displacement load elastic curve obtained for different types of boundary conditions at the shell edges are presented in Fig. 8. Numerical results obtained by Wood with para-linear three-dimensional shell elements [7] are also shown for comparison. Note the differences between the boundary conditions imposed by Wood and those used in the present analysis.

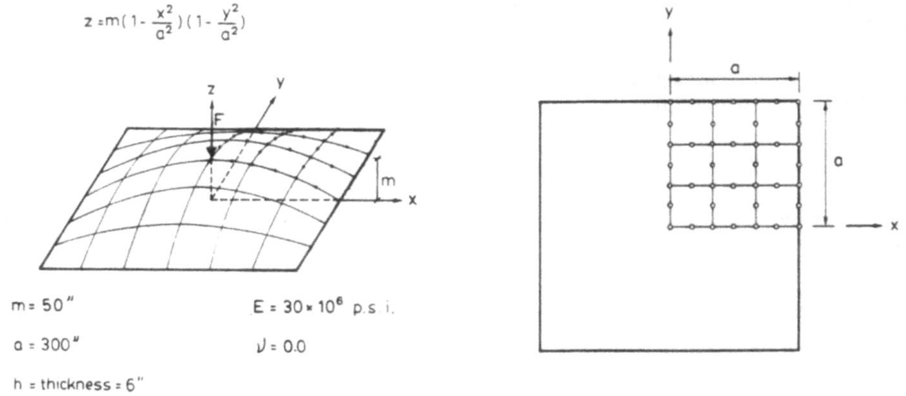

Fig. 7. Parabolic shell under central point load. Geometry, material properties and finite element mesh.

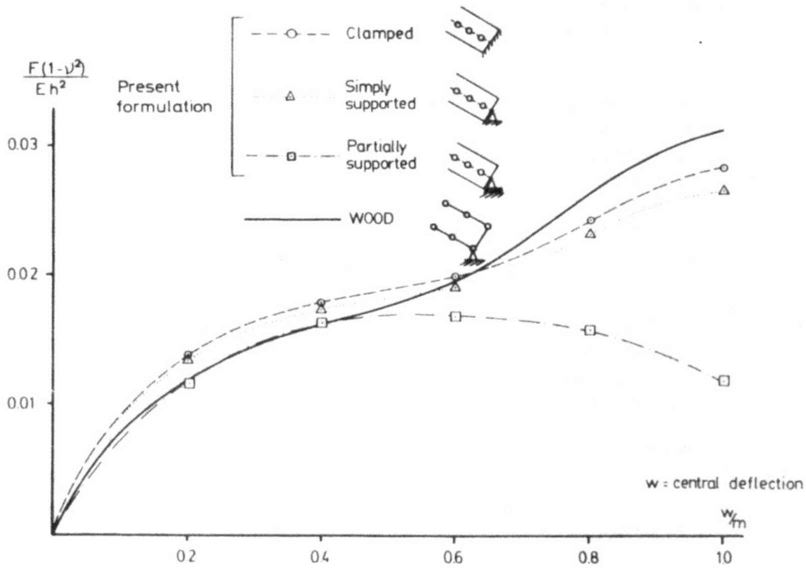

Fig 8. Parabolic shell. Central displacement versus load for different types of boundary conditions.

Example 2. Cylindrical shell.

The cylindrical shell studied can be seen in Fig. 9. Note that only four elements have been used to discretize one quarter of the structure (due to symmetry). The shell is assumed to be simply supported on its straight edges and free on the curved ones.

In the elastic analysis presented an increasing displacement, w, is prescribed in point A and the corresponding vertical reaction force, F, obtained. In Fig. 11 a plot of the value of F and the vertical displacement of point B, at the center of the free edge, versus w is presented for two different values of the shell thickness.

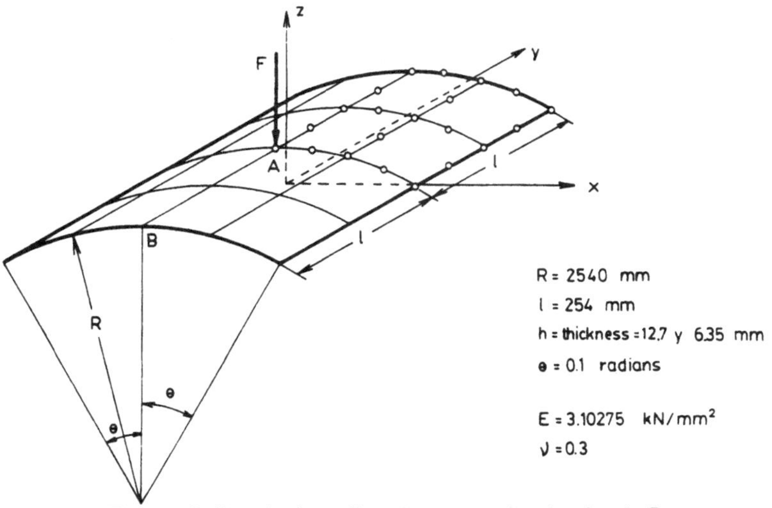

Fig. 9. Cylindrical shell under central point load. Geometry, material properties and finite element mesh.

Note the good agreement between the results obtained with the present formulation and those given by Surana [16] and Sabir and Lock [5] for the same problem.

Example 3. Spherical shell.

The last example is the analysis of a spherical shell with fully clamped edges under an increasing vertical point load acting in the center of the shell, as it can be see in Fig. 10.

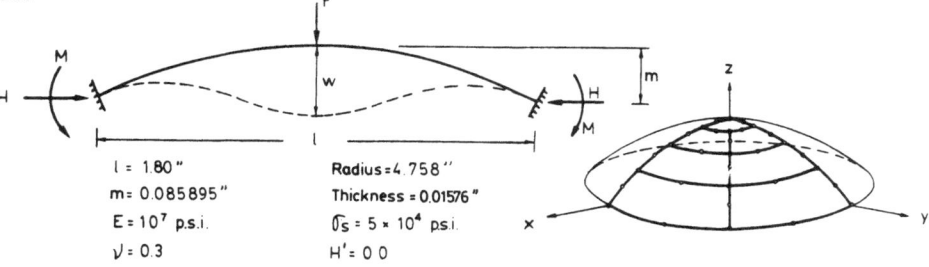

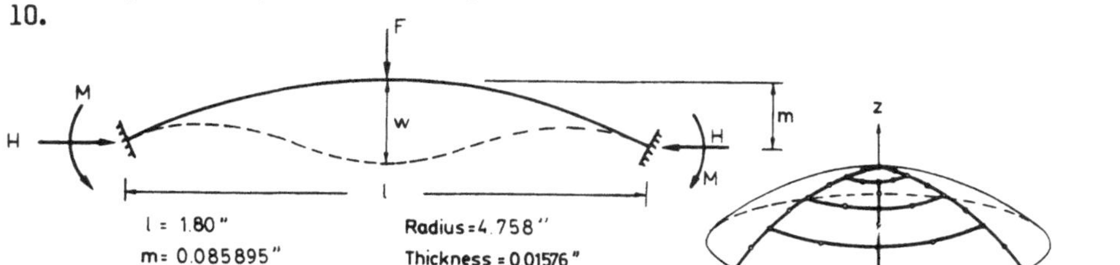

Fig. 10. Spherical shell under central point load. Geometry, material properties and finite element mesh.

Again, due to symmetry, only a quarter of the structure has been considered. A mesh of seven eight noded elements has been used. Results for the load/deflection at the center have been plotted in Fig. 12 and good agreement with results obtained by Wood [7] , with a two-dimensional axi-symmetric formulation, is obtained.

9. FINAL REMARKS

A total Lagrangian finite element formulation for the geometricaly non linear analysis of 3-D shells with large displacement and finite rotations has been presented. The formulation allows for shear deformation effects and large curvatures in the shell surface. Explicit forms of all finite element matrices have been obtained via the use of a local coordinate system based on the principal curvature directions. The

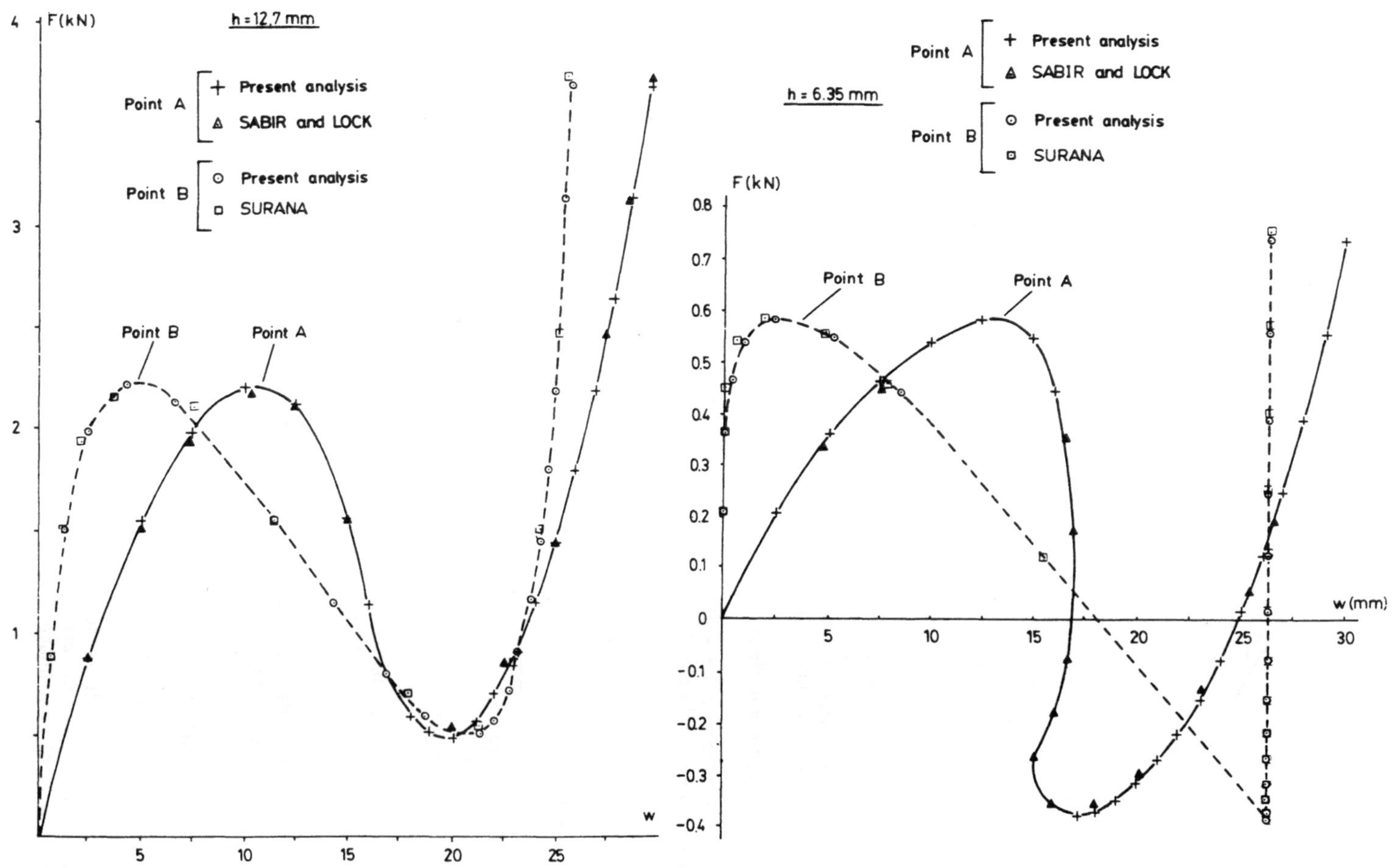

Fig. 11. Cylindrical shell. Deflections versus load for different thickness values.

accuracy of the formulation has been checked in a series of examples of large displacement analysis of shells.

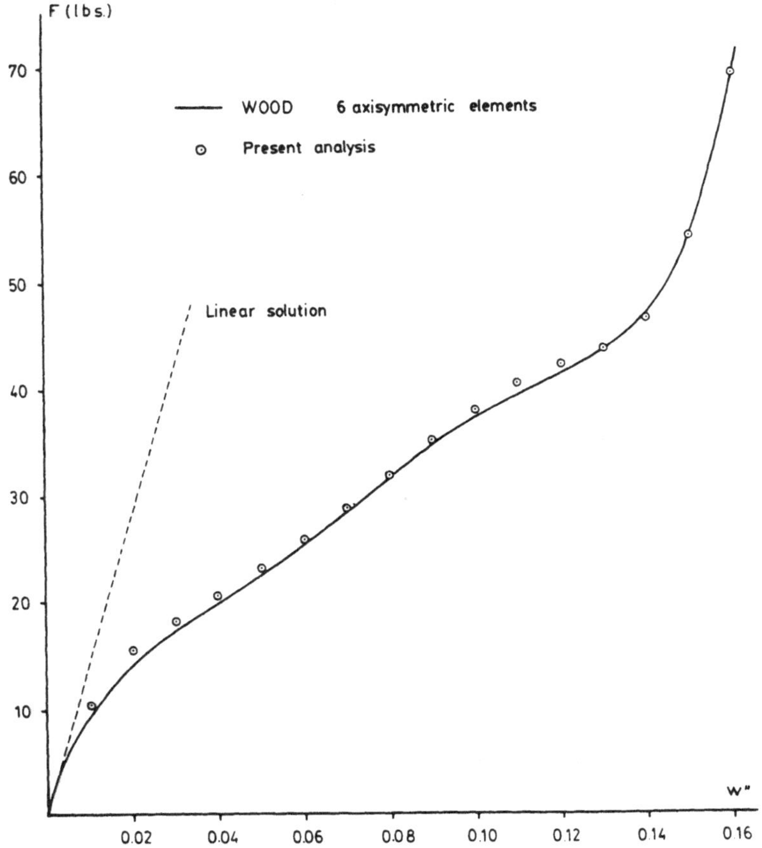

Fig. 12. Spherical shell. Central deflection versus load.

REFERENCES

1. ARGYRIS J.H., BALMER H., KLEIBER M., HINDENLANG U.
"Natural description of large inelastic deformations for shells of arbitrary shape-Application of trump element".
Journ. Computer Meth. Appl. Mech. and Eng., Vol. 22, pp. 361-389, 1980.

2. HUGHES J.R., LIU W.K.
"Non-linear finite element analysis of shells - Part I. Three-dimensional Shells".
Journ. of Comp. Methods in Appl. Mech. and Eng., Vol. 26, pp. 331-362, 1981.

3. NAYAK G.CH.
"Plasticity and large deformation problems by the finite element method".
University of Wales, Swansea, C/Ph/15/71.

4. RAMM E.
"A plate/shell element for large deflections and rotations".
Formulations and Computational Algorithms in Finite Element Analysis, Editors K.J. Bathe, J.T. Oden and W. Wunderlich, M.I.T. Press, 1977.

5. SABIR A.B., LOCK A.C.
"The application of finite elements to the large deflection geometrically non-linear behaviour of cylindrical shells".
Variational Methods in Engineering, Brebbia and Tottenham editors, Southampton Univ. Press, 7, pp. 66-75, 1973.

6. STRUIK D.J.
"Lectures on classical Differential Geometry".
Addison-Wesley Publishing Company Inc. Massachusets, 1961.

7. WOOD R.D.
"The application of finite element methods to geometrically non-linear structural analysis".
University of Wales, Swansea, C/Ph/20/73, 1973.

8. ZIENCKIEWICZ O.C., NAYAK G.C.
"A general approach to problems of large deformation and plasticity using isoparametric elements".
3rd. Conf. on Matrix Methods in Struct. Mech., A.F.I.T., Wright-Pattermon, Ohio, October 1971.

9. WASHIZU K.
"Variational Methods in elasticity and plasticity".
Pergamon Press, Oxford-New York, 2nd. ed., pp. 182-203, 1975.

10. ZIENCKIEWICZ O.C.
"The finite element method".
McGraw-Hill, New York, 1979.

11. OLIVER J.
"Una formulación cuasi-intrínseca para el estudio, por el método de los elementos finitos de vigas, arcos, placas y láminas sometidos a grandes corrimientos en régimen elastoplástico" (in spanish).
Ph.D. Thesis. E.T.S. Ingenieros de Caminos. Universidad Politécnica Barcelona, Spain, 1982.

12. OLIVER J., OÑATE E.
"A total lagrangian formulation for the geometrically non linear analysis of structures using finite elements. Part I. Two-dimensional problems: Shell and Plate structures".
To appear in Int. Jnal. Num. Meth. Eng.

13. KLEIBER M., KONIG J.A., SAWCZUK A.
"Studies on plastic structures: Stability, Anisotropic hardening, Cyclic loads".
Journ. Computer Methods in Appl. Mech. and Eng., Vol. 33, pp. 487-556, 1982.

14. CRISFIELD M.A.
"Incremental/Iterative solution procedure for non-linear structural analysis".
Proceedings of the International Conference on Numerical Methods for Non-Linear Problems, Swansea 1980 - Pineridge Press, Swansea, U.K.

15. PUGH E.D.L., HINTON E., ZIENCKIEWICZ O.C.
"A study of quadrilateral plate bending elements with reduced integration".
Int. Jnal. Num. Meth. Eng., Vol. 12, pp. 1059-79, 1978.

16. SURANA K.S.
"Geometrically non-linear formulation for the curved shell elements".
Int. Jnal. Num. Meth. Eng., Vol. 19, pp. 581-615, 1983.

Large Deformations of Elastic Conical Shells

W. HÜBNER

Hochschule der Bundeswehr München
Fachbereich Luft- und Raumfahrttechnik
Institut für Mechanik
Neubiberg, Federal Republic of Germany

Summary

Axisymmetric conical shells under axial forces are investigated. A geometrically nonlinear approach leads to the REISSNER-MEISSNER equations, which allow the calculation of large deformations. These two nonlinear second order equations have been integrated by a matrix method, suggested by E.L. AXELRAD. Another effective solution, suitable for a small computer, uses the RUNGE-KUTTA-integration combined with an iteration program for the unknown boundary values. Useful results for the practical design of conical springs have been published, which give slight corrections to the famous ALMEN-LASZLO-paper of 1936. While conical springs are rather flat and thick, the presented theory can be used for steep and thin shells. Then the "spring-characteristics", the force-deflection curves, become very complicated and their stability must be discussed. A simple criterion of stability is derived here out of the DIRICHLET definition. Presented as a problem of catastrophe theory, interesting curves in the parameter plane are obtained.

1. Derivation of the REISSNER-MEISSNER-Equations

The conical shells investigated in this paper and its most important notations are shown in Fig. 1.

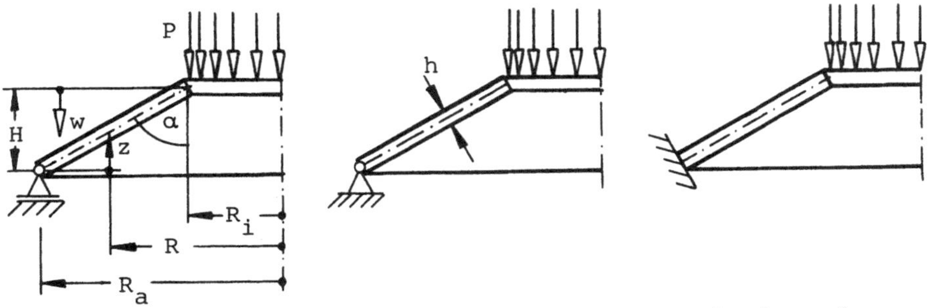

a) simply supported b) hinged c) clamped

Fig.1. Conical shells with different boundary conditions.

The geometrical displacement configuration is shown in Fig. 2. On the left-hand side the undeformed state is presented. The deformed state on the right is marked by an asterisk. The meridional tangent to the middle surface has the tilt angle α vs the vertical axis of symmetry.

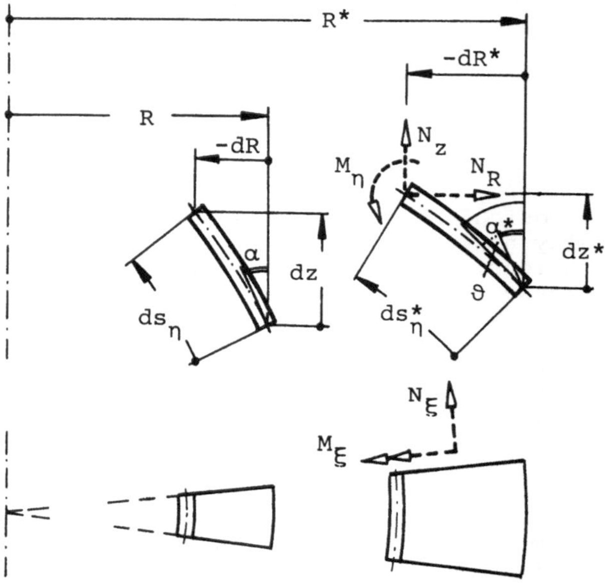

Fig.2. Shell element with deformation geometry, section forces, and section moments.

Important variables are the angle of tangent change $\vartheta = \alpha^* - \alpha$ as well as the curvature changes in circumferential and in meridional direction

$$\varkappa_\xi = -\frac{1}{R}(\cos\alpha - \cos\alpha^*) \ , \quad \varkappa_\eta = \frac{d\vartheta}{ds_2} = -\frac{d\vartheta}{dR}\sin\alpha \ . \quad (1),(2)$$

By differentiating the radial displacement relationship the compatibility condition is obtained

$$u_r = R\varepsilon_\xi = R^* - R \ , \quad \frac{d}{dR}(R\varepsilon_\xi) = (1+\varepsilon_\eta)\frac{\sin\alpha^*}{\sin\alpha} - 1 \ , \quad (3),(4)$$

where ε_ξ and ε_η are the hoop strain and the meridional strain of the middle surface. In the z-direction the deformation is given by

$$dz^* = -\frac{dR}{\sin\alpha}(1+\varepsilon_\eta)\cos\alpha^* . \qquad (5)$$

The section forces and moments N_R, N_z, M_ξ, M_η are defined per unit length of the shell's middle surface in the undeformed configuration and sketched in Fig. 2 at the upper section edge. As the presented problem is symmetric with respect to geometry and load, the shearing forces, the twisting moments and the transverse shear Q_ξ vanish.

The e q u i l i b r i u m e q u a t i o n s will be derived using the abbreviation

$$[\ ]' = \frac{d[\ ]}{dR} = -\frac{d[\ ]}{ds_\eta}\frac{1}{\sin\alpha} .$$

The stress function usually is introduced by $dV/ds_\eta = N_\xi$ or here

$$N_\xi = -V'\sin\alpha . \qquad (6)$$

The radial equilibrium without surface loads $(N_R R)' + N_\xi/\sin\alpha = 0$ together with (6) leads to

$$V = N_R R . \qquad (7)$$

In the direction of the symmetry-axis we get

$$(N_z R)' = 0 , \qquad \text{wherefrom} \quad N_z R = -\frac{P}{2\pi} = \text{const.} \qquad (8)$$

follows. Here P is the axisymmetric load applied at the inner edge. Assuming small displacements $1+\varepsilon_\eta \approx 1$, we obtain the bending moment equilibrium

$$(M_\eta R)'\sin\alpha - M_\xi\sin\alpha^* + V\cos\alpha^* - \frac{P}{2\pi}\sin\alpha^* = 0 . \qquad (9)$$

The specific normal force in meridional direction follows from

$$N_\eta = -N_R \sin\alpha^* + N_z \cos\alpha^* = -\frac{1}{R}(V\sin\alpha^* + \frac{P}{2\pi}\cos\alpha^*) \ . \qquad (10)$$

The elastic law of the shell is obtained by using the normal hypothesis (KIRCHHOFF-LOVE)

$$Eh\varepsilon_\xi = N_\xi - \nu N_\eta \ , \quad Eh\varepsilon_\eta = N_\eta - \nu N_\xi \ , \qquad (11)$$

$$M_\xi = D(\varkappa_\xi + \nu\varkappa_\eta) \ , \quad M_\eta = D(\varkappa_\eta + \nu\varkappa_\xi) \ , \qquad (12)$$

where E is the elastic modulus, ν the POISSON's ratio,

$$D = Eh^3/[12(1-\nu^2)] \qquad (13)$$

the flexural rigidity, and h the thickness of the shell. Using the above relationships, we get out of (11) and (12)

$$\begin{aligned} Eh\varepsilon_\xi &= -V'\sin\alpha + \frac{\nu}{R}(V\sin\alpha^* + \frac{P}{2\pi}\cos\alpha^*) \ , \\ Eh\varepsilon_\eta &= -\frac{1}{R}(V\sin\alpha^* + \frac{P}{2\pi}\cos\alpha^*) + \nu V'\sin\alpha \ , \end{aligned} \qquad (14)$$

$$\begin{aligned} M_\xi &= -\frac{D}{R}[(\cos\alpha - \cos\alpha^*) + \nu R\vartheta' \sin\alpha] \ , \\ M_\eta &= -\frac{D}{R}[R\vartheta'\sin\alpha + \nu(\cos\alpha - \cos\alpha^*)] \ . \end{aligned} \qquad (15)$$

The first of the REISSNER-MEISSNER-equations follows out of the nonlinear compatibility equation (4) by inserting the strains (14)

$$\begin{aligned} &-(RV')'\sin\alpha + \underline{\nu(V\cos\alpha^* - \frac{P}{2\pi}\sin\alpha^*)\vartheta'} - Eh(\frac{\sin\alpha^*}{\sin\alpha} - 1) + \\ &+ \frac{\sin\alpha^*}{\sin\alpha}\frac{1}{R}(V\sin\alpha^* + \underline{\frac{P}{2\pi}\cos\alpha^*}) = 0 \ . \end{aligned} \qquad (16)$$

The second equation is obtained from the bending moment equilibrium (9) with (15)

$$\begin{aligned} &-(R\vartheta')'\sin^2\alpha + \frac{1}{R}(\cos\alpha - \cos\alpha^*)\sin\alpha^* + \\ &+ \frac{V}{D}\cos\alpha^* - \frac{P}{2\pi D}\sin\alpha^* = 0 \ . \end{aligned} \qquad (17)$$

These are two nonlinear differential equations for the unknown variables ϑ and V. As $d\alpha/dR = 0$ has been used, they are valid for conical shells only. If flat conical shells are considered, the underdotted terms are numerically less important. Here however they cannot be dropped.

2. Solution Methods and Boundary Conditions

The integration of (16) and (17) is done by writing them as a system of first order differential equations. This system is completed by the equations for R* and z* from (3), (4), and (5), in order to get the deformed meridian. Therefore it is reasonable to use the approximation $1+\varepsilon_\eta \approx 1$. The system reads as follows

$$\begin{aligned}
\vartheta' &= (R\vartheta')/R \,, \\
(R\vartheta')' &= [\tfrac{1}{R}(\cos\alpha - \cos\alpha^*)\sin\alpha^* + \\
&\quad + \tfrac{V}{D}\cos\alpha^* - \tfrac{P}{2\pi D}\sin\alpha^*]/\sin^2\alpha \,, \\
V' &= (RV')/R \,, \\
(RV')' &= [\nu(V\cos\alpha^* - \tfrac{P}{2\pi}\sin\alpha^*)\vartheta'\sin\alpha + \\
&\quad + \tfrac{1}{R}(V\sin\alpha^* + \tfrac{P}{2\pi}\cos\alpha^*)\sin\alpha^* - \\
&\quad - Eh(\sin\alpha^* - \sin\alpha)]/\sin^2\alpha \,, \\
z^{*\prime} &= -\cos\alpha^*/\sin\alpha \,, \\
R^{*\prime} &= \sin\alpha^*/\sin\alpha \,.
\end{aligned} \tag{18}$$

The variables are regarded as components of the state vector

$$\mathbf{x}^T = [\vartheta,\ (R\vartheta'),\ V,\ (RV'),\ z^*,\ R^*] \,, \tag{19}$$

so that this system of differential equations has the form $x_j' = f_j(x_k)$, which is suitable for numerical integration, e.g. by the RUNGE-KUTTA-method.

Due to Fig. 1, three different boundary conditions for the outer edge are discussed. It is either simply supported (a) or hinged (b) or clamped (c).

In the simply supported case (a) the radial normal force $N_R(R_a)$ and the bending moment $M_\eta(R_a)$ vanish. Then we get from (7) and (15)

$$V_a = V(R_a) = 0 \; , \tag{20}$$

$$(R\vartheta')_a = - \frac{\nu}{\sin\alpha} [\cos\alpha - \cos(\alpha+\vartheta_a)] \; . \tag{21}$$

The radial displacement of the outer edge $u_a = R^*(R_a) - R_a$ is obtained out of (3) and (14)

$$u_a = \frac{1}{Eh} [-(RV')_a \sin\alpha + \nu V_a \sin(\alpha+\vartheta_a) + \frac{\nu P}{2\pi} \cos(\alpha+\vartheta_a)] . \tag{22}$$

The two terms ϑ_a and $(RV')_a$ are unknown.

In the hinged case (b) there is $u_a = 0$, and we get from (22) the equation

$$V_a = (RV')_a \frac{\sin\alpha}{\nu \sin(\alpha+\vartheta_a)} - \frac{P}{2\pi} \cot(\alpha+\vartheta_a) \; . \tag{23}$$

As in the simply supported case (a) equation (21) holds, and ϑ_a as well as $(RV')_a$ are unknown.

In the clamped case (c) besides $u_a = 0$, there is

$$\vartheta_a = 0 \; . \tag{24}$$

The value V_a follows from (23), while $(R\vartheta')_a$ and $(RV')_a$ are unknown.

In each of the three cases, the state vector at the outer edge $\mathbf{x}_a$ contains two unknown components. As $\mathbf{x}_a$ is the starting vector of the integration, the unknown variables have to be varied such, that at the end of the integration - at the inner edge - the radial normal force N_R and the bending moment M_η vanish, i.e.

$$V_i = V(R_i) = 0 \;, \qquad (25)$$

$$(R\vartheta')_i \sin\alpha + \nu[\cos\alpha - \cos(\alpha+\vartheta_i)] = 0 \;. \qquad (26)$$

For this problem, a computer program has been developped and successfully applied. In some domains, it has been necessary, to vary the load P instead of the geometric unknown ϑ_a in the cases (a) and (b), or the unknown $(R\vartheta')_a$ in case (c).

3. Spring Characteristics and Deformed Meridians

Integration of equations (18) leads to many interesting results, out of which the deflection at the inner edge

$$w = H - z^*(R_i) \;, \qquad (27)$$

dependent on the load P applied there, shall be discussed in detail. This dependence is presented - as usual in engineering - as a spring characteristic $P = f(w)$. Studying the influence of different parameters of flat, conical shells [3], it has been shown, that the form of the spring characteristics mainly depends on H/h, i.e. the relationship of initial height to shell thickness. This is valid also for the steeper shells, discussed here. They are much more dependent on H/h than on the angle α. Regarding also different thicknesses h, the shells are better characterized by the parameter $H/(h \sin\alpha)$, which for flat shells ($\alpha \to 90^\circ$) is very close to H/h. It is also useful to give the spring characteristics in a normed graph P/P_H versus w/H, as it is usually done with conical springs in [1] and [3]. The value P_H is the load where flat springs get "plates" (w=H). It has been given in [3] as $P_H = (4\pi D\cos\alpha)/(R_a+R_i)$. So the presentation in Fig. 3 is approximately valid for a variety of parameters, though it has been computed for $h = 0{,}32$ mm, $R_a = 50$ mm, $R_i = 25$ mm, $E = 4900\ N/mm^2$, $\nu = 0{,}3$. This figure, of course, is based on axial symmetry. If radial buckling is taken into account, the graphs will show befurcation, depending on the wall thickness h.

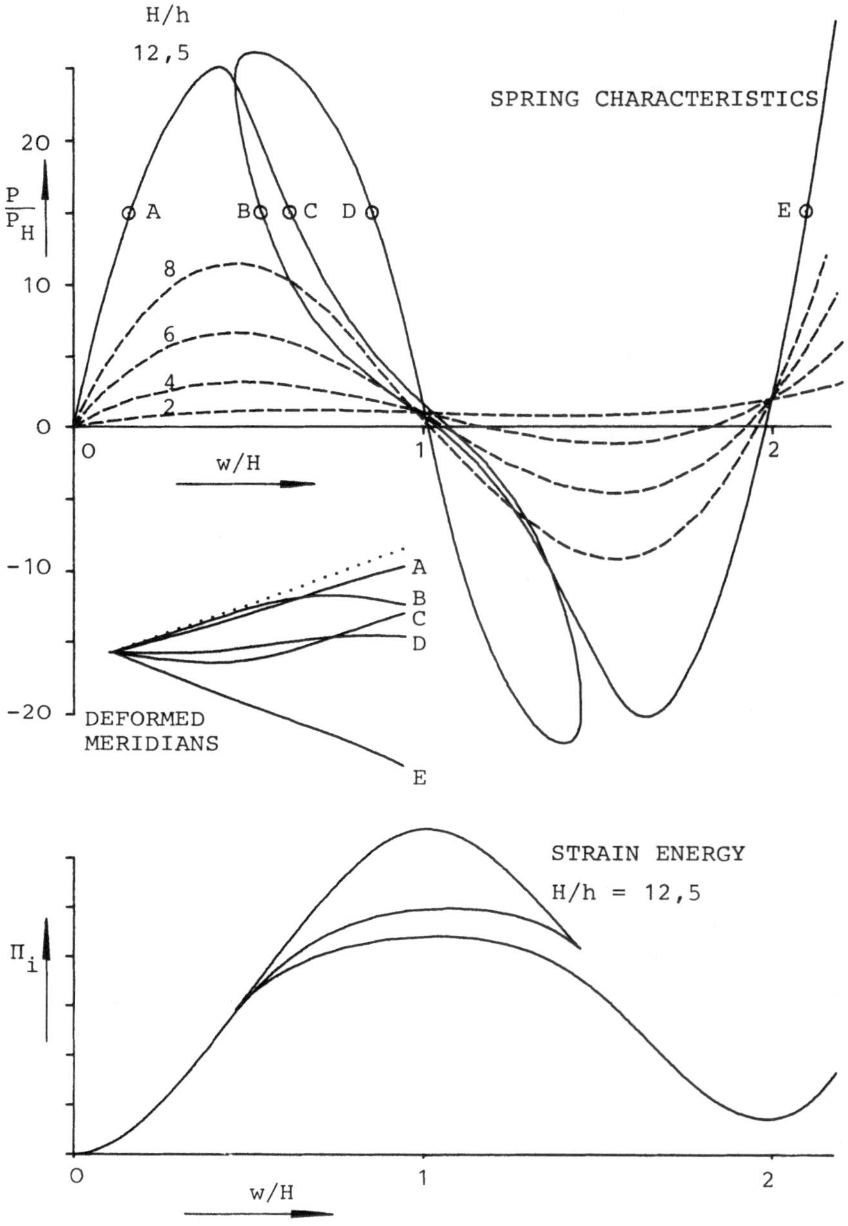

Fig.3. Force-deflection curves (spring characteristics) at different parameters H/h with some deformed meridians and strain energy (H/h = 12,5).

Fig. 3 shows the spring characteristics for the simply supported case (a). The flattest graph here (H/h=2) corresponds to the steepest of [4] and [7]. For bigger values H/h, the well known collapse characteristics are obtained, further increasing of H/h leads to more complicated graphs. As an example, H/h = 12,5 is plotted. Therefore the deformed meridians corresponding to the points A, B, C, D, E are shown as well as the strain energy Π_i.

For spring characteristics in the h i n g e d case (b) similar phenomena as discussed above can be observed, however at lower H/h-values. In the c l a m p e d case (c) there is no "snap through". As physically obvious, the springs of case (c) have high load capacities, however they are of less interest for stability discussion.

4. Potential Energy and Stability

Every spring characteristic in Fig. 3 shows equilibrium states, the stability of which cannot be directly evaluated. The best way to examine this stability is to use the energetic criterion of DIRICHLET. Its derivation is to be found in [2]. Therefore the potential energy Π is used. As the load P is conservative, it is

$$\Pi = \Pi_i + \Pi_o = \Pi_i - Pw \ , \tag{28}$$

where the deflection w at the inner edge is given by (27). The strain energy $\Pi_i = \int \sigma_{jk}\, \varepsilon_{jk}\, dV/2$ leads to

$$\Pi_i = \frac{1}{2} \int [N_\xi \varepsilon_\xi + N_\eta \varepsilon_\eta + M_\xi \varkappa_\xi + M_\eta \varkappa_\eta]\, dA \ , \tag{29}$$

and together with the material equations to

$$\Pi_i = \frac{\pi}{\sin\alpha} \int_{R_a}^{R_i} R[\frac{1}{Eh}\ (N_\xi^2 - 2\nu N_\xi N_\eta + N_\eta^2) + \\ + D(\varkappa_\xi^2 + 2\nu\varkappa_\xi\varkappa_\eta + \varkappa_\eta^2)]dR \ . \tag{30}$$

With (1), (2), (6), and (10), Π_i is obtained depending on the variables of the state vector. The computation is done by adding

$$\Pi_i' = \frac{\pi}{R}\left\{\frac{1}{Eh}\left[(RV')^2 \sin\alpha - 2\nu(RV')(V\sin\alpha^* + \frac{P}{2\pi}\cos\alpha^*) + \right.\right.$$
$$\left. + (V\sin\alpha^* + \frac{P}{2\pi}\cos\alpha^*)^2/\sin\alpha\right] +$$
$$+ D[(\cos\alpha-\cos\alpha^*)^2/\sin\alpha + 2\nu(R\vartheta')(\cos\alpha-\cos\alpha^*) +$$
$$\left. + (R\vartheta')^2 \sin\alpha]\right\} \tag{31}$$

to the system (18). Load P in (31) is used here instead of the variables ϑ, V and their derivatives, jet the function $P[\vartheta,(R\vartheta'),V,(RV')]$ is not known explicitely. Therefore

$$\partial\Pi_i/\partial P = 0 \ . \tag{32}$$

The criterion of DIRICHLET says that the equilibrium states are given by extrema of Π ($\delta\Pi = 0$); they are stable if Π has a minimum ($\delta^2\Pi > 0$). If the load at the edge P is conservative, and there are no other loads - as it is the case here - the presentation $\Pi(w)$ allows the following simple stability discussion: From (28) we get with P = const.

$$\delta\Pi = \frac{\partial\Pi}{\partial w}\,\delta w = (\frac{\partial\Pi_i}{\partial w} - P)\ \delta w \ , \tag{33}$$

and therefore for the equilibrium states

$$\frac{\partial\Pi_i}{\partial w} = P \ . \tag{34}$$

As this is valid for every point (P,w) of the spring characteristic, also for the total curve P = f(w)

$$\frac{\partial\Pi_i}{\partial w} = f(w) \tag{35}$$

holds. Therefore it is not necessary to compute $\Pi_i(w)$ by a labourous integration of (31), but it can be obtained from spring characteristic P = f(w), the points of which are given by the simple integration

$$\Pi_i(w) = \int_0^w f(\zeta)\,d\zeta \quad . \tag{36}$$

The numerical coincidence with the value computed by (31) has been shown. Differentiating (33), we get for a given equilibrium (P,w)

$$\delta^2\Pi = \frac{\partial^2\Pi}{\partial w^2}\,\delta w^2 = \left(\frac{\partial^2\Pi_i}{\partial w^2} - 0\right)\delta^2 w \ , \tag{37}$$

and thereform as a necessary and sufficient condition of stability

$$\frac{\partial^2\Pi_i}{\partial w^2} > 0 \ . \tag{38}$$

Because of (35) it is not necessary to differentiate Π_i twice with respect to w, but we get directly

$$\frac{\partial f(w)}{\partial w} > 0 \ . \tag{39}$$

This means, the slope of the spring characteristic $P = f(w)$ directly gives the stability criterion: sections with positive slope are stable, sections with negative slope are unstable, and points with horizontal tangent are indifferent.

5. Presentation as a Catastrophe Theory Problem

The mechanical examples of catastrophe theory are frequently of the typ of our problem [8]. The presentation in catastrophe theory is normally as follows: The spring characteristics are plotted in a thre dimensional graph (Fig. 4) with the initial slope - or here H/h - as the third parameter. The spring characteristics form the "equilibrium surface", which in our case is rather complicated. The projection of this surface in the "catastrophe map", i.e. the P-H/h-parameter plane, leads to traces, which characterize the type of catastrophe.

For the flat conical spring, the famous cusp - curve is

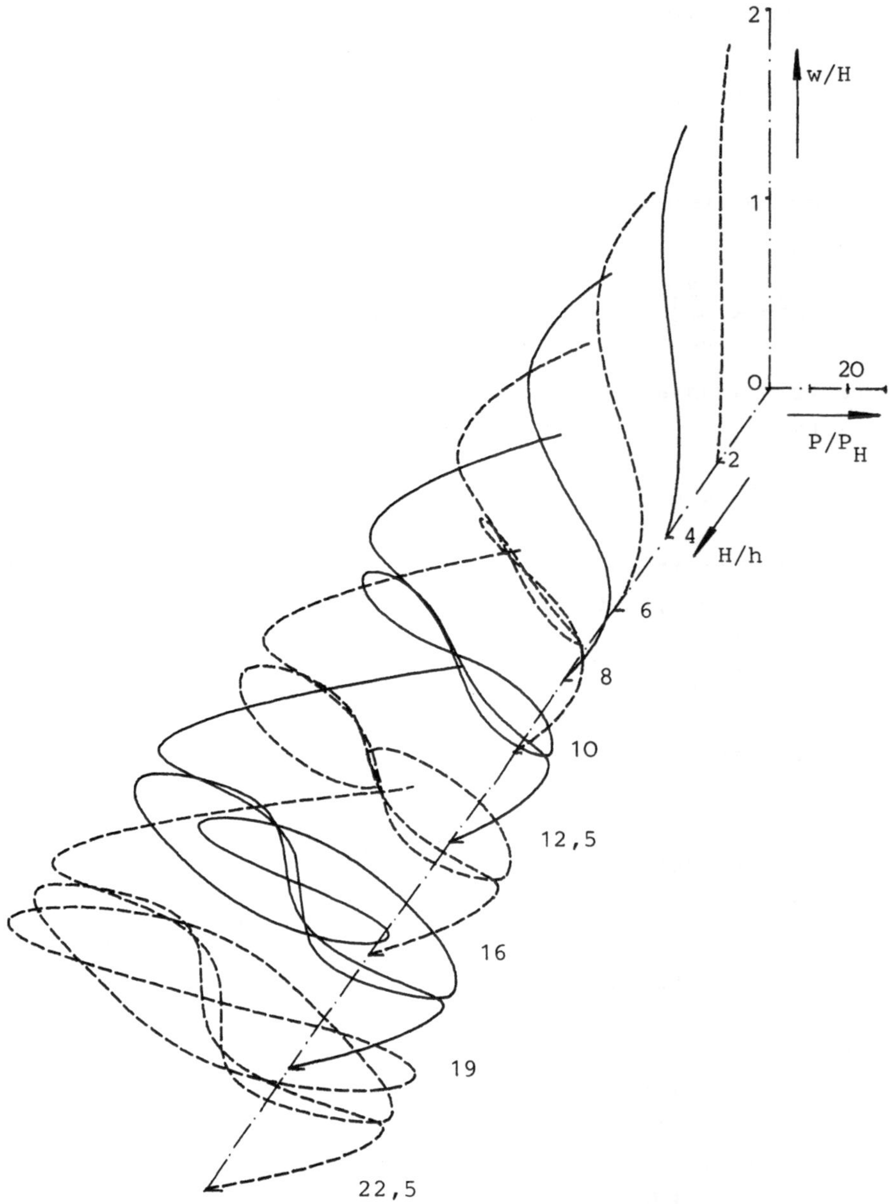

Fig.4. Spring characteristics with different H/h in a three-dimensional presentation.

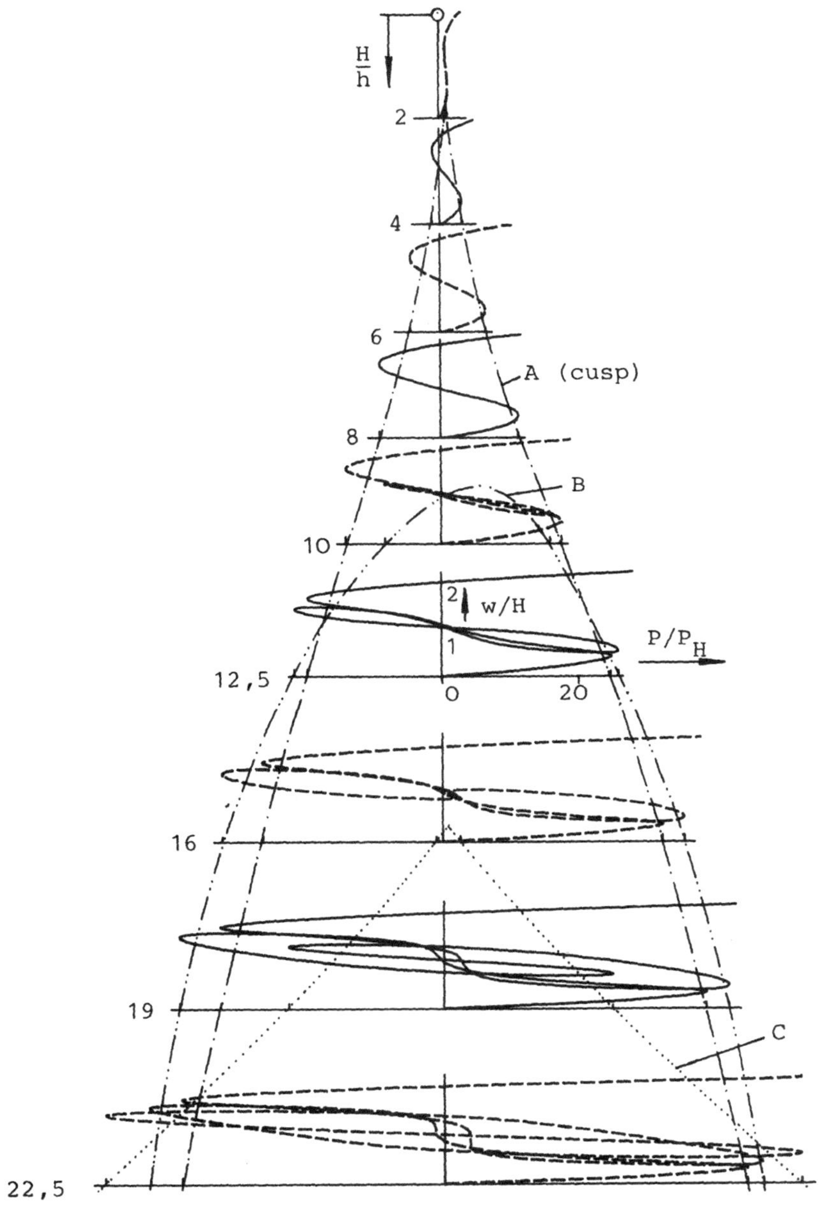

Fig.5. Catastrophe map with cusp A and two other traces B and C.

obtained by projecting the extremal P-values in the parameter plane. For steeper shells this operation becomes more clear in Fig. 5, where the third dimension is obtained by shifting the zero point of the spring characteristic downwards according to H/h. If the extrema are projected on the individual P-axis, we get from the first maximum and the last minimum the well known cusp. With increasing H/h, additional extrema of P arise, which form other traces. The diagram Fig. 5 gives a global presentation of our axisymmetric nonlinear problem. However it does not fit in the catalogue of catastrophe theory [8], where regionally linearized problems are treated.

6. References

1. Almen, J.O.; Laszlo, A.: The Uniform-Section Disk Spring. Trans. ASME 58(1936) 305-314.

2. Parkus, H.: Mechanik der festen Körper. Wien, New York: Springer 1966.

3. Axelrad, E.L.: Schalentheorie. Stuttgart: Teubner 1983.

4. DIN 2092: Tellerfedern; Berechnung (1978).

5. Emmerling, F.A.: Nichtlineare Biegung und Beulen von Zylindern und krummen Rohren bei Normaldruck. Ing. Archiv 51 (1982) 1-16.

6. Hübner, W.; Emmerling, F.A.: Axialsymmetrische große Deformationen einer elastischen Kegelschale. ZAMM 62 (1982) 408-411.

7. Hübner, W.: Deformationen und Spannungen bei Tellerfedern. Konstruktion 34(1982) 387 - 392.

8. Poston, T.; Stewart, I.: Catastrophe Theory and its Applications. London: Pitman 1978.

The Mechanics of Drape

D. W. LLOYD

Department of Textile Industries,
University of Leeds

Summary

The special difficulties of modelling the complex deformations of textile fabrics are discussed in the context of flexible shell theory, and the particular requirements of analysis methods in this area are described. Examples of previous attempts to model particular problems of complex fabric deformations are used to illustrate the advantages and disadvantages of the methods used. Finally, an outline is given of a computational approach which is under development at Leeds at the moment.

Introduction

"Drape" is normally used to describe the way in which a textile fabric forms elegant, doubly curved folds when deformed under its own weight. However, drape is increasingly being used to describe the wider range of complex deformations exhibited by fabrics, and this wider meaning of drape will be assumed here. Textile fabrics exhibit, in drape, deformation behaviour beyond that normally encountered in flexible shells. A sheet of paper, if held flat, will bend under its own weight; if bent about one axis it becomes stiffened, and will resist bending about a second, normal, axis. This behaviour belongs precisely to the class of shell behaviour defined at the start of the colloquium as "flexible". A fabric, in contrast, suffers no stiffening under these circumstances, and will bend through large displacements about two normal axes simultaneously without difficulty. This type of complex deformation can occur because fabrics are different to other materials, in that they exhibit a structure that is coarse on a macroscopic level.

These features of fabric deformations, the coarse physical structure, and the extremely large deformations possible, present unique difficulties of analysis in the context of flexible shell theory. The coarse structure enables a fabric to behave partly as a mechanism; this imparts exceptional mechanical properties to fabrics, and ensures that the mechanisms of deformation in fabrics are different to other materials. This may be seen by considering a doubly curved configuration imposed on different sheet materials. From differential geometry it is known that double curvature requires local changes in the area of the sheet. Rubber achieves this area change through a Poisson effect, by changes of thickness. Sheet metal, in car bodies for instance, is forced into permanent area changes through plastic strains. As neither mechanism is available in paper, paper is incapable of adopting double curvature; instead it exhibits regions of single curvature which meet at sharp points - regions of area change over an infinitesimal area - and exhibit a "crumpled" appearance. In a woven fabric, in contrast, the yarns can behave as elements of a lattice, with only frictional constraints at joints. The shear modulus is typically a factor of 10^5 less than the tensile moduli, so area changes are accommodated through easy shear.

Any approach at analysing the complex deformations of fabrics which is based on modelling the individual yarns in a fabric, with realistic yarn properties and all the boundary conditions at each interlacing specified, will lead to impossibly many degrees of freedom; so the fabric is normally replaced theoretically by an "equivalent" continuum. The consequences of the coarse structure, and its effects on fabric behaviour, lead to this continuum being represented as two-dimensional, as the thickness of the fabric does not play the same role as in a truly continuous material. This also corresponds to the way that the fabric mechanical properties and stresses are measured in practice, as, for example, forces per unit width instead of per unit cross-sectional area.

The need for theoretical analysis of very flexible fabric shells is two-fold. There is insufficient understanding of the drape behaviour of traditional textiles - clothes, furnishings, etc. Also, as fabrics find increasing use as primary structural elements in major engineering structures (for example, the technical limit of a fabric dome is about 350 m in diameter, compared to 45 m for masonry (1)), and as important components, accurate and reliable design and stress analysis techniques are required. This need is largely unfulfilled, as most of the textile mechanics literature is concerned with predicting fabric properties from structural models.

If a design and stress analysis technique is to be useful, it must be possible for the methods and results to be comprehended at a useful level by interested non-specialists, as the number of specialists in textile mechanics is very small.

Elastica Theory

An obvious approach to overcome the difficulties of analysis is to make simplifications in particular cases. In the case of fabrics bent about a single axis, or about parallel axes, the fabric cross-section may be modelled using planar elasticas. Although it is not difficult to include extension of the elastica(2), the method will be illustrated by an example of an inextensible elastica.

The example is that of the folding of heavy fabric sheets under their own weight, when fed vertically downwards.(3) This type of deformation occurs in a number of production processes in the fabric and clothing industries. It is convenient to define the so-called "bending length" of the fabric, b, as $(A/w)^{\frac{1}{3}}$, where A is the fabric bending rigidity and w the fabric weight per unit length. It is a simple matter to transform forces and distances to dimensionless form using b and w, to obtain the following expression of the differential equations of the elastica:

$$\frac{d^2\theta}{d\sigma^2} - \phi_x \sin\theta + \phi_y \cos\theta = 0$$

$$\frac{d\phi_x}{d\sigma} = 0$$

$$\frac{d\phi_y}{d\sigma} = 1$$

$$\frac{dX}{d\sigma} = \cos\theta$$

$$\frac{dY}{d\sigma} = \sin\theta$$

where σ is dimensionless arc length ϕ_x and ϕ_y are components of the dimensionless forces acting on the elastica in the X and Y directions, and θ is the angle between the tangent to the elastica and the X axis. The boundary conditions, and the stages in the deformation are illustrated in Figures 1 to 3, and the process is more fully described elsewhere.(3) An analysis, including extensibility, relevant to corrugated or folded plates subject to tension is described in Reference (2).

These analyses were carried out using the UMIST Bending Curve Package, which enables the integrations, specification of boundary conditions, initial curvatures, etc. to be done in a convenient manner.

Methods based on elastica theory have the advantages of simple application, and of the formulation of the problem, if not the details of the solution

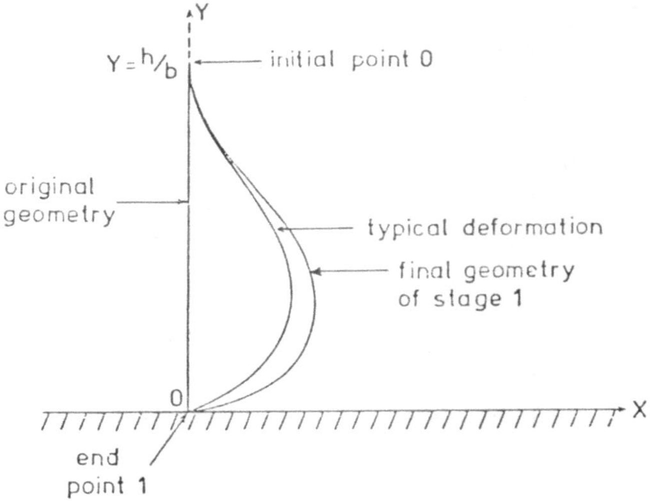

stage 1

point	0	1
σ	0	l/b *
$\emptyset_X$	?	
$\emptyset_Y$	?	
θ	$-\pi/2$	
$\frac{d\theta}{d\sigma}$	?	0
X	0	0
Y	h/b	0

* l/b increased incrementally to model physical process

Figure 1

Elastica Model of Deformed Fabric

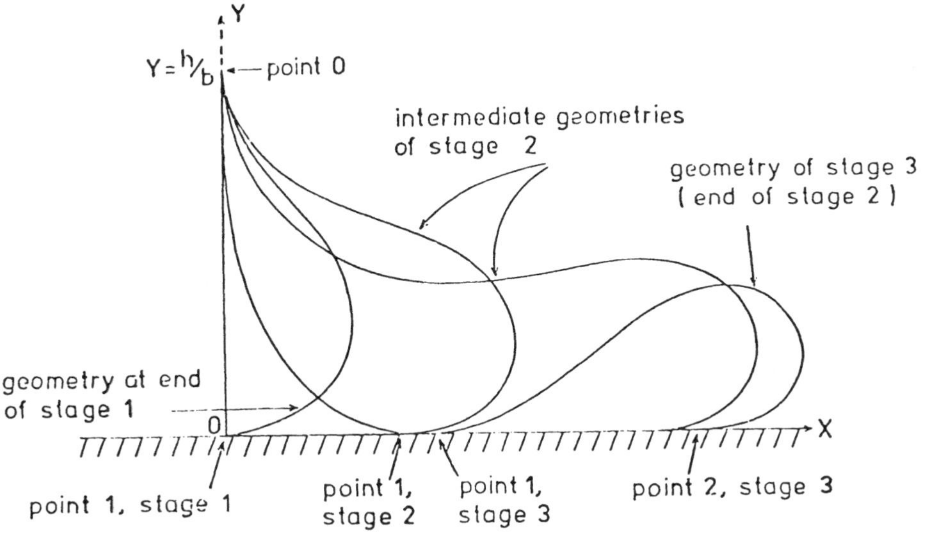

stage 2

point	0	1
σ	0	? (= L_1)
$\emptyset_X$	?	
$\emptyset_Y$	?	
θ	$-\pi/2$	$-\pi$
$\frac{d\theta}{d\sigma}$	?	0
X	0	$(l/b - L_1)$*
Y	h/b	0

* rolling condition as l/b increased

stage 3

point	0	1	2
σ	0	? [= L_{1c}]	? [= $L_{1c}+L_{2c}$]
$\emptyset_X$	?		
$\emptyset_Y$	?		
θ	$-\pi/2$	0	$-\pi$
$\frac{d\theta}{d\sigma}$	?		0
X	0	[= X_{1c}]	[= X_{2c}]
Y	h/b	0	0

values in square brackets are numerical constants recorded for later use

Figure 2

Elastic Model of Deformed Fabric (Continued)

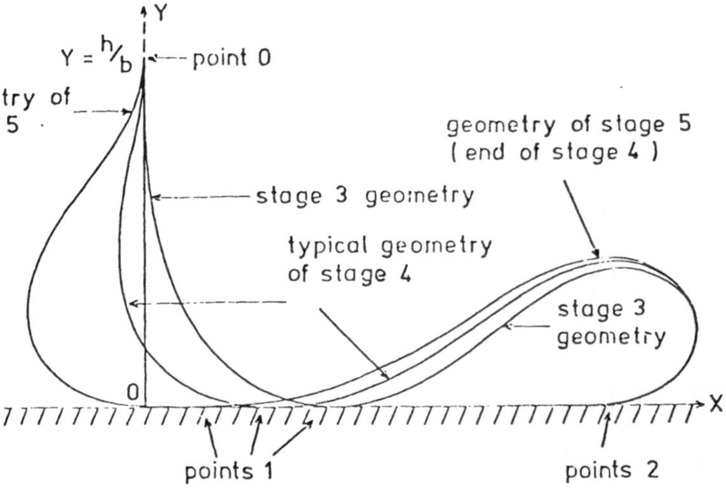

stage 4

point	0	1	2
σ	0	? $(=L_1)$	? $(=L_1+L_2)$
$\emptyset_X$	?	?	
$\emptyset_Y$	?	?	
θ	$-\pi/2$	0	$-\pi$
$\frac{d\theta}{d\sigma}$	?		0
X	0	$(X_{1c}-L_1-L_{1c})^*$	$(l/b-L_1-L_2)^*$
Y	h/b	0	0

* rolling conditions, l/b increased incrementally

stage 5

point	0	1	2
σ	0	? $(=L_1)$	? $(=L_1+L_2)$
$\emptyset_X$	?	?$^+$	
$\emptyset_Y$	?	?	
θ	$-\pi/2$	0	$-\pi$
$\frac{d\theta}{d\sigma}$	?	0	0^+
X	0	$[=X_F]$	$(X_{1c}+L_{2c}+X_{2c}-L_2-X_F)^*$
Y	h/b	0	0

$^+$ it can be proved analytically that the boundary value $(\frac{d\theta}{d\sigma})_2 = 0$ is satisfied if and only if $(\emptyset_x)_1 = 0$, but for the sake of completeness this boundary value - initial value pair are included

* combined rolling condition

Figure 3. Elastica Model of Deformed Fabric (Continued)

techniques, being easy to understand. This makes these methods readily explained to non-specialists.

Finite Element Models

Methods based on elastica theory are unsuitable for more complex deformation cases, so finite element methods were examined as a possible approach. Since very large deformations were to be expected, the NONSAP[4,5] finite element programs were used.

Cases of planar deformations of fabrics are clearly within the capabilities of finite element methods, but the author is unaware of any element which enables very large transverse displacements and very large membrane strains in anisotropic materials to be modelled. Simplifications are again necessary. Two example solutions are shown,[6] one of a "strip test" tensile test of a fabric, Figure 4, which exhibits the features expected of a real fabric and a model of the ballistic deformation of a knitted fabric, Figs. 5 - 7. A knitted fabric was chosen because these fabrics are more nearly isotropic than woven fabrics, and the experimental arrangement enabled an axisymmetric model to be used.

The problems associated with using finite element methods for modelling complex fabric deformations are again caused by the large displacements and extreme mechanical properties. The finite element method is most suited to stiff structures and small strains; flexible materials require increasing computing effort for decreasing returns. Large deformations can lead to ill-conditioning as elements become over-distorted, Figure 7, and the mechanical properties of many fabrics (for example, Poisson ratios close to unity) seem designed to cause computational problems.

Current Developments

The limitations of the techniques used to date have lead to the development at Leeds of a computational method for flexible shells similar in philosophy to the Bending Curve Package for elasticas. The approach, as yet incomplete, is to use the differential geometry of surfaces to represent the undeformed and deformed geometrics of the fabric, and hence to derive strain measures and compatibility equations for the fabric, and to integrate these numerically together with the equilibrium equations for particular sets of boundary conditions, with the associated initial value problem being solved by an iterative technique.

The method will only be summarized here, as it is not regarded as ready for

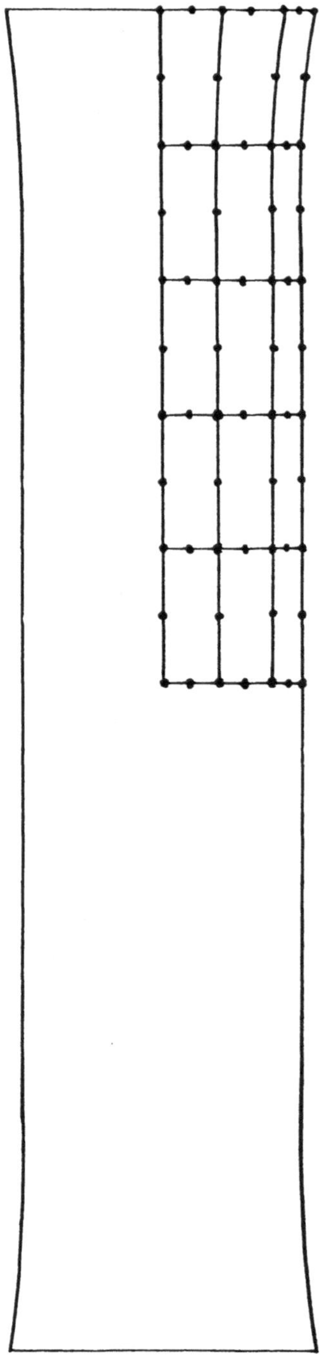

Figure 4
Finite Element Model of Fabric "Strip" Tensile Test

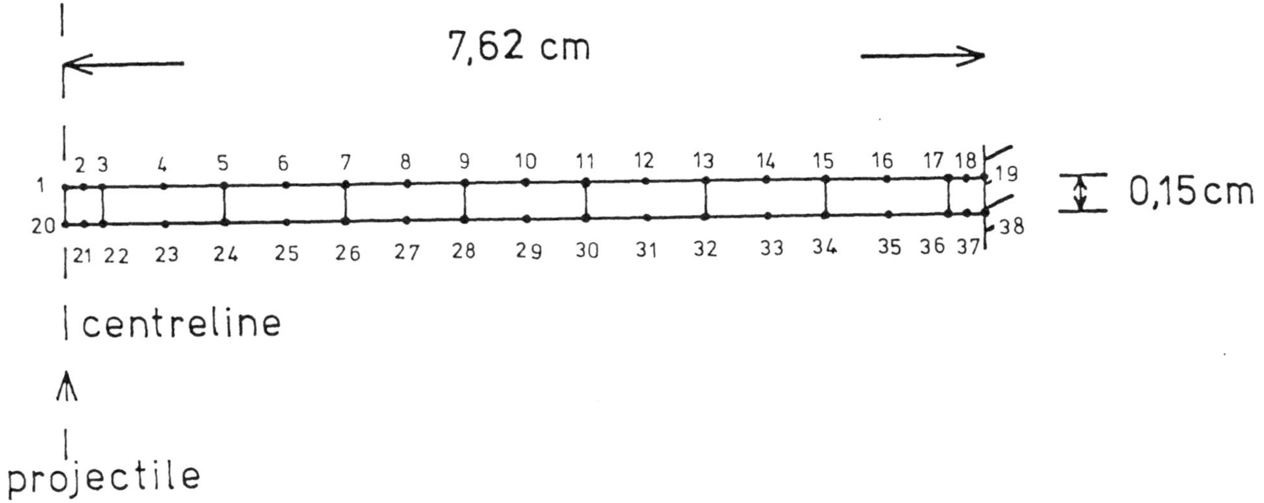

Figure 5

Finite Element Model of Ballistic Penetration of Knitted Fabric: Deformed Geometry

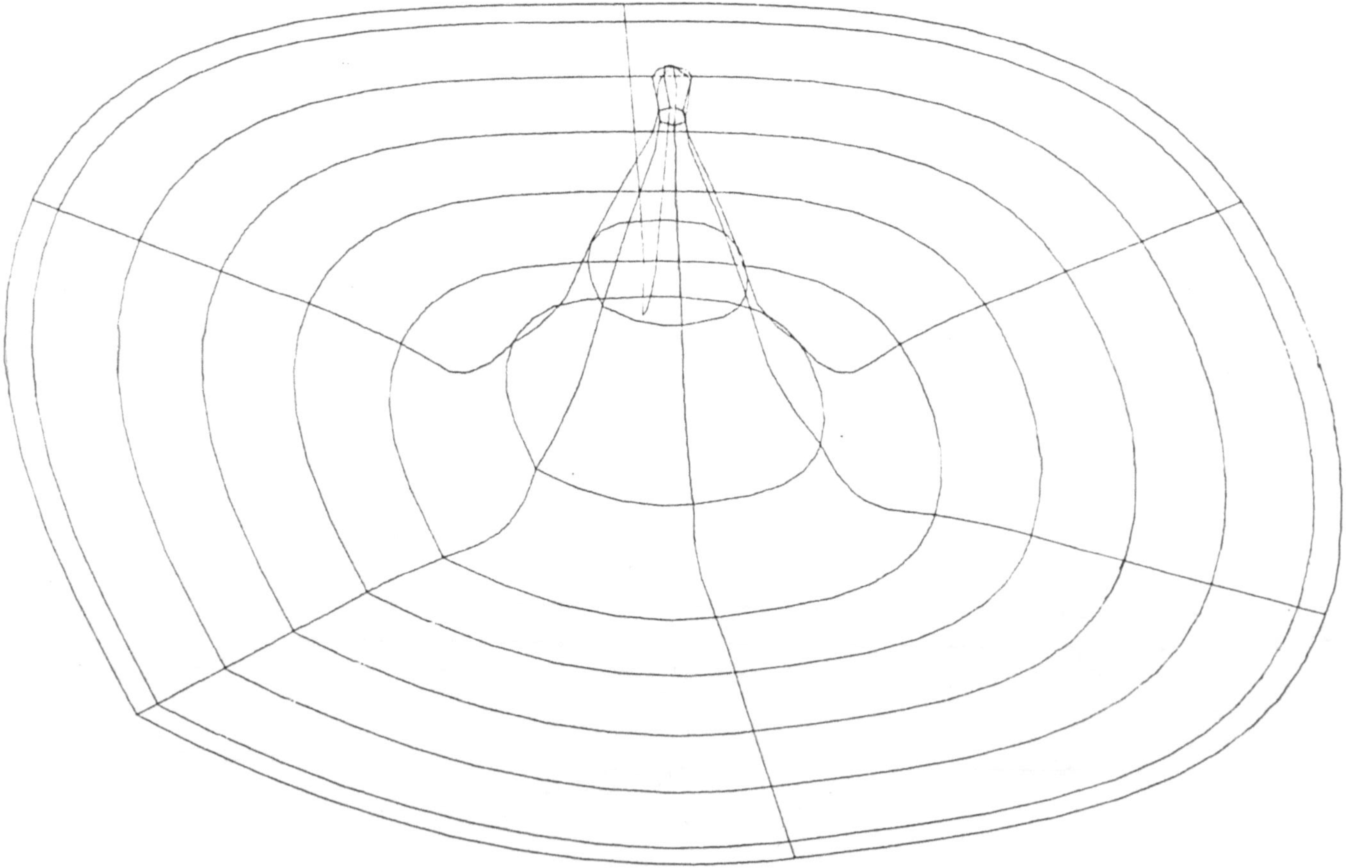

Figure 6
Finite Element Model of Ballistic Penetration of Knitted Fabric: Deformed Cross-section.

formal publication at present.

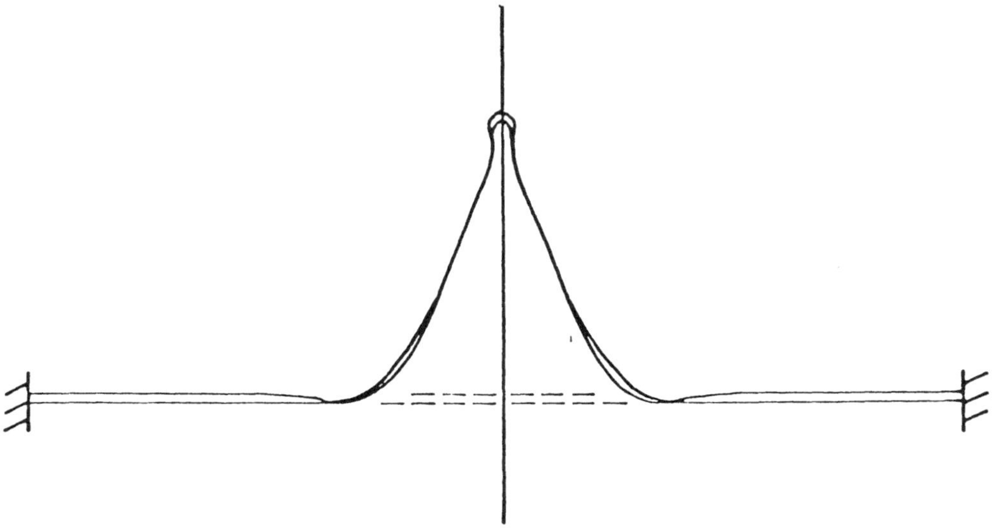

Figure 7. Finite Element Model of Ballistic Penetration of Knitted Fabric: Finite Element Mesh

The fabric in both deformed and undeformed states is represented by a surface embedded in Euclidean 3-space, and oriented by a unit normal vector field, U. A shape operator is defined in terms of the directional derivative of U with respect to tangent vectors in the surface. Scalar products of the shape operator with tangent vectors in the surface (closely related to the second fundamental form) are used to define normal curvatures and twists of the surface. The twist can be interpreted as the torsion of a geodesic through the point under consideration in the direction of the tangent vector, and the curvatures and twist exhibit tensor transformation behaviour.

Intrinsic, or convected, coordinates are used to describe the surfaces and the connection equations are expressed in terms of adapted frame fields on these coordinates. Membrane strains are calculated in terms of the difference in metric tensors (first fundamental forms) for the two states. Bending and twisting strains are calculated from the difference between the normal curvatures and twists for the strained and unstrained surfaces.

Compatibility equations are derived from the properties of the Lie bracket applied to certain derivatives of position vectors to the deformed surface, which gives particularly convenient expressions for computation. The equilibrium equations derive from the general expressions given by Ericksen and Truesdell,(7) by neglecting internal couples appropriate to polarized media, and expressing in component form. Finally the constitutive relationships are expressed as a 6 x 6 matrix of terms which relate particular stress-resultants and couple-resultants to the Green strains, and the curvatures and twist.(8) This is computationally convenient, as the terms may refer to complex empirical or theoretical constitutive equations.

This formulation, mainly in terms of vectors and components, is designed to be more easily comprehensible to non-specialists, if less elegant and concise than the more normal tensor formulation.

References

1. Blumberg, H., Building with Coated Fabrics: The Present Position and World-Wide Trends; in the Design of Textiles for Industrial Applications, pp. 137-158. Papers of the 61st Annual Conference of the Textile Institute, 1977, edited by P.W. Harrison, published by The Textile Institute, Manchester, 1977.

2. Buckley, C.P., Lloyd, D.W., Konopasek, M., "On the deformation of slender filaments with planar crimp: theory, numerical solution and applications to tendon collagen and textile materials", Proc.Roy.Soc., Lond. A 372, 33-64 (1980).

3. Lloyd, D.W., Shanahan, W.J., Konapasek, M., "The folding of heavy fabric sheets", Int. J. Mech. Sci., 20, 521-527 (1978).

4. Bathe, K.J., Wilson, E.L., Iding, R.H., "NONSAP - A Structural Analysis Program for Static and Dynamic Response of Non-linear Systems", Report No. UCSESM 74-3, University of California.

5. Bathe, K.J., Ozdemir, H., Wilson, E.L., "Static and Dynamic Geometric and Material Non-linear Analysis", Report No. UCSESM 74-4, University of California.

6. Lloyd, D.W., "The Analysis of Complex Fabric Deformations", in "Mechanics of Flexible Fibre Assemblies", Proc. of NATO Advanced Study Inst., Kilini, Greece, 1979, ed. J.W.S. Hearle, J.J. Thwaites, J. Amirbayat, NATO ASI Series E: Applied Sciences No. 38, Sijthoff & Noordhoff, 1980.

7. Ericksen, J.L., Truesdell, C., "Exact Theory of Stress and Strain in Rods and Shells", Archive for Rational Mechanics and Analysis, 1, 295-323 (1958).

8. Shanahan, W.J., Lloyd, D.W., Hearle, J.W.S., "Characterizing the Elastic Behaviour of Textile Fabrics in Complex Deformations", Textile Research Journal, 48 (9), 495-505 (1978).